The Handbook of Environmental Chemistry

Founded by Otto Hutzinger

Editors-in-Chief: Damià Barceló • Andrey G. Kostianoy

Volume 28

The Handbook of Environmental Chemistry

Recently Published and Forthcoming Volumes

The Turkmen Lake Altyn Asyr and Water Resources in Turkmenistan
Volume Editors: I.S. Zonn and A.G. Kostianoy
Vol. 28, 2014

Oil Pollution in the Baltic Sea
Volume Editors: A.G. Kostianoy and O.Yu. Lavrova
Vol. 27, 2014

Urban Air Quality in Europe
Volume Editor: M. Viana
Vol. 26, 2013

Climate Change and Water Resources
Volume Editors: T. Younos and C.A. Grady
Vol. 25, 2013

Emerging Organic Contaminants in Sludges: Analysis, Fate and Biological Treatment
Volume Editors: T. Vicent, G. Caminal, E. Eljarrat, and D. Barceló
Vol. 24, 2013

Global Risk-Based Management of Chemical Additives II: Risk-Based Assessment and Management Strategies
Volume Editors: B. Bilitewski, R.M. Darbra, and D. Barceló
Vol. 23, 2013

Chemical Structure of Pelagic Redox Interfaces: Observation and Modeling
Volume Editor: E.V. Yakushev
Vol. 22, 2013

The Llobregat: The Story of a Polluted Mediterranean River
Volume Editors: S. Sabater, A. Ginebreda, and D. Barceló
Vol. 21, 2012

Emerging Organic Contaminants and Human Health
Volume Editor: D. Barceló
Vol. 20, 2012

Emerging and Priority Pollutants in Rivers: Bringing Science into River Management Plans
Volume Editors: H. Guasch, A. Ginebreda, and A. Geiszinger
Vol. 19, 2012

Global Risk-Based Management of Chemical Additives I: Production, Usage and Environmental Occurrence
Volume Editors: B. Bilitewski, R.M. Darbra, and D. Barceló
Vol. 18, 2012

Polyfluorinated Chemicals and Transformation Products
Volume Editors: T.P. Knepper and F.T. Lange
Vol. 17, 2012

Brominated Flame Retardants
Volume Editors: E. Eljarrat and D. Barceló
Vol. 16, 2011

Effect-Directed Analysis of Complex Environmental Contamination
Volume Editor: W. Brack
Vol. 15, 2011

Waste Water Treatment and Reuse in the Mediterranean Region
Volume Editors: D. Barceló and M. Petrovic
Vol. 14, 2011

The Ebro River Basin
Volume Editors: D. Barceló and M. Petrovic
Vol. 13, 2011

Polymers – Opportunities and Risks II: Sustainability, Product Design and Processing
Volume Editors: P. Eyerer, M. Weller, and C. Hübner
Vol. 12, 2010

Polymers – Opportunities and Risks I: General and Environmental Aspects
Volume Editor: P. Eyerer
Vol. 11, 2010

The Handbook of Environmental Chemistry Also Available Electronically

The Handbook of Environmental Chemistry is included in Springer's eBook package *Earth and Environmental Science*. If a library does not opt for the whole package, the book series may be bought on a subscription basis.

For all customers who have a standing order to the print version of *The Handbook of Environmental Chemistry*, we offer free access to the electronic volumes of the Series published in the current year via SpringerLink. If you do not have access, you can still view the table of contents of each volume and the abstract of each article on SpringerLink (www.springerlink.com/content/110354/).

You will find information about the

- Editorial Board
- Aims and Scope
- Instructions for Authors
- Sample Contribution

at springer.com (www.springer.com/series/698).

All figures submitted in color are published in full color in the electronic version on SpringerLink.

Aims and Scope

Since 1980, *The Handbook of Environmental Chemistry* has provided sound and solid knowledge about environmental topics from a chemical perspective. Presenting a wide spectrum of viewpoints and approaches, the series now covers topics such as local and global changes of natural environment and climate; anthropogenic impact on the environment; water, air and soil pollution; remediation and waste characterization; environmental contaminants; biogeochemistry; geoecology; chemical reactions and processes; chemical and biological transformations as well as physical transport of chemicals in the environment; or environmental modeling. A particular focus of the series lies on methodological advances in environmental analytical chemistry.

Series Preface

With remarkable vision, Prof. Otto Hutzinger initiated *The Handbook of Environmental Chemistry* in 1980 and became the founding Editor-in-Chief. At that time, environmental chemistry was an emerging field, aiming at a complete description of the Earth's environment, encompassing the physical, chemical, biological, and geological transformations of chemical substances occurring on a local as well as a global scale. Environmental chemistry was intended to provide an account of the impact of man's activities on the natural environment by describing observed changes.

While a considerable amount of knowledge has been accumulated over the last three decades, as reflected in the more than 70 volumes of *The Handbook of Environmental Chemistry,* there are still many scientific and policy challenges ahead due to the complexity and interdisciplinary nature of the field. The series will therefore continue to provide compilations of current knowledge. Contributions are written by leading experts with practical experience in their fields. *The Handbook of Environmental Chemistry* grows with the increases in our scientific understanding, and provides a valuable source not only for scientists but also for environmental managers and decision makers. Today, the series covers a broad range of environmental topics from a chemical perspective, including methodological advances in environmental analytical chemistry.

In recent years, there has been a growing tendency to include subject matter of societal relevance in the broad view of environmental chemistry. Topics include life cycle analysis, environmental management, sustainable development, and socio-economic, legal and even political problems, among others. While these topics are of great importance for the development and acceptance of *The Handbook of Environmental Chemistry,* the publisher and Editors-in-Chief have decided to keep the handbook essentially a source of information on "hard sciences" with a particular emphasis on chemistry, but also covering biology, geology, hydrology and engineering as applied to environmental sciences.

The volumes of the series are written at an advanced level, addressing the needs of both researchers and graduate students, as well as of people outside the field of "pure" chemistry, including those in industry, business, government, research establishments, and public interest groups. It would be very satisfying to see these volumes used as a basis for graduate courses in environmental chemistry. With its high standards of scientific quality and clarity, *The Handbook of*

Environmental Chemistry provides a solid basis from which scientists can share their knowledge on the different aspects of environmental problems, presenting a wide spectrum of viewpoints and approaches.

The Handbook of Environmental Chemistry is available both in print and online via www.springerlink.com/content/110354/. Articles are published online as soon as they have been approved for publication. Authors, Volume Editors and Editors-in-Chief are rewarded by the broad acceptance of *The Handbook of Environmental Chemistry* by the scientific community, from whom suggestions for new topics to the Editors-in-Chief are always very welcome.

Damià Barceló
Andrey G. Kostianoy
Editors-in-Chief

Contents

Introduction

Igor S. Zonn and Andrey G. Kostianoy

Abstract This book presents a systematization and description of the knowledge on water resources in Turkmenistan and on the "Altyn Asyr Lake" water reclamation project. The publication is based on observational data, scientific literature mainly published in Russian editions, and long-standing experience of authors of the chapters in the scientific research in Turkmenistan. A special attention is paid to satellite monitoring of the Altyn Asyr Lake construction and satellite remote sensing of water resources in Turkmenistan. International team of authors from several countries combined their efforts and contributed to this book on the Altyn Asyr Lake Project, the first one published in Western Edition. This book is addressed to the specialists working in various fields of environmental problems and ecology, water resources and management, land reclamation and agriculture, and regional climate change in Turkmenistan and Central Asia.

Keywords Altyn Asyr Lake, Central Asia, Environment, Turkmenistan, Water Resources

This book is devoted to one of the largest hydro or water reclamation project, which is carried out in Turkmenistan, one of the countries in Central Asia – construction of the Turkmen lake "Altyn Asyr" ("Golden Age"), which is sometimes called "the Turkmen Sea" or "the Karakum Lake" [1].

The independence of the former Soviet republics of Central Asia (Fig. 1) has put before them a number of major and serious problems for the development of market economy.

I.S. Zonn
Engineering Research Production Center for Water Management, Land Reclamation and Ecology "Soyuzvodproject", 43/1, Baumanskaya ul., 105005 Moscow, Russia

A.G. Kostianoy (✉)
P.P. Shirshov Institute of Oceanology, Russian Academy of Sciences, 36, Nakhimovsky Pr., Moscow 117997, Russia
e-mail: kostianoy@gmail.com

I.S. Zonn and A.G. Kostianoy (eds.), *The Turkmen Lake Altyn Asyr and Water Resources in Turkmenistan*, Hdb Env Chem (2014) 28: 1–8, DOI 10.1007/698_2012_218,

Fig. 1 The map of Central Asia (http://www.lib.utexas.edu/maps/commonwealth/cis_central_asia_pol_95.jpg)

The background for this development is the specific geographical environment, made up mainly of desert areas with extreme climatic conditions, barren and sparsely populated. In 2011 Turkmenistan celebrated 20 years of independence. Since the establishment of a new state (1991) and declaration of its neutrality (1995), face of the country has changed significantly. In “a happy era of the mighty state,” proclaimed by President Gurbanguly Berdymukhamedov (Gurbanguly Berdimuhamedow), in water industry of the country, as in other sectors of economy, there are significant transformations. Water industry development has become one of the priorities of the state policy. In arid conditions, as a rule, water resources are very limited; therefore the conservation and management of water resources is a priority when discussing plans for economic and social development of the countries of Central Asia.

Water is vital for Turkmenistan, the country where 80% of territory is covered by one of the world’s largest deserts Karakums. Its availability is vital for the development of irrigated agriculture and water supply for population. The main source of surface water is the Amu Darya River, which through the Karakum Canal (also known as the Karakum River) brings together in a single complex the irrigation drainage systems of Lebap, Mary, Akhal, and Balkan velayats (provinces) (Fig. 2). In addition, the Amu Darya River meets the needs of Dashoguz Velayat. Most of the water resources of Turkmenistan are used to grow a variety of crops on irrigated

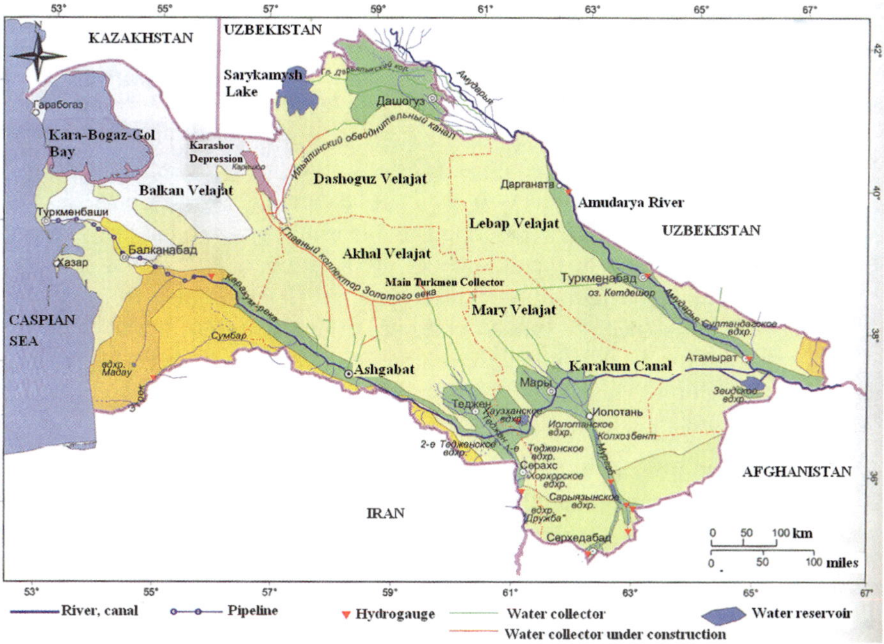

Fig. 2 Turkmenistan and its hydraulic network

lands – cotton, wheat, forage crops, vegetables, melons, etc. They are also used by industrial enterprises and utilities.

As a result of irrigation of fields for agricultural production, collector drainage waters (CDW) are formed. In order to maintain a good level of irrigated lands, it is required to drain water (usually mineralized) from irrigated fields beyond the agriculture zone. Deficit of irrigation water in agriculture can be compensated due to large stocks of mineralized groundwater and drainage water. The total volume of drainage water is estimated to be about 6 km^3/year, and taking into account the volume of drainage water generated on the territory of neighboring Republic of Uzbekistan, this figure can reach over 11 km^3/year. The total volume of drainage water discharged annually into Sarykamysh Depression (Sarykamysh Lake) and sands of the Karakum Desert is more than 6 km^3 with an average mineralization of 3–5 g/l. Such water could potentially be used for growing crops, flushing of saline soils, and irrigation of desert pastures.

Drainage waters forming on the irrigated lands of Turkmenistan and those coming from Uzbekistan cause a number of ecological problems for the environment:

1. Runoff of drainage waters from Turkmenistan and Uzbekistan back to the Amu Darya River, and, as a consequence, the deterioration of water quality in the lower reaches of the Amu Darya River.

2. The negative impact of the CDW extra discharge from the territory of the Republic of Uzbekistan on the irrigated lands of Dashoguz Velayat via the interstate Ozerny and Daryalyk collectors. This causes an increase in the water levels in the CDW collectors and reduction in the efficiency of the drainage systems in the area of their influence.
3. Elevation of the water table on the reclaimed lands, which, coupled with the deteriorating quality of irrigation water, speeds up salinization of the lands and a sharp decline in their fertility. It also eliminates the possibility of efficient desalination of lands as washing does not bring tangible results. As a result, irrigated lands become very strongly saline and cannot be used further for agricultural processes.
4. In the absence of receiving water reservoirs for CDW drained from irrigated fields in Mary, Akhal, and Balkan velayats of Turkmenistan, it was decided to temporarily dump CDW into the Karakum Desert, where there was flooding of lands and pastures. This decision led to a negative impact on the ecology of the flooded pasture lands.

As can be seen from all of the above-mentioned environmental problems related to CDW, Turkmenistan alone is faced not only with domestic water management issues, such as preventing the shrinking of grazing areas and protection of the Central Karakum from flooding by CDW, but also with interstate issues related to prevention of CDW discharge into the Amu Darya River. Currently CDW are discharged into the Amu Darya River, Sarykamysh Lake, and low places in the desert.

To solve this problem with CDW discharge, in 1960s creation of the Trans-Caspian collector was suggested, which had to collect all the drainage water from the Murghab, Tedzhen, and other irrigated areas and forward them to the Caspian Sea. In 1970s, a new project was elaborated which was aimed to discharge CDW into the closed Karashor Depression (Fig. 2). It was planned to forward CDW from Tashauz (now Dashoguz) to Karashor via ancient Uzboy river bed beside the Sarykamysh Lake. Second, the Main Turkmen Collector 720 km long was planned to build from Chardzhou (now Turkmenabad) through the center of the Karakum Desert along the river bed of the ancient Amudarya (along Unguz salt marshes) (Fig. 2). It was supposed to extend the Murghab and Tedzhen river beds to the Main Collector. Another collector was planned from the Kopetdag Plain to Karashor Depression. In the early 1980s construction of this hydraulic network began, but then was stopped due to the collapse of the USSR.

The National Program "Strategy of economic, political and cultural development of Turkmenistan until 2020" shows the construction of the Turkmen lake "Altyn Asyr" as one of the priorities in the country. It will allow:

1. To prevent the discharge of CDW in the Amu Darya River, improve water quality in the river, and, thus, radically change the current negative environmental situation in the lower reaches of the river.
2. To gather into a single stream all CDW from irrigated lands of Dashoguz, Lebap, Mary, Akhal, and Balkan velayats and direct them into the Karashor Depression – future Altyn Asyr Lake.
3. To use 400,000 ha of flooded desert pastures in agriculture.

4. To decrease by 2 m water level in the Ozerny and by 1.2 m in the Daryalyk collectors, ensuring normal operation of drainage systems in Dashoguz Velayat.
5. To use the supplementary drainage collectors for irrigation of pastures, which will increase their water supply and productivity in the area of 1.3 million hectares.
6. To reduce the discharge of CDW into Lake Sarykamysh and decrease the lake level to a stable point – −9 m, which will reduce the water area of the lake and shrink large areas of shallow water.
7. To reduce the threat of destruction of transport communications (gas pipelines, road bridges, aqueducts, etc.) in Dashoguz Velayat.

According to the general plan, the total length of the collectors of the Turkmen lake "Altyn Asyr" is over 2,650 km, the area of influence of collectors is around 2.5 million hectares, the construction period – about 10 years [1]. Construction of the Turkmen lake began in 2000. By 2008 Dashoguz collector of 385 km long was built. Also two water dams of 600 m and 22 km long were constructed. The confluence of two collectors will be connected with Karashor Depression with a canal of 74 km long. The first phase of the hydraulic network was put into operation in 2009. Work is underway on the supply Murghab, Tedzhen, and Dzharsky collectors.

The construction is designed in three phases. Upon completion of the second and third stages, in Turkmenistan a single drainage system will be constructed. The total length of the main and secondary collectors will reach 2,654 km. The length of the Turkmen lake "Altyn Asyr" is planned to be 103 km, width – 18.6 km, average depth – 69 m, capacity – 132 km^3, and area – about 2,000 km^2. Up to 10 km^3 of CDW is planned to forward to the lake. It is expected to withdraw CDW from Dashoguz Velayat with a rate of 210 m^3/s, and from the Lebap, Mary, Akhal, and Balkan provinces another 240 m^3/s. Filling the lake would take about 15 years and $4.5 billion. Realization of this ambitious project will solve many social, ecological, and economic problems in the nearest future [1].

In June 2012, speaking at the Academy of Sciences of Turkmenistan, President of the country Gurbanguly Berdymukhamedov has highlighted the importance of one of the tasks of the Project: "the development of technology for effective use of the Altyn Asyr Lake, its tributaries and adjacent lands in agriculture and socioeconomic area." In his publication "Economic strategy of Turkmenistan: Relying on the people, for the people," President G. Berdymukhamedov writes "...Our trump card in the set of solutions to this problem is water." From the ancient times, Turkmens cherish water as the most precious wealth, measuring every drop of water on weight of gold. Our ancestors created a unique irrigation system with a unique "kyariz" method of irrigation. Today, based on the folk traditions to lead a deal to its logical conclusion, we have to solve the problem of water supply in a complex with commissioning of fallow lands and ecological safety on a regional scale. The Turkmen Lake "Altyn Asyr," already stretching for hundreds of acres in the center of the Karakum Desert, will serve to this important goal.

In a sense, Turkmenistan is a country of the experiment – scientific and engineering experiment devoted to delivery of water to the areas where it is needed to solve the agricultural and socioeconomic problems. The first experiment in 1950s was the construction of the Karakum Canal (now called the Karakum River) with a length of about 1,450 km. Its route was twice subject to change. Construction of the canal had an important impact on the overall economic development and improvement of social conditions of the population in Turkmenistan.

As in the case with the construction of the Karakum Canal, the world practice has a little experience in CDW transport for hundreds of miles. We can mention the project to the drain CDW from the San Joaquin Valley in California, USA to the San Francisco Bay via 302 km long collector. Therefore, such a big project like Turkmen Lake "Altyn Asyr" attracts a great interest of specialists, scientists, and ecologists from different countries. It acts as a natural model, the study of which will provide theoretical and practical bases for the use of mineralized water in the desert development taking into account the environmental safety.

Implementation of this huge water project, which has no analogues in the world, faces difficult problems today and even more problems may occur during its operation, as it was the case with the Karakum Canal. Today, some of the foreign press, specialists from different countries, and international organizations show criticism in relation to implementation of the "Altyn Asyr" Project [2]. The main concern of specialists is about environmental consequences of the project. It seems that it is a bit premature, as the project is still underway. By the end of 2012 there are still hundreds of kilometers of collectors to be constructed and the Karashor Depression is still empty. It will take about 15 years to fill it.

In this book we would like to inform the international scientific community about this great project in Turkmenistan, describe its details, show the results of the ongoing research, especially in satellite monitoring [3–6], and invite for cooperative research. To understand the idea of the project and of the construction of the Altyn Asyr Lake, it is required to analyze the country's water resources, and to focus attention on those that are used or could be used for the development of agriculture. Thus, the book includes chapters on general environmental conditions, Karakum Desert, water bodies and resources in Turkmenistan, groundwaters, and regional climate change. Special chapter is devoted to international cooperation of Turkmenistan in water sector. Previous water projects implemented in Turkmenistan – the Karakum Canal and Sarykamysh Lake – are described in detail. The Altyn Asyr Lake issues are in the focus of the book: from new morphometric characteristics of the lake, computed based on the three-dimensional digital elevation model of the Karashor Depression, to forecasts for its water and salt balance, future ichthyofauna and waterbirds, which will live in the hydrographic network of the Altyn Asyr Lake. Special attention is given to the ongoing satellite monitoring of water resources in Turkmenistan and the construction of the hydrographic network in the vicinity of the Karashor Depression, which we are working on since 2009.

In this book we combined efforts and expertise of an international team of specialists in different fields of natural, social, and political sciences from Turkmenistan, Russia, Ukraine, France, USA, and Israel. The publication is based on observational data, scientific literature mainly published in Russian editions, and long-standing experience of some of the authors in the scientific research in

Turkmenistan. The authors did not try to achieve unanimity on the Altyn Asyr Lake Project, and sought to reflect personal professional opinion to the issues discussed.

This is the first book about the Altyn Asyr Lake published in Western Edition, thanks to Springer-Verlag Publishers who supported our idea. Undoubtedly, it will be followed by others, upon the progress in the implementation of the project and filling the Karashor Depression with CDW. We hope the book will become a milestone in the realization of the project, which in July 2014 will celebrate 5 years since the beginning of filling the new hydrographic network by CDW. This book is addressed to the specialists working in various fields of environmental problems and ecology, water resources and management, land reclamation and agriculture, and regional climate change in Turkmenistan and Central Asia.

This book may be regarded as a follow-up volume to our previous books on Central Asia published in Springer-Verlag: "The Caspian Sea Environment" (2005), "The Aral Sea Encyclopedia" (2009), "The Aral Sea Environment" (2010), and "The Caspian Sea Encyclopedia" (2010) [7–10]. On behalf of the authors, we would like to thank Springer-Verlag Publishers for their timely and steady interest in the environment of the Central Asia and their support of our publications.

Finally, we would like to note that in this book readers will find the same geographical terms written in a different way. For example, a capital of the country is written as Ashkhabad, Ashgabat, and Ashgabad. Amu Darya River sometimes is written as Amudarya, etc. This is explained by two reasons. First, after declaration of independence of Turkmenistan in 1991, there was a change in many geographical names and names of cities in the country. Thus, Chardzhou town became Turkmenabat, Nebitdag – Balkanabat, Krasnovodsk – Turkmenbashi (as well as Turkmenbashi Bay and Peninsula), Tashauz – Dashoguz, Kazandzhik – Bereket, Karakum Canal – Karakum River, etc. Second, in some cases this is due to different transliteration from old and new names in Russian (Cyrillic letters), in other cases – from Turkmen Language (Latin letters). Standardization of spelling of the Turkmen names and terms in English is a big task, which is outside of the scope of this book; moreover all the names and terms used in this book are today encountered together in Turkmenistan. We hope this will not lead a reader to confusion.

References

1. Turkmen Lake (2010) "Altyn Asyr", Ashgabat, p 104 (in Turkmen, Russian and English)
2. Stone R (2008) A new great lake – or dead sea? Science, v. 320
3. Kostianoy AG, Lebedev SA, Solovyov DM (2011) Satellite monitoring of water resources in Turkmenistan. Int Water Technol J 1(1):4–15
4. Kostianoy AG, Lebedev SA, Zonn IS, Lavrova OYu, Solovyov DM (2011) Satellite monitoring of Turkmenistan, Moscow, Signal, p 16 (in Russian and English)
5. Kostianoy AG, Zonn IS, Solovyov DM (2012) Construction of the Turkmen lake «Altyn Asyr» and the results of its satellite monitoring in 2009-2011. Int Water Technol J 2(1):59–66

6. Kostianoy AG, Zonn IS, Solovyov DM (2012) The Turkmen Lake "Altyn Asyr": satellite monitoring of its construction. Modern problems of remote sensing of the Earth from space. 9(1):219–227 (in Russian)
7. Kostianoy AG, Kosarev AN (eds) (2005) "The Caspian Sea Environment". The handbook of environmental chemistry. Vol.5: water pollution, Part 5P. Springer, Heidelberg, p 271
8. Zonn IS, Glantz M, Kostianoy AG, Kosarev AN (2009) "The Aral Sea Encyclopedia". Springer, Heidelberg, p 292
9. Kostianoy AG, Kosarev AN (eds) (2010) "The Aral Sea Environment". The handbook of environmental chemistry. Springer, Berlin, V7, p 332
10. Zonn IS, Kostianoy AG, Kosarev AN, Glantz M (2010) "The Caspian Sea Encyclopedia". Springer, Berlin, p 527

Turkmenistan: Landscape–Geographical Features, Biodiversity, and Ecosystems

Eldar A. Rustamov

Abstract This chapter describes in brief the geopolitical location of Turkmenistan and physiographical features of its territory, background (deserts and mountains), and intrazonal (river valleys and oases) landscapes; provides specific features of its biodiversity and basic ecosystems; focuses on susceptibility of ecosystems to the effect of anthropogenic factors, in particular in view of extensive land development, and redistribution of water resources.

Keywords Amudarya and Karakumdarya (Karakum Canal) rivers, Biodiversity, Ecosystems, Karakum Desert, Kopetdag Mountains, Landscapes, Turkmenistan, Types of deserts, Wetlands

Contents

1 General Information

Turkmenistan is one of the Central Asian republics locating in its southwestern part. The republic lies between latitudes 35°08′ and 43°48′N, and longitudes 52°27′ and 67°41′E. Its area (without the Caspian offshore area) is 491.2 thousand square

E.A. Rustamov (✉)
Ministry for Nature Conservation of Turkmenistan, Ashgabat, Stepnaya, 21, 744000, Ashgabat, Turkmenistan
e-mail: elldaru@mail.ru

I.S. Zonn and A.G. Kostianoy (eds.), *The Turkmen Lake Altyn Asyr and Water Resources in Turkmenistan*, Hdb Env Chem (2014) 28: 9–22, DOI 10.1007/698_2012_179,

kilometer. The territory extends for 1,100 km from west to east and for 650 km from north to south. To the east and north Turkmenistan borders Uzbekistan, to the northwest – Kazakhstan, to the west over the Caspian Sea – Azerbaijan, to the south – Iran, and to the southeast – Afghanistan.

Administratively the territory of the country is divided into 5 velajates (provinces) and 53 etraps (districts) with 21 cities (including capital – Ashgabat) and 77 settlements. The population of Turkmenistan is about six million growing annually by 5% on the average. More than half of the population (55%) lives in rural areas.

The territory of Turkmenistan covers the southwestern part of the Turan Plain extending in the arid zone of Central Asia. The nature and biodiversity of the country correspond to its southern position in the moderate climate belt. The modern natural processes are determined by open desert landscapes, generally drainless territory, climate aridity, and looseness of surface soils. At the same time, in Turkmenistan and, in fact, in the whole Central Asian region the anthropogenic factors acquire the growing significance which influences essentially the natural environment, in general, as well as the transformation and dynamics of ecosystems and their biodiversity.

The climate here is sharply continental and arid. The climate continentality is revealed in frequent and considerable fluctuations of the daily and annual meteorological characteristics, while aridity – in very low precipitations, slight cloudiness, and higher dryness of air. On the Caspian coast and in mountains the continental climate becomes somewhat milder. The average annual air temperature varies from 21°C to 34°C, however, on the plains the temperature ranges from 15°C to 18°C in the southeast to 11–13°C in the north, in mountain regions it is 10–15°C and at still higher altitudes (1.5–2.0 thousand meter) – 7–10°C.

The soil cover is rather diverse due to orographic peculiarities of the territory, hydrothermal factors, and soil-forming rocks. In the flat part of Turkmenistan the soils are extending following the latitudinal-zonal regularity, while in mountains – the vertical-belt.

By the order of decrease in their natural fertility the soils may be arranged as follows: light gray soils, meadow soils of ancient irrigation, meadow–gray, and meadow–takyr-like soils. They are all used in irrigated farming, while takyr, gray–brown, and sandy-desert soils are used as grazing lands [1,2].

2 Landscape and Environmental Peculiarities of the Territory

In physiographical and landscape–environmental terms (Fig. 1) Turkmenistan may be divided into two unequal parts: flat northern part and piedmont-mountain southern part. The flat lands account for 85% of the whole territory of the country with dominating deserts (73%) and only 12% with oases and water surfaces. The rest of the territory (15%) is covered by mountains and foothills. The lowest altitudes are recorded in the north in the Karakum Desert – 92 m below the sea

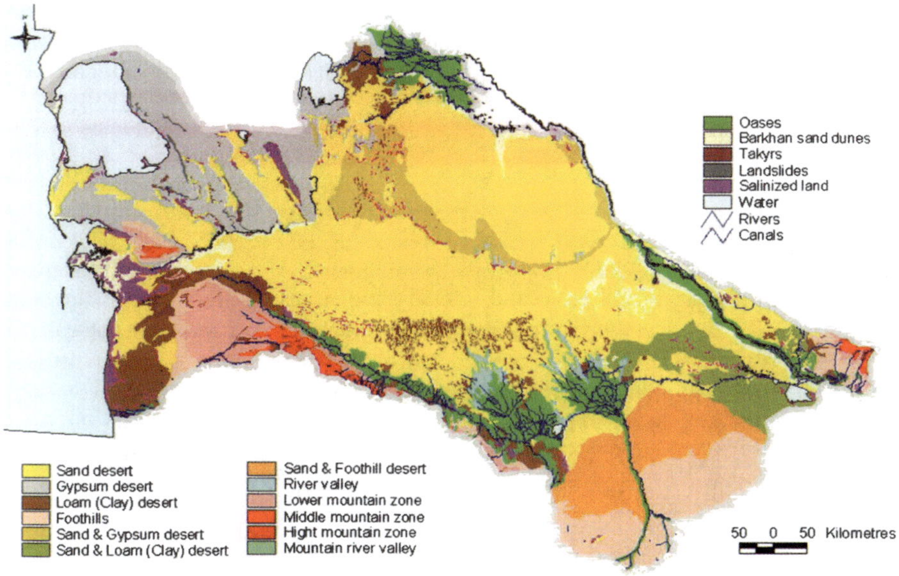

Fig. 1 Landscape and ecosystem division of Turkmenistan territory

level (Akcha-gaya Depression), the highest areas are found in the south in Kopetdag (Chopan Mountain – 2,889 m, Shakhshakh Mountain – 2,912 m) and in the far east in Koitendag (Airy-baba Mountains – 3,139 m).

The plains are generally inclined to the west – to the Caspian Sea and to the north – to Sarykamysh. They are subdivided into lowland and highland plains.

Lowland plains of Turkmenistan include the Karakum Desert (Central, Zaunguz, and Southeastern), Sarykamysh Depression, seaside lowland and river valleys of Amudarya, Karakumdarya (Karakum Canal), Murgab, and Tedjen dissected by a dense network of artificial irrigation canals. The Central (or lowland) Karakums stretch from west to east from the ancient bed of Western Uzboy to Kelif Uzboy (occupied at present by the Karakumdarya River valley in its upper reaches), while from north to south from the Unguz Depression to the midstream of Karakumdarya. The Zaunguz Karakums are limited in the west and south by a narrow chain of small Unguz depressions, in the east – by the Amudarya River and in the north – by the Sarykamysh Depression. This is an extensive area covered with overgrown large sand ridges, narrow kyr, and crushed stone ridges. The Southwestern Karakums stretch southward from the Middle Karakumdarya smoothly passing into the hills of Badkhyz and Karabil. Some researchers identify the Eastern Karakums as an independent area.

The Sarykamysh Depression has smooth slopes and a flat bottom; at present as well as in the quaternary period it was occupied by a vast lake with the same name (3.47 thousand square kilometer) surrounded by clay-solonchak areas. The maritime lowland is located among the western spurs of the Kopetdag Mountains, southern margins of the Krasnovodsk Plateau, Caspian Sea and in the south – the valley of the Atrek River (in its lower reaches). The surface is mostly covered with clays, while in the marginal area solonchaks and sands prevail. The river valleys in plains have a few mildly revealed terraces (to three of them). The Amudarya delta is composed of old alluvial and modern sandy and sandy-clay sediments. The deltaic plain of the river is sloping smoothly northward (to the Aral) and also westward (to Sarykamysh). Quite recently it was developed for agricultural use.

High plains of Turkmenistan were formed from Tertiary plateaus that adjoin in the north and northwest the lowland plains of Turkmenistan. These include Krasnovodsk Plateau, Southern Usturt Plateau, and related smaller residual uplands – Butentau, Duzgur, Tarimgaya, Kangakyr, Zengibaba, and Eshchekankrenkyr (Akchagai). The Krasnovodsk Plateau covers the peninsula bearing the same name and extends to the northwest to the Oktumkum sands, to the north to Kara-Bogaz-Gol Depression (with the bay of the same name), and to the southeast to Chilmamedkum. In the northeast it passes into Tuarkyr folded area with the "mountains" Ersarybaba and Babashi and chinks Begendjalykyr, Tyuverkyr, Akkyr, Tekedjik, and Koimatdag stretching for considerable distances. The Southern Usturt Plateau locates in the central part of Northern Turkmenistan and includes broken uplands (Chelyungkyr, Kaplankyr) alternating with depressions filled with the sands of Uchtagankum and Kumsebshen and solochaks of Kazakhlyshor and Karashor (for the project – the cup of Turkmen Lake Altyn Asyr).

3 Deserts and Their Biodiversity

In Turkmenistan the sandy, clay, crushed-stone, and solonchak deserts can be found. They have different soils, vegetation, and wildlife. All desert areas are characterized by a long dry summer and poor vegetation not forming the closed canopy that may develop only during a short spring.

Sandy deserts occupy both the lowland and highland plains and the largest of them is the Karakum Desert (350 thousand square kilometer). Quite naturally, the deserts leave their "imprint" on the natural environment of the country, in general, determining its soil conditions and biodiversity. Deserts are characterized by environmentally specific combination of contrasting factors, some of which are in excess (abundance of heat), while others in deficit (water insufficiency). Deserts represent a life domain with highly specific biodiversity having deep and century-old genealogical links with the arid conditions. In the course of evolution its components, such as plants and animals, managed to develop and acquire a sum of morphophysiological and ecological–ethologic adaptations enabling them to survive in such extreme environment. The conditions for existence of biodiversity in sandy deserts (not on drifting barkhans, but in hummocky-ridge deserts) are more favorable than in clay and other types of deserts. Sand is capable to keep water at great depths. And water gives life to plants, while availability of rather rich vegetation with its fodder potential and protective properties creates, in its turn, the favorable conditions for animals [3].

Flora of deserts, in particular of the Karakums, comprises some 700 varieties. On barkhans the vegetation is scattered widely or is completely absent. Quite seldom one can find here *Stipogrostis* sp., some varieties of *Calligonum* sp., Saltwarts (*Salsola* sp.), *Ammodendron conollyi*, and some others. The vegetation on the hummocky and ridge-hummocky sands is represented largely by the groups of White Saxaul (*Haloxylon persicum*) with a grass cover of *Carex physodes*, rarely Bulbous Meadow Grass (*Poa bulbosa*) and other grasses. Several varieties of *Calligonum*, saltwarts (*Salsola* sp.) are also met here, in particular *Salsola richteri*, Milk-vetch (*Astragalus* sp.), and *A. conollyi*. Perennial grasses are also growing here, such as Wild Onions (*Allium* sp.), Bindweeds (*Convolvulus* sp.), Iris (*Iris* sp.), Desert Candle (*Eremurus* sp.), and others. The groups of Black Saxaul (*Haloxylon aphyllum*) including several dozens of plant varieties are usually growing in lowlands. The Black Saxaul (*H. aphyllum*), unlike White Saxaul (*H. persicum*), may form thickets that comprise also *Calligonum*, Milk-vetchs (*Astragalus* sp.), Saltwarts (*Salsola* sp.), Jointfir (*Ephedra* sp.) with the grass cover made largely of annual grasses. Vast areas overgrown with well-preserved Black Saxaul (*H. aphyllum*) forming in some places "desert forests" are not so extensive as it was in the past century [4].

Fauna of the Karakums comprises 2 species of amphibian, 40 species of reptiles, 238 of birds, and 50 species of mammals [5,6], while the invertebrates living here are not adequately studied and comprise about five to six thousands of species [7]. Out of reptiles the stenotopic species in the sandy areas are Common Wonder (*Teratoscincus scincus*) and Comb-toed (*Crossobamon eversmanni*) Geckos, Lichtenstein's Toadhead (*Phrynocephalus interscapularis*) and Secret Toadhead (*Phrynocephalus mystaceus*) Agamas, Reticulate Racerunner (*Eremias grammica*), and Rapid Racerunner (*Eremias scripta*). In areas with compacted substrate the Even-fingered (*Alsophylax pipiens*) and Southern Even-fingered (*Alsophylax laevis*) Geckos, Sunwatcher Toadhead Agama (*Phrynocephalus helioscopus*) and *Phrynocephalus raddei*, *Eremias intermedia*, and *Eremias lineolata* occur. Out of the eurytopic reptile species you can meet Horsfield's Tortoise (*Agrionemys horsfieldii*), Steppe Agama (*Trapelus sanguinolentus*), Grey Monitor (*Varanus griseus*), Iranian Saw-scaled Viper (*Echis multisquamatus*), and Steppe Ribbon Racer (*Psammophis lineolatus*). They can live both in sands and on clay and crushed-stone surfaces. The areas composed of compact soils may become the habitat for such birds as Black-bellied Sandgrouse (*Pterocles orientalis*), Eurasian Stone-curlew (*Burhinus oedicnemus*), Houbara Bustard (*Chlamydotis undulata*), and Greater Short-toed Lark (*Calandrella brachydactyla*). Shrubs growing in sandy areas become the nestling places for Turkestan Ground-Jay (*Podoces panderi*), Great Grey Shrike (*Lanius pallidirostris*), and some Old World warblers (*Sylviidae*). Among the eurytopic birds we can name Little Owl (*Athene noctua*), Eurasian Eagle-Owl (*Bubo bubo*), Crested Lark (*Galerida cristata*), Isabelline Wheatear (*Oenanthe isabellina*), and Desert Finch (*Rhodospiza obsoleta*). Among the mammals the most typical stenotopic species connected with sand substrate are Piebald shrew (*Diplomesodon pulchellum*), Long-clawed Ground Squirrel (*Spermophilopsis leptodactylus*), several species of jerboa (Northern Three-toed

Jerboa *Dipus sagitta*, Comb-toed Jerboa *Paradipus ctenodactylus*, and others), Sand Cat (*Felis margarita*). You can also meet such stenotypes as Asiatic Wild Ass (*Equus hemionus*), jerboa (Thick-tailed Three-toed Jerboa *Stylodipus telum* and Blanford's Jerboa *Jaculus blanfordi*), and some other mammals, but their habitat is usually clay plains. Wolves (*Canis lupus*), foxes (*Vulpes vulpes*), Long-eared Hedgehogs (*Hemiechinus auritus*), and Desert Hares (*Lepus capensis*) are widespread and eurytopic in the Turkmen desert, they may also be found beyond its borders. In solonchak and stony deserts the composition of animal species is much poorer as their existence in such tough living conditions compared to sandy and clay areas is very hard. Therefore, in the deserts the overgrown sands may boast of the richer biodiversity. Somewhat inferior in this respect are clay deserts. And barkhan sands, stony and, in particular, solonchak sandy areas are most poor in organic life [3,4].

4 Mountains and Their Biodiversity

The mountains in Turkmenistan include ranges referring to the Turkmen–Khorasan system in the southwest of the country, piedmonts Parapamiz (Badhyz and Karabil) in the south and Koytendag or Kugitangtau Mountains, the outspurs of the Hissar Mountains in the far east. The mountains in Turkmenistan are low and medium height. All have no snow caps. They are affected by the desert landscape, thus, they do not have enormous water reserves. The maximum height of the Great Balkhan Mountains is 1,880 m (Mount Aplan), Badhyz – 1,267 m, Karabil – 984 m, Koytendag – 3,139 m, Kopetdag – 2,889 m (Mount Chopan in Turkmenistan), and 2,942 m (Mount Rizeh in Iran).

The main range in the Turkmen–Khorasan Mountains is Kopetdag making their northern part and ending in the northwest as small Kurendag Range and further on as Small and Great Balkhan Ranges separated from Kurendag with a sandy-clay plain. The Kopetdag Range is not high, but it is rather elongated stretching from northwest to southeast for 500 km and further beyond Turkmenistan, but near the borders of Iran and Afghanistan it passes into Parapamiz Range. Its piedmonts are made of two systems of smoothly rolling uplands: Badkhyz in the south of the Tedjen–Murghab interfluve and Karabil eastward of the Murghab River valley.

Kopetdag includes some parallel running ranges and plateaus (Garaul-Nokhur, Desht-Tuman plateaus, and others). It is intersected by numerous rocky gorges and deep valleys over which rivers, not so water abundant as plain rivers, flow. In the north Kopetdag neighbors on the piedmont plain (to 20 m and in some places to 40 m wide) sloping northward some areas of which are used for growing agricultural crops.

The Great Balkhans are the isolated mountains stretching from west to east for 35 km. In the west the Great Blakhans border on the Krasnovodsk Plateau, in the north and northeast – the Chilmammetkum sands, in the east – the western margins of the Karakums, and in the south they are separated with the solonchak-clay strip

Kelkor from not high Small Balkhan Ranges (779 m) belonging already to the Kopetdag Mountains. The Koytendag Ranges are rising quite apart from others on the Tukmen–Uzbek border; they are a part of the Pamir-Alay Mountains. The Koytendag is heavily cut with canyon-like gorges and has rather smooth slopes.

Vegetation cover in mountains of Turkmenistan varies from belt to belt, but there is no clear-cut difference between the belts. Above the altitude of 1,000 m Turkmen juniper (*Juniperus turcomanica* – in Kopetdag and Balkhans, *Juniperus zeravschanica* – in Koytendag) is widespread. The northern slopes of Kopetdag have no coniferous forests. The shrubs and trees are mostly found in moist mountain gorges. For the Badkhyz most typical are pistachio thin forests (*Pistacia vera*) and giant Ferula Badrakema (*Ferula badrakema*).

In the mountains the most widespread fauna is as follows: reptiles – Himalayan Agama (*Laudakia chernovi*), Turkestan Rock Agama (*Laudakia lehmanni*), Caucasian Agama (*Laudakia caucasia*), and Khorasan agama (*Laudakia erythrogastra*) as well as several species of *Eremias*; snakes – Eurasian Worm Snake (*Typhlops vermicularis*), Dahl's Whip Snake (*Coluber najadum*), and Striped Dwarf Racer (*Eirenis medus*). The Caucasian (*L. caucasia*) and Turkestan (*L. lehmanni*) agamas are stenotopic and live in the rocky and stony areas. Such species as Central Asian Cobra (*Naja oxiana*), Horsfield's Tortoise (*A. horsfieldii*), and Grey Monitor (*V. griseus*) are eurytopic and are met not only in mountains, but on plains, too. The most typical mountain birds are Lammergeier (*Gypaetus barbatus*), Caspian Snowcock (*Tetraogallus caspius*), Red-billed Chough (*Pyrrhocorax pyrrhocorax*), White-winged Grosbeak (*Mycerobas carnipes*), Eastern Rock Nuthatch (*Sitta tephronota*), Wallcreeper (*Tichodroma muraria*), Song Thrush (*Turdus philomelos*), Ring Ouzel (*Turdus torquatus*), and Rock Thrush (*Monticola saxatilis*). These species do not leave mountains. Thus, White-winged Grosbeak (*M. carnipes*) prefer living in juniper stands, while Wallcreeper (*T. muraria*) – on bare rocks and cliffs. In the mountains some other bird species may be found, for example, Eurasian Magpie (*Pica pica*), Golden Eagle (*Aquila chrysaetos*), Little Owl (*A. noctua*), Rock Dove (*Columba livia*), Chukar (*Alectoris chukar*), Common Quail (*Coturnix coturnix*), but they are eurytopic and may migrate deep into the plain. Out of mountain mammals there are Wild Goat (*Capra aegagrus*) and Markhor (*Capra falconeri*), Urial (*Ovis vignei*), Afghan pika (*Ochotona rufescens*) Treecreeper, Meadow vole (*Microtus* sp.), and rare Ural Field Mouse (*Sylvaemus uralensis*). Night bats prefer, in fact, living in mountains, but some species when the conditions are suitable live on the plains [3].

5 Oases and Their Biodiversity

Turkmenistan, like whole Central Asia, is a closed drainless area with very poor river network. Not high mountains are dry compared to other Central Asian mountain systems. Precipitations are greater in the mountains, than on the plains, but they are mostly absorbed by soft soils and lost to evaporation. Their smaller amount flows down the mountain slopes or appears in the form of springs over the ground surface. Oases are connected with the river valleys where agriculture is developed.

All large rivers take their origin beyond the borders of the country. These are such transborder rivers as Amudarya, Murghab with tributaries Kushka and Kashan, Tedjen, Atrek with tributaries Chendyr, and Sumbar as well as small rivers flowing down the Kopetdag slopes and also the Koytendarya River in Koytendag. The Amudarya River 1,409 km long with only 799 km within Turkmenistan borders accounts for 90% of all surface water resources in the country. At the same time, the network of artificial canals and headers is well-developed, in particular in the zone of Karakum Canal (or Karakumdarya River, 1,380 km long) and Turkmendarya River (180 km) that take their feeding from the Amudarya River basin. The total length of irrigation canals is over 40 thousand kilometer, while of the collecting and drainage network – over 35 thousand kilometer. The surface waters of flat Turkmenistan, apart from rivers and canals, include artificial reservoirs (16 in total) and lakes. The Karakumdarya River valley and the dense network of artificial canals in the flat part of the country appeared in the past century as a result of irrigation construction.

The Murghab and Tedjen rivers formed blind deltas that at present are included into the collecting and drainage system of the Turkmen Lake of Altyn Asyr.

The rivers, except Karakumdarya, are not overgrown due to high water turbidity. But the lakes forming oxbows in river floodplains are overgrown heavily. The coastal aqueous vegetation in lakes is represented by Common Reed (*Phragmites australis*), some varieties of Cattail (*Typha* sp.) and Club-rush (*Scirpus* sp.) River valleys are characterized by quite specific wildlife. In fresh waters, mostly in Amudarya and Karakumdarya rivers, many fish species may be found, such as Catfish (*Silurus glanis*), Bream (*Abramis brama*), Common Carp (*Cyprinus carpio*), Turkestan barbel (*Barbus conocephalus*), Snakehead (*Channa argus*) (invader), Asp (*Aspius aspius taeiniatus*), as well as Bighead Carp (*Aristichthys nobilis*) and Grass Carp (*Ctenopharyngodon idella*) that were acclimatized here still 50 years ago; out of amphibian – Lake Frog (*Rana ridibunda*) and European Green Toad (*Bufo viridis*); out of reptiles – Caspian Turtle (*Mauremys caspica*) and Europeanpond Turtle (*Emys orbicularis*), Grass Snake (*Natrix natrix*), and Tesselated Water Snake (*Natrix tessellata*). The birds living in riparian tugai include Common Pheasant (*Phasianus colchicus*), Shikra (*Accipiter badius*), Stock Dove (*Columba oenas*), Turtle Dove (*Streptopelia turtur*), Common Cuckoo (*Cuculus canorus*), Pallid Scops Owl (*Otus brucei*), Common Nightingale (*Luscinia megarhynchos*), and some other small sparrow species. Water basins are the habitat for dubbling ducks, Common Coot (*Fulica atra*), Black-winged Stilt (*Himantopus himantopus*), Pratincoles (*Glareolidae*), *Stemidae*, and others. Among mammals there are found otters living in Tedjen, Murghab, Sumbara, and Etrek water areas. In the floodplains of the same rivers Wild Boars (*Sus scrofa*) may be met, while in sea waters – rare Caspian Seal (*Pusa caspica*).

Describing environmental conditions of biodiversity in oases we should mention such habitats as tugai in river valleys, cane thickets near lakes and reservoirs, and anthropogenic landscapes. Tugai forests are represented by rather original and specialty tree and shrub communities. They include, first of all, two varieties of poplars – *Populus euphratica* and *Populus pruinosa* combined with Eastern Oleaster (*Elaeagnus orientalis*) and different varieties of Tamarisk (*Tamarix* sp.) interweaved with lianas. By now the tugai have been destroyed in many places and are used for growing agricultural crops. Some tugai areas survived only near the Amudarya and small areas near Murghab and Tedjen rivers [4].

Agricultural lands are used for cultivation of cotton, grain and fodder crops, vegetables, and melon crops. They are also occupied by orchards, vines, etc. The total area of agricultural lands in Turkmenistan is over 402 thousand square kilometer of which 385 thousand square kilometer are grazing lands, while 17 thousand square kilometer are irrigated lands. At the same time, the areas suitable for irrigation and drainage development in the future make about 177 thousand square kilometer [8]. The productivity of arid grazing lands used for free-range animal husbandry is low and depends greatly on the seasonal climatic conditions of a particular year or several years. Moreover, arid grazing lands are affected by deflation processes, in particular, in the technogenic zones during construction of roads and gas pipelines. In the agrarian sector the priority crops are wheat and cotton accounting for 49% and 42% of the sown areas, respectively.

The anthropogenic complex of species is being formed in the river valleys and oases with agricultural lands, residential buildings, and other structures needed by a man, in orchards and parks of settlements and cities, in field and road protective shelterbelts. Habitats appearing as a result of human activities in Turkmenistan tend to broaden. The classical example here may be the changes in the desert landscapes and habitats in the zone of the Karakumdarya that were observed after construction of this artificial waterway in the mid-twentieth century that led to formation of the intrazonal anthropogenic landscape along its route. We can name here the most typical representatives of the fauna found here: mammals – Common Pipistrelle (*Pipestrellus pipistrellus*), Short-tailed Bandicoot Rat (*Nesokia indica*), Gray Dwarf Hamster (*Cricetulus migratorius*), House Mouse (*Mus musculus*), Common jackal (*Canis aureus*); birds – Laughing Dove (*Streptopelia senegalensis*), Common Swift (*Apus apus*), Eurasian Hoopoe (*Upupa epops*), Common Myna (*Acridotheres tristis*), Common Blackbird (*Turdus merula*), Eurasian Tree Sparrow (*Passer montanus*), and others; reptiles – several species of gecko (*Gekkonidae*); amphibian – Lake Frog (*R. ridibunda*) and European Green Toad (*B. viridis*).

6 Ecosystems

On the territory of Turkmenistan the specific ecosystems were developed and exist corresponding to its division into the plain and mountain parts. On the plains the desert ecosystems mostly prevail, while in the mountains with the clear-cut altitudinal belts some specific ecosystems are found. Wetland and anthropogenic ecosystems occur mostly on plains and much less in mountains. Desert ecosystems presently cover 79% of the country's territory (without the Caspian offshore zone) of which 73% are flat-desert ecosystems and 6% are piedmont-desert. Other ecosystems account for 2–9% of the territory. They are developed quite unevenly (Fig. 1) and their contours are usually not clear-cut; they have smooth transitional zones among them called ecotones.

Desert ecosystems. For the plains the most typical are sandy desert (overgrown and half-overgrown) ecosystems that compared to others possess more diverse plant and animal life, both as concerns the range of available species and their quantity. In this respect, the ecosystems of clay deserts take the second place, and there is an ecotone existing between these two types – sandy-clay deserts. The most poor are crushed-stone and solonchak ecosystems as well as the ecosystem of drifting barkhans. In addition to what was said above, we should stress their low productivity. The main limiting factors here are moisture deficit and high air temperatures. Due to water shortage the plants (main producers) grow sparsely and do not form closed stands. Their small leaves are adapted to save water, while thorns protect them from being eaten by animals as a source of moisture. As is known, the photosynthesis, breathing, and growth processes in life organisms go on quicker at temperatures from 20°C to 40°C, while the shade air temperatures in Turkmen deserts rise higher than 40°C, and soil may be heated to 70°C and more. Accordingly, the high temperatures do not only cause intensive evaporation, but also slow down the vital life processes.

From the mid-twentieth century, as a result of unceasing development of the plain and desert ecosystems their transformation has become practically irreversible. Disposal of great quantities of highly saline drainage waters into the depressions in desert led to waterlogging of arid lands, including grazing lands.

Mountain ecosystems. These ecosystems cover the south of the country and contribute to climate formation there, although they themselves are affected by the nearby desert plains. The main specific feature with these ecosystems is their high biodiversity compared to plains which may be attributed to their vertical zoning and varying environmental conditions on slopes of different exposure. The wildlife here comprises many endemics characterized by small areas of extension. For example, in Kopetdag possessing 1,700 plant varieties the endemics account for 19.5% (332 varieties), and this is the highest percentage for mountain regions of Central Asia. Mountain ecosystems of Turkmenistan, in particular in Southwestern Kopetdag, reveal low stability to the economic use regime, including recreation. Thus, overgrazing may lead to soil erosion and formation of landslides. The soil cover may also degrade as a result of plowing of slopes, destruction of the tree and shrub vegetation. The mountain ecosystems in Turkmenistan are believed to be the depository of biodiversity and, to some extent, of fresh water. However, they are not connected, in fact, with the wetland ecosystems of plains. Mountain ecosystems (in Southwestern Kopetdag) belong to one of the centers of cultural plant origin. These ecosystems require milder regime of use, in particular in the zones of their use for recreation purposes, as development of tourism may bring irreversible changes in them [8]. The living conditions for wild animals in mountains are more diverse, therefore, these ecosystems are more abundant than in deserts. The life in mountain ecosystems dictates certain requirements to components of biodiversity, but they are less tough than in desert ecosystems [3].

Wetland ecosystems. Wetlands ecosystems of Turkmenistan are divided [9] into coastal–marine and valley–oasis ones which, in their turn, are broken down into lacustrine, river, tugai, and oasis ecosystems, i.e., the excessively wetted territories of the natural and anthropogenic origin. At the same time, a rather unique "mosaic" of wetland ecosystems may be found on plains of Turkmenistan.

In the offshore zone with depths to 200 m in the Turkmen sector of the Caspian Sea we can identify the coastal–marine ecosystem making an integral part of the single ecosystem of the Caspian Sea. This synergism is supported by active horizontal and vertical water circulation and migration of fish and seals over the whole water area of the sea. The most vivid example of synergism of this ecosystem is its reaction to invasion of Sea Walnut (*Mnemiopsis leidyi*). This species hibernates in the south, near the Iranian coast, while in summer with the currents it has enough time to expand over the whole water area as far as the offshore area. The ecosystem of the Caspian Sea, regardless of its integral nature, consists of several subordinated ecosystems. One of the main dividing factors is the bottom relief, the so-called sills, controlling the rate of water exchange: inside each part of the sea it is greater than among nearby parts. The top level in the ecological hierarchy is taken by the sea ecosystem in general, that is followed by the ecosystems of the Northern, Middle, and Southern Caspian, still lower levels are taken by the subordinated ecosystems. The greatest diversity can be found in the offshore ecosystems among which the upwelling ecosystem may be distinguished near the eastern coast of Middle Caspian [10] to which the Turkmen sector of the sea belongs.

Disturbance and development of the coastal–marine ecosystem in disregard of environmental requirements and expertise lead to pollution of water and coastal ecosystems, reduction of biodiversity, and deterioration of the social and economic conditions. At the same time, the Turkmen sector of the Caspian Sea is considered to be the purest, and this is the result of the targeted efforts on protection of the natural environment.

Out of all components of biodiversity in the coastal–marine ecosystem the wetland birds are directly linked with the mainland water bodies and in the future with Turkmen Lake of Altyn Asyr because the river and, in particular, large lake ecosystems are the places of concentration of water fowl on their transcontinental migration routes (Afro-Eurasian and Central Asian), while the Turkmen Caspian and inland water bodies being wetland ecosystems are the environmentally favorable places for passage and hibernation of these birds.

River and lake ecosystems (including reservoirs as artificial lakes). The whole diversity of river ecosystems may be grouped into three main types of habitat: open shallow water areas, cane thickets and, in particular, tugai.

Tugai (fringe woodlands) is a specialty relict type of the ecosystem – the floodplain tree-shrub and grass communities developed in floodplains of rivers in Central and Western Asia. The tugai ecosystem includes two types of vegetation – tugai and cane where hygrophytes (*P. australis*, *Typha* sp., *Scirpus* sp., and *Carex* sp.) are growing on meadow and meadow-alluvial soils. Perennial thickets of cane form plavni in some places. The critical condition of the tugai ecosystems in river valleys in Turkmenistan (except the Amudarya nature reserve) as in the whole Central Asian region requires their comprehensive and transboundary monitoring and protection [11]. The fertilizers applied amply and washed from fields, animal husbandry wastes and sewage waters produce their negative effect on river ecosystems of Turkmenistan.

Lake ecosystems are confined mostly to river floodplains. But such lakes (and their floodplains) may be also formed as a result of water discharge from irrigation systems and their existence depends on the scale and condition of land irrigation. In the past century artificial lakes – reservoirs were constructed in Turkmenistan for seasonal flow regulation. We should also mention here that the waterlogged areas in the river valleys and near-oasis zone as due to drainage of irrigated lands and disposal of wastewaters beyond the borders of oases some desert ecosystems got transformed into the lakes. Small water basins filled with brackish waters in which salinity grows with every passing year have appeared. Unfortunately, this leads to soil salinization and formation of small solonchaks. Diversion of saline waters beyond irrigated zones is the fact as well as the actions on wastewater desalination and its utilization. The water levels in Turkmen lakes are subject to significant seasonal and annual variations and this depends on climatic conditions in a particular year and the rate of silt deposition, in particular, in reservoirs. Water level fluctuations in lakes change the water salinity in them. On desert plains some small lakes and their floodplains exist only intermittently: in spring during rainfalls they get filled with water and in summer they dry out turning into solonchaks with the surface salt crust.

We should stress here the great significance of the wetland ecosystems for a man. They accumulate and store fresh water that is cleaned of pollutants there. The wetlands are the habitat of many plants and animals, including those put on the Red Book of Turkmenistan (2011). These ecosystems are required for the life of local communities and provide possibilities for tourism development. However, excessive pollution affects the ecological equilibrium and is able to interfere with the capacity of water bodies for self-purification.

Among the natural valley–oasis ecosystems that suffered from intense anthropogenic transformation there is the Amudarya delta the southwestern part of which is located in Turkmenistan. As is known, in the past 50 years due to uncontrolled use of water resources for irrigation of agricultural lands, the average river flow into the Aral Sea (initially 56 cubic kilometer a year) was reduced to more than tenfold. And the grave disturbance of ecosystems and the major ecological disaster in the Central Asian region led to drying out of the Aral Sea with all consequences for the natural environment.

Against the deficit of water resources, climate aridization, drop of the water level in the Aral Sea, and significant water level fluctuations in the Caspian Sea the wetland ecosystems in Turkmenistan face also potential anthropogenic threats, both direct and indirect. The direct threats include actions resulting in the reduction of biodiversity and, in general, degradation of ecosystems, such as destruction of tugai, overgrazing, poaching. The indirect threats produce their negative effects through pollution from drainage waters, industrial (the Caspian is also polluted with oil and oil products) and municipal and domestic wastewaters. As a result, we have salinization, deflation, and desertification of the wetland ecosystems being combined with ineffective systems of irrigation and land use and transborder "misuse" of river flows. This gives rise to reduction of areas, still greater degradation of wetland ecosystems leading to loss of biological diversity [12].

Anthropogenic ecosystems. Anthropogenic ecosystems are believed to be secondary towards the natural ones. They have appeared on agricultural and urbanized territories. Their share accounts for over 9% of the Turkmenistan area. Beginning from the second half of the past century the irrigated lands in the country has increased multiply. On the virgin lands the new anthropogenic ecosystems – oases, such as Khauzkhan, Akhal, Shasenem, and others of smaller size were formed. These ecosystems are distinguished by their intrazonality and mosaic nature. In the oases one can find the attending and "interacting" elements of wild and cultural plants. Of course, the endeavor to maximize agricultural production reduces sharply the biodiversity, the animals and, in particular, plants are subject to artificial selection. In this context, the weed vegetation also plays an important role as concerns preservation of biodiversity and integrity of phytocenosis in the anthropogenic ecosystem. In the arid environment the existence of various components of biodiversity is impeded, thus, many species get adapted to the conditions of the anthropogenic ecosystem as they find them more optimal. Therefore, the fauna in such ecosystem represents an assembly, and the diversity, say, of birds, and their number here are higher than in nearby natural, in particular, desert ecosystems. This is most well visible in arid plains as irrigation and any other man's activities lead to complication of the structure of the ecosystem which, in its turn, attracts here the respective species of wild fauna, thus, increasing their population.

In conclusion it should be stressed that the structure and dynamics of biodiversity are dependent intimately on the condition of ecosystems and ecological, geographic, and climatic peculiarities. The ecosystems of Turkmenistan are strongly affected by negative anthropogenic factors, including transborder. The greatest changes are observed in the desert and wetland ecosystems that were affected by wide-scale development of lands with accompanying redistribution of water resources. This in full measure may be referred to the whole Central Asian region. This is why the project "Turkmen Lake Altyn Asyr" was included into the regional program of action on improving the ecological, social, and economic situation in the Aral Sea area for 2002–2010 [13]. The natural ecosystems have the invaluable importance to support the life of the growing population of the country. In view of restricted possibilities for extending the cultivated lands and water shortage and in view of fragility of desert and mountain ecosystems, the greater attention should be focused on their study and monitoring until the well-balanced evaluation of their economic use is obtained.

References

1. Statistical Yearbook of Turkmenistan: 1997–2001 (2002) Turkmenmillikhasabat, Ashgabad, 265 p. (in Turkmen)
2. Development scheme of land reclamation and water management in the USSR for the period till 2000 in the Turkmen SSR (1983) vol. 7, book 5. Ashgabad, 174 p. (in Russian)

3. Rustamov AK (2011) Animal world of Turkmenistan and its protection (on the example of the vertebrate). Ylym, Ashgabad, 296 p. (in Russian)
4. Rustamov A, Rustamov E (2007) Biodiversity conservation in Central Asia: on the example of Turkmenistan. NEF, Tokyo, 273 p
5. Rustamov AK (1992) Animal world of the Karakums. In: The Karakum Desert and the Thar Desert. Ylym, Ashgabad, pp 146–156 (in Russian)
6. Rustamov AK (1994) Ecology of birds in the Karakum Desert. In: Biogeography and ecology of Turkmenistan. Kluwer Academic, Dordrecht/Boston/London, pp 247–264
7. Tokgaev T, Rustamov EA (1994) Changes in biodiversity of the Karakum Fauna caused by desertification. Problems Desert N 4–5, pp 95–101 (in Russian)
8. NPAEP – National Plan of Action of President of Turkmenistan on Environment Protection (2002) Ashgabad, 233 p. (in Russian)
9. Rustamov EA (2002) Strategic aspects of wetland ecosystems of Turkmenistan as a habitat of birds. Problems of Ornithology of Turkmenistan, Moscow, pp 80–90 (in Russian)
10. Monakhov GA (2005) On unity and hierarchy of Caspian ecosystems. In: Transactions of XVI international conference "Lomonosov-2009", CaspMNIC, Astrakhan, pp 10–11 (in Russian)
11. Treshkin SE (2011) Tugai degradation in Central Asia and possibilities for their restoration. Thesis of Doctor's Dissertation, Volgograd, 45 p. (in Russian)
12. Saparov UB, Esenov PE (2006) Analysis of water ecosystem condition in Turkmenistan. In: Problems of preservation of inland water conservation in Central Asia and Southern Caucasus. Report of GVP Program for CAiK, Almaty-Tashkent, pp 122–127 (in Russian)
13. Turkmen Lake "AltynAsyr" (2010) TDNG, Ashgabad, 99 p

The Karakum Desert

Igor S. Zonn and Paltamet E. Esenov

Abstract The Karakum Desert is one of the largest deserts worldwide. It locates within one Central Asian country – Turkmenistan. Covering about 80% of its territory it is vital for the economic development of the country. It holds in storage the oil, gas, and other mineral deposits. Distant-range cattle husbandry is practiced here; cotton growing, feed production, melon crop cultivation, and horticulture are developed on irrigated lands.

Keywords Climate, Desert, Economy, Relief forms, Soils, Vegetation, Water resources

Contents

I.S. Zonn (✉)
Engineering Research Production Center for Water Management, Land Reclamation and Ecology "Soyuzvodproject", 43/1 Baumanskaya Str., 105005 Moscow, Russia
e-mail: igorzonn@yandex.ru

P.E. Esenov
National Institute of Deserts, Flora and Fauna, 15, Bitarap Turkmenistan Street, Ashgabat 744000, Turkmeinstan

I.S. Zonn and A.G. Kostianoy (eds.), *The Turkmen Lake Altyn Asyr and Water Resources in Turkmenistan*, Hdb Env Chem (2014) 28: 23–38, DOI 10.1007/698_2012_192,

1 Introduction

Creation of the Turkmen Lake “Altyn Asyr” occupying one of the depressions in the Karakum Desert required construction here of the collector and drainage network and infrastructure that later on should support performance of this magnificent project. Accordingly, it seems appropriate to open this chapter with the description of the Karakum Desert, its components, and ecological and geographical characteristics.

Turkmenistan is one of the leading countries in Central Asia. Its territory locates in the drainless basins of the Caspian and Aral seas. It belongs to the zone of extra-tropical deserts of the Northern Hemisphere. The largest sandy desert Karakum covers the northern and central parts of Turkmenistan (about 80% of the country’s territory) (Fig. 1).

The Karakum Desert locates in the arid zone with 80–300 mm of precipitations a year, thinned and scanty vegetation represented by perennial and annual succulents. This is the zone of nomadic cattle rising; irrigated farming is impossible here.

2 Landforms

The Karakum Desert (in Turkmen “Gara Gum” meaning black or overgrown sand) is one of the Asian deserts. It occupies the Turanian Plain east of the Caspian Sea, with the Aral Sea to the north and the Amudarya River and the Kyzyl Kum Desert to the northeast (Fig. 1). It is characterized by symmetrical location to the left of the large regional river – Amudarya, a vast flat “descend” to the seaside of the Caspian Sea (Kara-Bogaz-Gol Bay) confined in the south and southeast with a medley of low plateau (Kopetdag).

One of the most ancient deserts of the world – the Karakum is an inland natural sandy aggradation desert. It has clear-cut geographical borders. The northern border is formed by the Amudarya deltaic flatland with modern and ancient irrigated lands. The northeastern border runs over the Amudarya valley, while the eastern – over the state border with Afghanistan. In the southeast the Karakum Desert is limited by on spurs of the Barabil and Batkhyz uplands, while in the south and southwest it gradually merges with the hills and piedmonts of medium-height Kopetdag Mountains. The Sarykamysh Depression and the ancient dry bed of Western Uzboi are regarded as the northwestern and western borders of the desert. Sometimes the Precaspian Plain of Turkmenistan is also referred to the Karakum Desert calling it the Southwestern Karakums. The Karakums Desert in the above borders is about 360 thou km^2 [1].

The desert has a generally flat surface being the lowest in Central Asia. During its geological history the territory of the desert suffered more than once the sea transgressions. After the sea recess the exposed area was affected by wind and flowing waters. At the same time, the territory was subject to tectonic impacts that

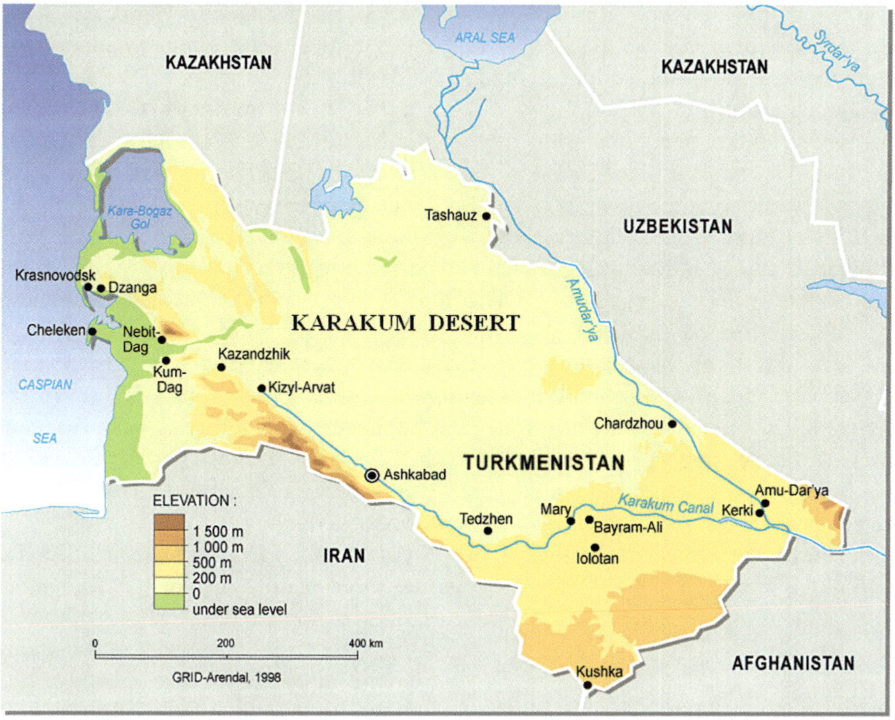

Fig. 1 Map of Karakum Desert

formed plates with an area of hundreds and thousands of square kilometers limited by gentle and steep slopes and, in some places, by abrupt cliffs (chinks) to 50–100 m high.

The present-day surface of the Karakum Desert was formed in the recent 30–40 million years. During this time the mountain formation processes were underway in the territories neighboring the Karakums in the south and southeast. Intensive eolian processes destroyed the formed mountains turning them into a drift area. The Upper-Tertiary seas found on the territory of the modern Karakums were gradually shrinking in size and receding westward. From the Late Pliocene the Karakums were devoid of all seas. The pra-Amudarya River flowing down from mountains and reaching the northern part of the modern Low Karakums contributed much to the surface formation here. The flatland over which the river was meandering was filled with the material brought with its waters. About one million years ago, the pra-Amudarya affected by tectonic processes deviated westwards. The debris material brought not only with the waters of the pra-Amudarya, but also its tributaries Tedzhen, Murghab and others was deposited on the flatland of the Lower Karakums. The alluvial deposits filled the Central Karakums depression and their thickness exceeded 500 m in some places.

During this time the northern part of the desert – Zaunguz Karakums were slowly uprising. The Southwestern Karakums representing an inclined flatland were formed due to the action of ancient water streams.

The desert aridity was amplified with changing of the Amudarya (Oxa) riverbed that about 70,000 years ago went beyond the Low Karakums confines and turned northward to the Aral Sea. From this time on the Central Karakums were left without surface water recharge source. After the Amudarya River turned to the north the Tedzhen, Murghab, and other rivers flowing down from the Kopetdag stopped being tributaries although they brought waters to the southernmost part of the desert, i.e. they formed blind subaerial deltas. These deposits in the south of the Karakum sandy-clay flatland formed numerous takyrs and solonchaks. After turn of the Amudarya in the Low Karakums the eolian transformation of the upper layers of the Karakum series with the growing climate aridization became the prevailing process. In the Neogene-Quaternary Period the warm and humid phases alternating with dry cool periods was observed permanently across the Karakums territory which affected its landforms. Only in the recent 6,000 years we may identify the active deflation phases, each consisting of 15 century-wise rhythms, on the average [2].

Therefore, through its historical development and in the course of climatic variations in the Neogene-Quaternary Period three major parts of the Karakums and its modern relief were formed:

1. *Inclined flatland Zaunguz Karakums or Northern Karakums* (they are often called Zaunguz Plateau) covering the smaller northern part of the desert represent the ancient alluvial, deeply broken high flatland where alluvium is overlain with eolian sands composed largely of sands and carbonate sandstones. The relief of the flatland is mostly takyr with a mantle of eolian deposits to 30 m thick (10–20 m on the average) over marine deposits. The half-overgrown sands form long steeply sloping ridges to 40–60 m high. In the inter-ridge valleys the outcrops of clays and solonchaks are rather frequently met. The lowest point of the country (81 m below the sea level) is the Akdjakaya Depression in the Zaunguz Karakums.
2. *Low or Central Karakums* are a sandy desert on the ancient alluvial re-deflated flatland concave in its central part and gently sloping towards north and south, accumulating the detritus drifted from the nearby territories and divided by a chain of the Unguz depressions. It is composed largely of the sandy-clay deposits of the pra-Amudarya which are overlain with the deltaic sediments of the Tedzhen and Murghab rivers. Therefore, the whole territory is covered by laminated gray micaceous sands of very homogeneous composition, with thin interbeds or lenses of chocolate clays of the Tertiary and Quaternary age of water origin 500 m thick [3].
3. *Southeastern Karakums* locate in the interfluve of the Murghab and Amudarya rivers. This is an inclined flatland composed of the Quarternary alluvial-deltaic sediments from 0 m (in the south) to 100 m (in the north) thick. It includes deeply broken sands adjoining Karabilyu in the north, the sandy-loamy flatland with low sands and solonchaks of the so-called Obruchev steppe and the pre-Amudarya barkhan belt.

The mountains surrounding the desert were the source of the materials brought with rivers. The same circumstance determined the differences in the lithology of the surface. In the Karakums the deposits carried from the mountains by the pra-Amudarya and Amudarya prevail.

As concerns the relief, the Karakums are a small-hummocky, sandy, heavily rugged flatland generally sloping from east to west, with ridgy sands in the north and hummocky (small-hummocky) sands with flat clay areas, takyrs in the south.

Much of the Karakums surface is covered with eolian sandy deposits. In the Neogene-Quaternary Period of the geological history of the Karakums their territory received no less than 70 thou km^3 of sediments, out of which no less than 7 thou km^3 were subject to eolian transformation [4]. The sands are mostly fine-grained here. The fractions sizing 0.015–0.150 mm and 0.150–0.210 mm dominate. Due to remoteness of the main part of the Karakums from the areas of drifting the coarser fractions in the sand composition are quite rare.

The petrographic composition is dependent on the composition of rocks in drift source areas and on the work of water streams, eolian drift during which the less strong minerals were destroyed. The key role in the Karakums sandy deposits is given to quartz, of less significance is feldspar, debris of magmatic and carbonaceous rocks.

The sands of Zaunguz are distinguished by their light-yellow and red-yellow coloring due to abundant presence of quartz, feldspars, and granitoids. The sands of the Lower Karakums thanks to the presence in considerable amounts of magmatic rock debris (basic effusive rocks and acid gabbroids, diabases) have greenish-gray and steel-gray coloring.

Long-time development of the surface, frequent transgressions and regressions of the Caspian Sea, migration of river systems, climatic changes and other natural and anthropogenic factors combined to form specific types of the eolian relief of the Karakums.

Here we can find the following types of sand forms: hummocky, ridge, ridge-hummocky, cellular, barkhan-hummocky, barkhan, barkhan chains. The ridges occupying 60% of its territory dominate here.

In the Zaunguz Karakums the ridges are made of compact parent rocks overlain with sands. This distinguishes them from the ridge relief of the Lower Karakums. In both cases the sand ridges are extending nearly meridionally. In the Zaunguz they can be to 70 km long, 0.2–2 km wide and 5–30 m high. In the Lower Karakums their height ranges from 15 to 20 m and width 200–300 m.

The inter-ridge valleys have smaller ridges running parallel to the larger ones; they are composed of loose sands. Near the Kopetdag foothill plain the sand ridges are characterized by larger dimensions. In the inter-ridge valleys the takyr and takyr-like soils occur.

The central and southern parts of the Zaunguz Karakums abound in kyr ridges. Kyr is a long and narrow relief form composed of Zaunguz parent rocks. The width of the kyr strip may reach several dozen kilometers. Its length varies from 5 to 8 km reaching at times 20 km. Its crest width is to 1 km and the height ranges from 30–40 m to 75 m. The spacing among kyrs is 1–4 km. The kyrs tend to gradually lowering northwards being overlain with sands and smoothly transfer into sand ridges which, in their turn, sink from 25 to 7 m near Amudarya oases. Their length is in no way inferior to the kyrs, while their crest width is no more than 200 m with spacing in inter-ridge valleys being 1.5–2 km.

Hummocky sands represent a broadly extending form of eolian relief covering 30% of the Karakums territory. They are also found in various combinations with other eolian relief forms (barkhan-hummocky, ridge-hummocky and others). The hummocky forms are usually immobile, however, in the areas transitional from the barkhan relief to the hummocky one and back they acquire some mobility. The height of these forms is 1–2 m.

The desert-specific relief features include dry beds and drainless depressions that may be rather large. Some of them have a basin-like shape (e.g., Sarykamysh), others are linear – Unguz, Western and Kelif Uzboy and some smaller ones. Unguz extends latitudinally from the Amudarua to the Ekedje and Dodur shafts for nearly 400 km. its width is 15–20 km. Separating the Zaunguz Karakums and Lower Karakums, Unguz consists of a chain of individual depressions crossed with dykes. The bottom of small depressions is composed of shors and takyrs, and in the eastern part – of sands.

The Kelif Uzboy also represents a linear chain of depressions in the Southeastern Karakums. But unlike Unguz, these depressions have the clear-cut signs of being influenced by pra-Amudarya locating on one of its meanders across the Karakums. Not very deep depressions are divided by not high sandy dykes. Some depressions of Kelif Uzboy are filled with waters of the Karakum Canal.

Dry beds are the significant phenomena in the relief, river network, and watersheds of the desert. They are often very long and remind of rivers left without water, such as Western Uzboy and Kelif Uzboy in the Karakums. The Western Uzboy was once a channel diverting Amudarya waters from Sarykamysh to the Caspian Sea. This was the river bed. Kelif Uzboy is the trace of pre-Amudarya meandering when it ran along the piedmonts of Parapamiz and Kopetdag into the Caspian.

The Western Uzboy is the largest of other dry beds. It may be clearly tracked from the southern bank of Sarykamysh to the Kelkor solonchak and along the dry bed of Aktam as far as the Caspian Sea. The total length of Western Uzboy is 500 km, the maximum width of its valley is 3 km, its greatest depth is −40 m. In its riverbed one can see many strip-like "dry" salt lakes extending in the riverbed for 1–5 km. Their width is usually no more than 50–100 m. In the southern part there are several freshwater lakes as well as shors, takyrs, and salt ridges. Uzboy diverts excessive water from Sarykamysh.

3 Climate

The climate of the Karakums may be classified as the climate of deserts of temperate latitudes. It is sharply continental and very arid which may result from specific atmospheric circulation, location of the Karakums deep in the huge

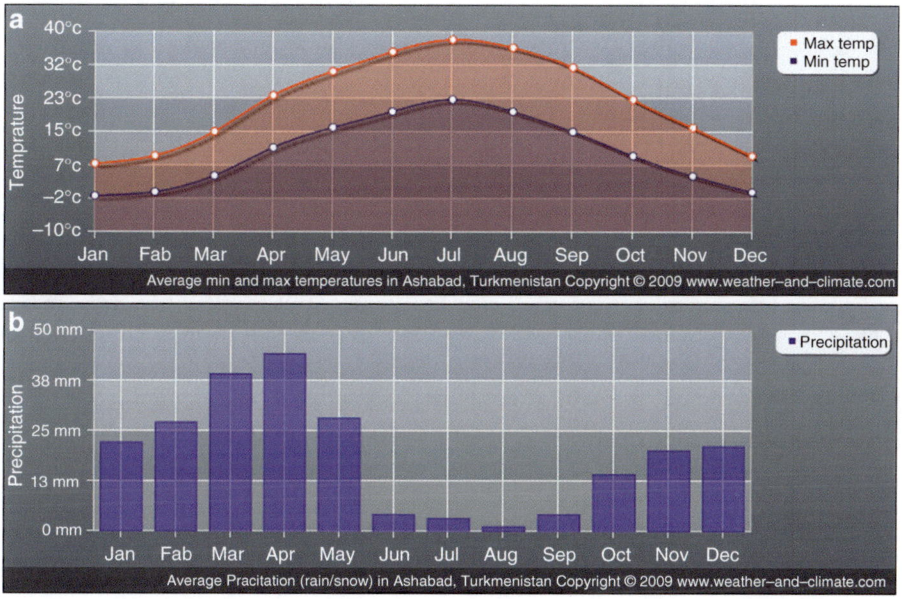

Fig. 2 The monthly mean minimum and maximum daily temperature in Ashkhabad (**a**). The monthly mean precipitation, including rain, snow, hail in Ashkhabad (**b**)

continent, its southern location, the nature of underlying surface and mountain systems in the southeast and south. The climate here is characterized by very hot cloudless and long summer, soft rainy spring, warm dry autumn, and frosty winter with frequent thawing periods.

The average annual air temperatures across the desert are positive varying from 11°C to 13°C in the north to 15–18°C in the southeast.

Winter is mild with low snow. The coldest month is January with the average temperature varying from −6°C in the northeast of the country to +3°C to +5°C in the south (Fig. 2). In some years, invaded by Siberian anticyclone the winters became very severe. Frosts with temperatures from −30°C to −35°C may last for long. The frost-free period is 230–250 days. Summer is very hot and dry. The maximum air temperature in July may rise above +30°C. The absolute maximum of +48°C to +50°C is registered in the Central and Southeastern Karakums. The daily amplitudes of air temperatures are very high reaching 50°, while on the ground surface they may be even 80°.

The distinguishing feature of the climate in Karakums is not only high air temperatures, but also a long sunshine period that may reach 2,800–3,100 h a year, on the average. The number of sunshine days in a year varies from 120 to 185 decreasing from south to north.

The Karakums refer to the zone of insufficient wetting. The average annual precipitations here vary from 80 to 300 mm (Fig. 2) and their quantity is growing from north to south. About 50–60% of precipitations fall in February and April.

High summer temperatures, insufficiency of atmospheric precipitations, and lack of surface flow contribute to formation of the air humidity regime. The relative humidity reaches its maximum in January making 70–78% and in June through September, the most arid period, it drops to 22–25%.

Low air humidity results in intensive evaporation from the water surface. The annual evaporation varies from 1,400 to 2,300 mm which is 15–20-fold more than annual atmospheric precipitations [5].

The permanent feature of deserts is winds. Slight winds (2–3 m/s) are blowing every day. At a speed of 4 m/s they form the wind-sand flow that triggers movement of unfixed sands and formation or rearrangement of sand relief. In the Central Karakums strong winds blow for 10 days a year, on the average, while in the Southeastern Karakums – to 50 days. Mists are usually observed in the southeastern areas from November through March for 10–20 days a year near water bodies. Thunderstorms occur ten times a year, in May, in the Zaunguz and Central Karakums.

4 Water Resources

As the book has a special chapter devoted to the water resources of Turkmenistan, we will provide here only a brief description of water resources in the Karakum Desert.

Location of Turkmenistan inside Eurasia, far from oceans, in the zone of deserts of temperate belt explains its low water supply. This situation is aggravated by the relief conditions. The greater part of the country has no surface waters. In the deserts the surface flow is formed only on small takyr and takyr-like watersheds composed of clay deposits. After rainfalls over 5 mm the surface flow is formed as very short-term floods. The size of the surface flow is dependent on the quantity of precipitations, their intensity and, to a great extent, on the size of a watershed and wetting before rainfall. The flow is accumulated in lower parts of takyrs and is quickly lost to evaporation and distributed seepage. The average many-year surface flow from 1 km^2 of takyrs varied from 5 to 20 km^3 a year. This flow and also limited groundwater reserves some of which are fed with mountain waters represent, in fact, the water resources of the Karakums. Their total volume is insignificant, hundred times smaller than the flow of Amudarya, Murghab, and Tedzhen rivers heading to the desert.

Ground waters occur nearly across the whole territory of Karakums. Closer to the recharge areas (waters of Amudarya, Murghab and Tedzhen rivers, and Kopetdag rivers seeping and being lost in sands) the ground waters become slightly saline. In the central part of the deserts their salinity is high. More than 10,000 dug pits are dispersed over the Karakums. Many desert regions have no freshwaters at all, but they have hundreds of pits (chirle) collecting rain waters from takyrs and directing them into deeper sandy horizons where they form lenses "on top of saline ground waters."

Table 1 Water resources and natural wetting of the Karakum Desert [6]

Description	Amount (km^3)
Amudarya River flow	12.3–13.6
Murghab River flow	1.46
Tedzhen River flow	0.730
Flow from takyrs	0.225
Flow from takyr-like watersheds	0.016
Static reserves of large fresh-water lenses	80
Sub-takyr fresh-water lenses on saline ground waters	0.0003[a]
Water of atmospheric precipitations without adjustments	42.4
The same with adjustments for wetting of the precipitation metering cylinder and inadequate account of winds	51.4

[a]Provided their reserves are replenished regularly.

Therefore, by rough estimates the water resources of the Karakums formed by the local surface flow (potential water resources) make presently 244 mln m^3, including flow from takyrs 225 mln m^3 and flow from takyr-like watersheds 15.7 mln m^3. Regardless of their insignificant amounts these water resources are important for watering of pasturelands although only a small part of them are used.

The summed up characteristics of water resources and natural wetting of the Karakums are presented in Table 1 below.

5 Soils

Low precipitations and very high temperatures in the vegetation period interfere with the development of biological and soil processes making them specific. Thus, the Karakums soils feature low thickness, meager humus content, poor structure, and nearly overall salinity.

The main types of soils developed here are sandy desert, gray-brown, takyr-like, takyrs, solonchaks, residual meadow and cultural irrigated soils of oases [1].

Sandy desert soils occur everywhere across sandy territories, in particular on the surfaces fixed with grass vegetation. They are not saline, but poor in nutritive substances and low-productive without application of respective organic and mineral fertilizers. These soils are used very effectively for cattle grazing (Fig. 3).

Gray-brown soils spread over the vast territory of Western and Northern Karakums in Zaunguzie. They develop on saline parent rocks. At the same time their long washing with atmospheric precipitations decreases the level of water soluble salts in the top soil layer. This is also facilitated by a relatively light texture of deposits making up these soils. Gray-brown soils as well as other soils of deserts are poor in humus which is explained by intensive mineralization of organic matter in soils in conditions of the arid desert climate. In some regions the gray-brown soils by their agrochemical properties are suitable for development of irrigated farming here.

Fig. 3 Sands in the Karakum Desert

These soils develop in automorphic conditions of ancient deltas, upper river terraces, and piedmont inclined plains composed of alluvial and proluvial deposits. In the regions of ancient farming the sediments generated by irrigated farming make their contribution into their formation.

Takyr-like soils, mostly of the transitional desert-oasis strip of Karakums, are a part of the arable land stock. Irrigation increases their humus content, improves structure and microaggregation permitting to receive high yields of various crops, including cotton.

Takyrs are developed broadly over the ancient deltaic and piedmont plains often combining with takyr-like soils and also in the inter-ridge depressions. They are formed in layered, mostly clay and loamy, alluvial and proalluvial deposits (Fig. 4).

Takyrs formed on outcrops of saline clay lenses and also on clay deluvium in inter-ridge depressions may be found quite often in the Karakums.

Residual meadow soils may be found on alluvial plains of the Amudarya, in the northwestern margins of the Murghab and Tedzhen deltas, in the Sarykamysh Depression, in the Western Uzboy valley, and in Southwestern Turkmenistan. They occur spot-like alternating with various sandy surfaces. These soils are generally slightly and medium saline, but non-saline ones are also found. These are highly fertile lands of the desert plain.

Solonchaks (shors) are developed among parent rocks forming deep (hundreds of meters) vast depressions mostly of the tectonic origin over eluvium of parent rocks. But more often shors appear among sands and in this case the depth of depressions may reach 20–40 m, the length – several kilometers and the width – hundreds of meters.

Extensive areas of solonchaks (shors) may be found on ancient deltaic plains of the Amudarya. They are formed on alluvial saline deposits at shallow groundwater occurrence (1–3 m).

Fig. 4 Takyr in the Karakum Desert

Depending on the salt composition and depth of groundwater occurrence all solochaks are divided into puffed, crust, crust-puffed, wet, and others. In many cases their whole profile is wetted in various degree as ground waters occur at a depth of 1–2 m. Takyrs often have spots of takyr solonchaks. Solonchaks are largely of the secondary origin and they are widely met as individual spots in oasis where ground waters are stagnant and occur close to the surface.

The cultivated-irrigated or irrigated soils are spreading within ancient and modern oasis. As a result of long irrigation a kind of cultured soils appeared here. They have a zonal nature and, at the same time, the anthropogenic origin.

6 Vegetation

About 700 varieties of higher plants may be found in the Karakums. The desert-specific climate, heavily saline ground waters, prevailing sandy substrate – all these factors, all together and individually – affect the plants obliging them to be heat-, cold-, salt-resistant, adapted to drifting sands, strong winds and dust storms.

Many plants of the desert refer to xerophytes, succulents, halophytes and demonstrate high adaptation to local conditions thanks to their morphological and physiological features. Desert plants manage to find sufficient water even in waterless and low-water areas. They possess special mechanisms protecting them from excessive heating and desiccation. Their evaporation is minimal. This becomes possible due to their deep root system (in saxaul it reaches 14 m) or development of horizontal roots satisfying their "thirst" with ground waters of the topsoil. Some plants have small leaves and no leaves at all. They assimilate with the

Fig. 5 Saxaul in the Karakum Desert

help of green offshoots (white and black saxaul) many of which fall off in summer (Fig. 5). Others have woolly leaves, waxy or glossy leaves (e.g., sandhill wattle) [1].

Desert vegetation is very sparse not forming closed canopy. Due to this fact and lack of leaves in large shrubs the deserts have no forests and, as a result, no shadow, underbrush and grass peculiar of forests.

In spring the whole territory of the Karakums, except barkhan sands, gets covered with a green carpet of ephemers and ephemeroids that scorch out in late April – early May. They mostly consist of sandy sedge grass (ilek) making good feed for cattle throughout a year. Out of shrubs the white and black saxaul prevail. Moving southwards the shrubs disappear giving place to grasses.

In the Karakums apart from higher plants the mosses and lichens (karakharsang) suppressing grass vegetation, in particular in Zaunguzie, are widespread. Barkhan sands are overgrown with selin and sparsely occurring shrubs – sandhill wattle and one-two varieties of kalligonums (kandym). The vegetation on barkhan sands is very sparse.

The desert vegetation consists of the following varieties: desert-woody thickets (white and black saxaul in combination with cherkez, kandym, sandhill wattle, and others); psammophytic shrubs (tamarisk, kandym, singren, sandhill wattle, and others); slightly overgrown bare sands – barkhans (wormwood varieties, exrophyte semishrubs – keurek, tansy and others); succulent thistle vegetation (sarsazan, glasswort, saltwort, and others) on solonetz soils and solonchaks; blue-green algae and lichens on takyrs; tugai (Asiatic poplar, oleaster, willow varieties, tamarisk, liana, cane, and others). In the river floodplains apart from background plants the perennial and annual grasses are growing on sands.

Availability of large shrubs, small shrubs and grassy vegetation creates a multi-layered pattern. It is most clearly visible in the sandy desert. It determines the possibility to use pastures in different seasons of a year. Grass pastures should be used for cattle grazing preferably in spring, summer, and autumn, while the multi-layered pastures may be used the year round.

7 Animals

The fauna of the Karakums is very specific. It is distinguished by high assimilation to the desert conditions, protective coloring of animals, rather poor species composition compared to other zones, and prevailing nocturnal animals.

The very important factor for animals is the desert climate, in particular the long warm period of a year, a short, usually warm, winter and availability of fodder in all seasons of a year (but not equally abundant in all seasons). The fauna of the Central Asian deserts is typical of the Turan Depression, and it has many common species with Middle Asia, North and Central Africa. The quantity and composition of species in different parts of the desert depend on difference of natural environment. Very specific is the fauna of river valleys where apart from the desert species you can find animals peculiar of dense tugai thickets and water bodies. There is also the fauna of oasis and settlements. Thus, the Amudarya valley numbers 211 bird species, while the Western Karakums – 118 species. The river valleys are the habitat for over hundreds of nestling birds, the desert – 20–30 species [1].

The animal world has two main complexes: fauna of sandy areas and fauna of areas with compact soils – clays and crushed stone. The animal world of sandy shrub deserts is most rich. The takyrs and, in particular, solonchaks are nearly lifeless.

The most typical mammals found here are: roofed – gazelle (everywhere); predatory – corsac, fox, wolf, dune cat (endemic), steppe cat, caracal; rodents – sand eel (greater sand eel, midday gerbil, Libyna jird, in particular the first one), thin-toed ground squirrel (in sandy areas), yellow gopher (in clay deserts), numerous jerboa – smaller, hairy-footed, comb-toed (endemic) and others; insectivorous – sand shrew and long-eared hedgehog; bats are distributed sporadically. The reptile population is rich and quite specific: steppe tortoise; snakes – carpet viper (poisonous), arrow snake, sand boa and many others; lizards – toad agama (long-eared, sand, takyr). Among the insects the most extensively represented are the colepterous (darkling bettle, leaf-horned and others), flies, ants, termites (several species). Quite frequently met are solifugae and scorpions. The fauna of ticks is quite diverse.

8 Economy

The Karakum Desert possesses significant and diverse natural resource potential: fuel-power (oil, gas, solar and wind energy), chemical raw materials (potassium and table salt, mirabilite, sulfur, and others), construction materials, agro-climatic resources (the longest vegetation frost-free period permits cultivation of warmth-loving crops), curative, fresh and saline ground waters, fertile (when irrigated) lands, pastures that, due to their forage and climatic conditions, offer the year-round grazing for cattle.

At present the surface and outlook of the Karakum Desert is formed by anthropogenic activities, although the exogenous factors should not be neglected.

Different volumes and conditions of water use led to appearance of different types of farms, different modes of local resource management. Depending on the supply of water, its quality either irrigated farming or cattle grazing farming, small or large irrigated areas, cultivation of selected crops appeared in the desert. Availability of water also controls the concentration of population, sizes of settlements, comfort of living, and life conditions.

Large oil and gas deposits were discovered in the Karakum Desert on the basis of which the oil and gas processing (Turkmenbashi) and chemical industries were developed. Gas and oil are transported via large main pipelines within the country and abroad: Korpedhe (Turkmenistan) – Kurt Kui (Iran) 200 km long; Dovletabad (Turkmenistan) – Serakhs – Khangeran (Iran) and others. Gas is used at large gas-turbine power plants in Turkmenbashi, Marakh, and Balkanabad.

Chemical industry uses the raw materials extracted in the Kara Bogaz Gol Bay in the west of the Karakum Desert. In Gaurdak the sulfur deposits are developed that provide raw materials to the chemical plant in Turkmenabat producing fertilizers for agriculture out of Gaurdak sulfur and Karatau phosphorites.

Very important is also the development of the transport-communication system in the desert where the railroads are of key significance. In the recent years the railroads Tedzhen – Serakhs – Meshkhed (300 km long), Ashkhabad – Karakumy – Dashoguz, Turkmenabat – Atamyrat (230 km long) were constructed with the further exit abroad. The internal automobile roads Turkmenbashi-Ashkhabad-Mary (980 km long), Ashkhabad-Mary-Turkmenabad (600 km long), Ashkhabad-Karakumy-Dashoguz (along the railroad), and others ensuring internal cargo transportation are improved to satisfy international standards.

Historically two types of irrigated farming have been established in the desert, they are small oasis and large oasis farming. The first type developed on piedmont plains near small water sources and on a border with the desert in flooded areas, in areas of ground water discharge on temporary water streams. The large oasis farming appeared in deltas and valleys of large and medium rivers and also near large artificial canals for inter-basin and inner-basin water transfers, first of all, the Karakum Canal.

The farming in Turkmenistan is completely dependent on artificial irrigation. Out of the total consumption of water resources 90% of water is used annually for irrigation purposes. As the water resources are limited only 2 mln ha of land may be put under irrigation annually. The key agricultural crops are cotton and wheat.

The main use of the Karakum Desert territory is for animal husbandry based on desert pasturelands. And their watering plays a key role in their development. The animal products produced in the desert are less costly than in the steppe zone. Sheep and camel rearing is practiced most extensively, and recently the horse breeding was added here.

Breeding of Karakul sheep distinguished by high adaption to the specific water-fodder situation in the desert has acquired great importance. The share of Karakul sheep in Turkmenistan makes 70% of the total flock. The greatest sheep stock is found in the Central and Southeastern Karakums. In the southwest and north of the desert, their quantity is much less due to scanty pastures.

For agriculture development the construction of the Turkmen lake Altyn Asyr in the center of the Karakums is in the focus of attention. Its construction will permit to use more wisely the collector-drainage waters presently disposed into the desert, the Sarykamysh Lake and partially into the Amudarya River as well as to improve the condition of irrigated lands and to increase the water resources of the country.

References

1. Babaev AG, Zonn IS, Drozdov NN, Freikin ZG (1986) Deserts. Mysl, Moscow (in Russian)
2. Kasianov KS (1976) Dynamics of soil wind erosion processes. Nauka, Moscow (in Russian)
3. Arangeldyev A, Zhumashov AP (1992) Peculiarities of relief formation and development in the Karakums. In: Karakum Desert and Thar Desert. Ylym, Ashgabat (in Russian)
4. Petrov MP (1973) Deserts of the world. Nauka, Leningrad (in Russian)
5. Orlovsky NS (1992) Climatic resources of the Karakums. In: Karakum Desert and Thar Desert. Ylym, Ashgabat (in Russian)
6. Kirsta VT (1992) Water resources of the Karakum Desert. In: Karakum Desert and Thar Desert. Ylym, Ashgabat (in Russian)

Climate Change in Turkmenistan

Elena Lioubimtseva, Jahan Kariyeva, and Geoffrey M. Henebry

Abstract More than 80% of Turkmenistan is desert; thus, key environmental issues are associated with redistribution and supply of limited water resources. Turkmenistan is projected to become warmer and probably drier during the coming decades. Aridity is expected to increase in all republics of Central Asia, but especially in the western part of Turkmenistan. The temperature increases are predicted to be particularly high in summer and fall but lower in winter. Especially significant decrease in precipitation is predicted in summer and fall, while a modest increase or no change in precipitation is expected in winter months. These seasonal climatic shifts are likely to have profound implications for agriculture, particularly in western Turkmenistan and Uzbekistan, where frequent droughts are likely to negatively affect cotton, cereals, and forage production, increase already extremely high water demands for irrigation, exacerbate the already existing water crisis, and accelerate human-induced desertification. The Amudarya is the most water-bearing river in Central Asia; its endorheic drainage basin includes the territories of Afghanistan, Tajikistan, Uzbekistan, and Turkmenistan. Fed by seasonal snowmelt of snowpacks and glaciers, the flow of the Amudarya may increase due to intensified melting of the glaciers and snowpacks under a warming climate, which could further contribute to expansion of agricultural land use at the expense of converted natural areas. During the last few decades, Turkmenistan has experienced widespread changes in land cover and land use following the socioeconomic

E. Lioubimtseva (✉)
Geography and Planning Department and Environmental Studies Program, Grand Valley State University, B-4-202 MAK, 1 Campus Drive, Allendale, MI 49401-9403, USA
e-mail: lioubime@gvsu.edu

J. Kariyeva
Office of Arid Lands Studies, The University of Arizona, Tucson, AZ 85721, USA

G.M. Henebry
Geographic Information Science Center of Excellence, South Dakota State University, Brookings, SD 57007, USA

I.S. Zonn and A.G. Kostianoy (eds.), *The Turkmen Lake Altyn Asyr and Water Resources in Turkmenistan*, Hdb Env Chem (2014) 28: 39–58, DOI 10.1007/698_2012_175,

and institutional changes in the wake of the disintegration of the USSR in 1991, and subsequently followed by a decade of drought and steadily increasing temperatures. These changes in the vegetated landscape are sufficiently broad to be detectable from orbital sensors at multiple scales.

Keywords Arid environments, Central Asia, Climate change, Deserts, Drylands, Hydrology, Land cover, Land use change

Contents

1 Introduction

Located in Central Asia Turkmenistan is a largely desert country that lies between 35°08′ and 42°48′N and between 52°27′ and 66°41′E, north of the Kopetdag mountains, between the Caspian Sea in the west and Amudarya River in the east. Turkmenistan territory expands 650 km from north to south and 1,100 km from west to east. To its north Turkmenistan borders Kazakhstan, to east and northeast, Uzbekistan, to south, the Islamic Republic of Iran, and to southeast, Afghanistan. On the west, Turkmenistan abuts on the Caspian Sea. The economy of Turkmenistan is linked with the use of land and water resources and continues to rely on agriculture that is dependent on irrigation. Major freshwater sources for Turkmenistan are the Amudarya that flows into the Aral Sea and is the main irrigation source for Turkmenistan [1], Murghab and Tedjen Rivers that disappear in the Karakum Desert, and the Karakum Canal – an unlined canal, which diverts water from the Amudarya [2]. More than 80% of Turkmenistan is desert. Turkmenistan occupies an area of 491,210 km^2 and had a population of 5,041,995 in 2010 [3]. The Karakum Desert comprises 80% of the area of the country, while 17% either mountainous or various types of clay, loess, and stony deserts. Only about 3% of the area is suitable for agriculture [4]. Turkmenistan has some fertile land and modest water resources located primarily in the eastern part of the country.

Because farming is the main industry in the Central Asian region [5], key environmental issues are associated with redistribution and supply of water resources and the limited water availability throughout the territory. Redistribution of surface water resources of the region is as follows: Tajikistan uses 44% of the total regional river flow, Kyrgyzstan – 26%, Uzbekistan – 10%, Kazakhstan – 3%, and Turkmenistan – 2% [6]. About 15% of the flow belongs to Afghanistan.

Despite its extensive oil and gas resources, Turkmenistan remains a poor, predominantly rural country with the majority of population relying on intensive agriculture in irrigated oases and is extremely vulnerable to climate change. During the last few decades, Turkmenistan has experienced widespread changes in land cover and land use following the socioeconomic and institutional transformations of the region catalyzed by the USSR collapse in 1991. The decade-long drought events and steadily increasing temperature regimes in the region came on top of these institutional transformations, affecting vegetation–climate feedbacks at multiple spatial and temporal scales.

According to the Intergovernmental Panel on Climate Change's (IPCC) Fourth Assessment Report (AR4), developing nations and arid regions of the world are particularly vulnerable to climate change and climate variability [7, 8]. Climate change and variability affect arid ecosystems and their productivity through the changing patterns in temperature and precipitation, droughts, floods, heavy winds, and other extreme events. Water availability and food security of arid and semiarid zones has been always unstable due to their low natural productivity and high variability in both rainfall amounts and intensities. The increasing pressures caused by the global climate change on livelihoods deteriorate the human vulnerability to the on-going desertification processes and natural climatic variability. The impacts of climate change in desert countries, such as Turkmenistan, are likely to lead to still larger populations being affected by water scarcity and the risk of declining crop yields and increase the risk of environmental migrations and political conflicts caused by the decline of resources that are important to sustain livelihoods [9].

The goal of this chapter is to explore the interconnections between the climatic changes and variability that might be linked to the dynamics of the global atmospheric system and the regional-scale climatic and land cover changes. Particular attention will be given to the discussion of changing trends and future scenarios of climatic and environmental changes in Central Asia and Turkmenistan in particular. This chapter consists of the introduction, three major sections, and conclusions. The first section discusses past and present climate change and variability in Turkmenistan and a broader region of Central Asia. The second section provides analyses of climate change scenarios until the end of the next of the century and their potential implication for water resources and food security. The third section examines land use and land cover changes based on vegetation indices derived from remote sensing data. The final concluding section draws preliminary recommendations for adaptation measures to impacts of climate change.

2 Past and Present Climate and Water Resources

2.1 Climate and Physical Environments

Turkmenistan has a distinctive continental climate with average annual air temperature ranging from 12–17°C in the north to 15–18°C in the southeast and the absolute maximum temperature of 48–50°C in the Central and South-East Karakum Desert [10]. Geographic landscape of Turkmenistan is separated in two unequally divided ecoregions: the larger (80% of the area) ecoregion is represented by desert plains, primarily the Karakum (Black Sands) Desert, with limited precipitation [11], and the smaller (20%) one by mountains and their foothills. Water is of critical importance to Turkmenistan that is characterized by dry and continental climate and limited river network and fresh water resources [11]. Turkmenistan's climate is classified by the Köppen climate classification as *Bwk*, which is used to designate mid-latitude arid areas. Summers are long (from May to September), hot, and dry, with the average July temperature reaching more than 30°C with an absolute maximum of 52°C in the eastern Karakum [12]. Winters are relatively cold and moist in the north with a mean annual temperature of the coldest month, January, ranging from −10°C to 0°C. In the south, along the Kopetdag Mountains winters are warm and moist (with a January temperature typically 0°C or above). Very high daily temperature variability occurs with frequent sand storms and intense sunshine [13]. As in many other arid and semiarid regions, climate of Turkmenistan is highly variable. Throughout Turkmenistan, annual wind speeds are weak to moderate (0–5 m/s). Higher wind speeds (6–9 m/s) can be found along the northern slopes of the Kopetdag and the Caspian shore [13, 14].

The highest amount of precipitation is observed in the mountains – up to 398 mm in Koyne Kesir, the smallest – less than 80 mm above the Kara-Bogaz-Gol Bay. Temporal variability of precipitation is very high and precipitation has a distinctive spring maximum throughout most of the region. Precipitation during the cold season of the year is two to three times higher than during the warm months [14].

The major controls on precipitation change include latitudinal shifts of the westerly cyclonic circulation and position of the Siberian high [12]. The North Atlantic Oscillation (NAO) exerts an important control over the pattern of wintertime atmospheric circulation variability over arid and semiarid zones of Central Asia. Over the past four decades, the pattern captured in the NAO index has altered gradually from the most extreme and persistent negative phase in the 1960s to the most extreme positive phase during the late 1980s and early 1990s. Additionally, the patterns of precipitation as well as drought conditions in Central Asia have been linked to El Niño – Southern Oscillation (ENSO) phases [15]. Cold ENSO phases generally result in drought conditions in the region, while warm ENSO phases result in an increased precipitation [15, 16] and intensified vegetation response in nonagricultural area [17].

The processes in the Caspian Sea area and its level have a major impact on the balance of moisture and heat in the boundary layer atmosphere of western Turkmenistan. The Caspian Sea level correlates with climatic variability over

European Russia, because it depends entirely on climatic conditions and runoff in the basins of the Volga and Ural Rivers. After several decades of a steady raise, the level of the Caspian Sea has been going down since during the past 7 years [18]. On the other hand, the hydrology of Turkmenistan is determined by the runoff of the Amudarya, Tedjen, and Murghab Rivers and the major man-made "river" – the Karakum Canal.

2.2 Hydrological Network

Due to its arid climate more than 80% of Turkmenistan's territory lacks a constant source of surface water flow. The Amudarya is the most water-bearing river in Central Asia. Its endorheic drainage basin includes the territories of Afghanistan, Tajikistan, Uzbekistan, and Turkmenistan and occupies the area of about 465,000 km^2. Its upper part – known as the Pianj River – flows from the Lake Zorkul in Pamir Mountains and is called the Pamir River before it is joined by Vakhandaria in Afghanistan. From east to west, the following big tributaries flow into the Pianj: Gunt, Bartang, Yazgulem, Vanch, Kyzylsu, and Vakhsh. After joining Vakhsh, the Pianj changes its name to Amudarya. It received more tributaries in Tajikistan: Kafirnigan, Khanaka, Karatag, and Surkhandarya. All these rivers start in the Pamir and their runoff is entirely determined by the regime of the mountain glaciers. The source of the Amudarya or Pianj and its several tributaries starts in the Vrevskiy glacier on the north slope of the Hindukush at a height of about 4,900 m above sea level. Already between the confluence of the Pianj and the Vahsh to the town Kerki about 40% of the river runoff is used for irrigation, with considerable amounts being lost for evaporation and infiltration. The riverbed here is divided into multiple channels and forms the numerous islands. Most irrigated lands along the Amudarya valley stretch from Kerki in Eastern Turkmenistan to Nukus in Uzbekistan.

In the past and as recently as in the Middle Ages, the flow of Amudarya changed several times diverting water to the Sarykamysh hollow and further down to the Caspian Sea along the ancient Uzboy River [19]. The flow of water along the Uzboy continued almost to the end of the sixteenth century. The Amudarya changes its riverbed many times, eroding the right bank most of all and constantly displacing to the east. The Amudarya has two flow peaks: one in April–May, caused by the maximum of precipitation in the mountains and snow melting on the low mountain slopes. The second peak flow takes place in summer, in June and July, due to melting of the glaciers in the Pamir and Hindukush Mountains.

The Amudarya is fed by seasonal snowmelt of snowpacks and glaciers and intensified melting of the glaciers and snowpacks occurs under conditions of warming climate [20], causing temporary increase in river runoff, which further contributes to expansion of agricultural land use at the expense of converted natural areas. Unpredictability in runoff dynamics results from a simultaneous interplay of two dynamic processes: climate change impact coupled with transformations in socioeconomic priorities in newly independent countries. This unpredictability introduces a new concern related to changes in land use and land cover.

The nonagricultural ecosystems of the Amudarya composed of *tugai* woodlands, shrubs, and reeds were once continuous and typical ecosystems used to cover the floodplains of the Amudarya. These riparian woodland communities are comprised of the poplars, willows, tamarix, and honeysuckle woods, which alternate with meadows that occupied by reeds, cattails, and licorice [21, 22]. As a result of extensive land reclamation and clearance projects in the Soviet Union, most of the Amudarya *tugai* ecosystems were destroyed to promote irrigated agriculture and only 10% of the native vegetation remains [21]. *Tugai* forests represent a habitat for numerous endangered and valuable flora and fauna species, which are severely threatened through habitat destruction with virtually no virgin forest remaining today [21]. The total area of the Amudarya riparian woodlands now been reduced from about 500 km^2 to around 50 km^2, of which less than 10–15 km^2 is considered to be in relatively healthy condition [21]. Located in regularly flooded river valleys, the *tugai* forest areas represent the most favorable agricultural lands of the region owing to soil quality and more favorable moisture regime [23].

The Amudarya is a transboundary river crossing Turkmenistan, Uzbekistan, Tajikistan, and Afghanistan, but it is chiefly Turkmenistan and Uzbekistan, who share its waters for their agricultural needs. In January of 1996, Turkmenistan and Uzbekistan signed the "Agreement between Turkmenistan and the Republic of Uzbekistan on Cooperation in Water Use" that regulates the amount of water withdrawn from the river to 22 billion cubic meters (BCM) of the annual water flow of 54–68 BCM, for each country [1].

Two other major rivers in Turkmenistan are Murghab and Tedjen. Murghab starts in the Paropamisus Mountains in Afghanistan. After receiving its tributary Abykaisara it enters Turkmenistan. The river valley up to the delta is divided into two sections: ancient part, formed mainly by sand, which was exposed to considerable weathering, and the younger section with clayey deposits. It ends in a dry delta in the Karakum Desert about 100 km below the town Mary.

The Tedjen River starts at a junction of the Paropamisus and the Hindukush mountains on the border of Iran and Afghanistan. After passing the Herat valley, the river turns north and flows about 300 km in Turkmenistan. In summer, Tedjen dries up below the Pulihatum village because of the water withdrawal for irrigation. The Tedjen delta is a sandy–clayey plain occupied by irrigated croplands and reed marshes.

The Karakum Canal is the largest man-made river in Turkmenistan and also one of the largest irrigation and water supply canals in the world. Started in 1954 and completed in 1988, it is navigable over much of its 1,375 km length; it carries 13 cubic km of water annually from the Amudarya River across the Karakum Desert. The canal opened up huge new tracts of land to agriculture, especially to cotton monoculture heavily promoted by the Soviet Union, and supplying Ashgabat, the capital of Turkmenistan, with a major source of water. Prior to construction of the Karakum Canal, the total irrigated area in Turkmenistan was about 166,000 ha. In western Turkmenistan and the Pre-Kopetdag regions, water was insufficient not only for irrigation but also for domestic use [4]. Unfortunately, the canal allows almost 50% of the water to escape en route, creating lakes and ponds along the canal, and a rise in groundwater leading to widespread soil salinization problems [5].

2.3 Climate Change in the Past

Palaeoclimatic and archaeological data indicate that climate of Turkmenistan has experienced many natural fluctuations in the past. Its landforms carry relict features both of relatively short humid intervals with runoff higher than nowadays, and long arid periods [24–26]. Pollen and archaeological data from Central Asia suggest that climate change was followed by significant ecosystem changes [27]. Significant cyclical variations of regional climate and sea level during this period resulted from major changes in river discharge into it caused by climatic variability and several natural diversions of the Amudarya River away from the Aral Sea [19, 24, 28, 29]. Environmental reconstructions based on pollen and archaeological data from several locations in Central Asia suggest that climate of this region was featured by colder winter temperatures, cooler summers, and greater aridity during the Late Glacial Maximum (around 20–18,000 years before present) and again in the Younger Dryas (12,800 and 11,500 years before present) [24, 30, 31]. The so-called Djanak arid phase of the Younger Dryas was followed by an increase in temperatures and precipitation during the Early and Mid-Holocene. A trend toward greater humidity during the Holocene culminated around 6,000 years ago, a phase known in Uzbekistan and Turkmenistan as the Liavliakan pluvial [29, 32]. According to pollen reconstructions, mean annual precipitation in the deserts of Central Asia could be three times higher than at present in the middle of Holocene, when desert landscapes were probably entirely replaced by mesophytic steppes, with well-developed forest vegetation along the river valleys [33]. A general trend of aridization that started approximately 5,000 years ago was interrupted by multiple minor climatic fluctuations in this region at a much finer temporal scale [26].

Historical records available from the weather stations show a steady increase of annual and winter temperatures in this region since the middle of the twentieth century. The precipitation trends, however, are highly variable across the region, both spatially and temporally, reflecting the great natural rainfall variability and landform diversity. Precipitation records available in this region since the end of the nineteenth century show a slight decrease in the western part of the region, little or no changes throughout most of the region, and relatively significant increase in precipitation recorded by the stations surrounded by irrigated lands during the past 50–60 years [34]. This precipitation decrease in the area between the Caspian and Aral Sea mainly occurred since 1960 and coincides with the Aral Sea desiccation. Both the degradation of the Aral Sea and the dramatic fluctuations of the Kara-Bogaz-Gol Bay caused by the construction in 1980 of the Caspian-Kara-Bogaz-Gol Dam (followed by its demolition in 1992) have caused significant changes in albedo, hydrological cycle, and mesoclimatic changes throughout western parts of Kazakhstan, Uzbekistan, and Turkmenistan [27].

Meteorological data reveal an increase of annual and winter temperatures in Turkmenistan since the beginning of the past century. The mean annual temperature has increased by 0.6°C in the northern part of the country and by 0.4°C in the south since 1931. At the same time, the number of days with temperature higher than 40°C has increased since 1983 [35].

Durdyev [35] used records from 24 meteorological stations as a reference base to examine natural climate variability and demonstrate the temperature fluctuations from 1950 to 2004 in Turkmenistan. The study revealed that period from 1950 to 1970 was relatively stable period in temperature fluctuations, with obvious peak in temperature regimes 1971 [35]. These findings were supported by similar results that were found for Uzbekistan where temperature reference data from 40 meteorological stations were used to assess climate variability from 1930 to 1995 [23]. The results of Uzbekistan study had longer period of observation and were able to reveal two eras of variation/circulation patterns: 1931–1960 and 1961–1990 [23].

3 Future Climate Change Projections

Warming of the global climate system is unequivocal, as is now evident from observations of increases in global average air and ocean temperatures, widespread melting of snow and ice and rising global average sea level. Changes in atmospheric concentrations of greenhouse gases (GHGs) and aerosols and land cover alter the energy balance of the climate system. Global GHG emissions due to human activities have grown since preindustrial times, with an increase of 70% between 1970 and 2004 [20].

Atmosphere Ocean Global Climate Models (AOGCMs) representing physical processes in the atmosphere, ocean, cryosphere, and land surface are the most advanced tools currently available for simulating the responses of the global climate system to increasing GHS concentrations. Climate models predict temperature increase in Turkmenistan of 3–4°C by the middle and by more than 5°C by the end of this century. Precipitation projections are highly uncertain but given the exiting aridity and high interannual and interseasonal variability of climate of Turkmenistan, even a slight temperature increase is likely to deepen the existing water stress in the region [10].

The rates of the projected changes significantly differ across seasons, with much higher temperature changes generally expected during the winter months. Despite significant differences in the ranges of change among the scenarios, the majority of the recent AOGCM experiments tend to agree that precipitation is likely to increase both northward (European Russia and Central Siberia) and southward (Northern India, Iran, and Pakistan) from Central Asia [26]. For the Central Asian plains, however, the expectation is for increasingly dry conditions with a slight increase in winter rainfall but decreases particularly in spring and summer. This trend toward higher aridity is projected to be more significant west from 70°E to 72°E. The AR4 supports these findings, pointing out that Central Asia, particularly its western parts, is very likely to become drier during the coming decades [20]. The AOGCM scenarios appear to be consistent with the observed temperature and precipitation trend over the past decades in most of the region. However, it is uncertain the extent to which the observed and projected trends result primarily from the global restructuring of atmospheric circulation and changes in the teleconnections controlling macroclimatic conditions over Central Asia versus mesoclimatic changes induced by regional land use change [34].

We have conducted several assessments of the annual, seasonal, and monthly climate change scenarios for Central Asia produced by the AOGCMs used in the IPCC Third [36] and Fourth Assessment Reports [20]. Detailed discussion of the climate change scenarios for Central Asia can be found in Lioubimtseva [30], Lioubimtseva et al. [26], Lioubimtseva [37], and Lioubimtseva and Henebry [34]. AOGCM scenarios indicate a generally good agreement that the current trend of temperature increase in arid Central Asia is likely to continue. The ranges of precipitation projections are still uncertain. The majority of climate models project a slight decrease in precipitation rate over most of the region with a stronger decrease in the west and southwest and a very slight increase in the north and east of this region [37]. However, given the low absolute amounts of precipitation and high interannual, seasonal, and spatial patterns of precipitation across the region, the changes in precipitation rate projected by climate models should be treated with caution. It is the change in the temporal and spatial variability of precipitation and its seasonal distribution – rather than absolute precipitation values – that are more important for the assessment of human vulnerability of this arid region, but they are also more difficult to assess and project.

We have used the Model for the Assessment of Greenhouse-induced Climate Change (MAGICC) and Global and Regional Climate Change Scenario Generator (SCENGEN), version 5.3.2 [38] to generate regional climate change scenarios for Turkmenistan for the time intervals, centered around 2025 and 2050. MAGICC/SCENGEN software has been developed by the National Center for Atmospheric Research (NCAR). MAGICC model consists of a suite of coupled gas-cycle, climate, and ice-melt models integrated into a single software package. The software allows the user to determine changes in GHG concentrations, global-mean surface air temperature, and sea level resulting from anthropogenic emissions. SCENGEN constructs a range of geographically explicit climate change projections for the globe using the results from MAGICC together with AOGCM climate change information from the CMIP3/AR4 archive. We have constructed climate change scenarios for Turkmenistan centered on 2025 and 2050, assuming policy changes based on the A1B–AIM SRES marker scenario [36, 39]. MMD assessment involved scenarios from the AOGCMs developed by the following organizations: Canadian Centre for Climate Modelling and Analysis (CCCma), Centre National de Recherches Météorologiques, Météo France, National Center for Atmospheric Research, CCSR/NIES/FRCGC, Japan, CSIRO, Australia, and Hadley Centre for Climate Prediction and Research, Met Office United Kingdom (Tables 1 and 2).

A1 family scenarios assume rapid global economic growth that leads to high energy demand and hence to a steep increase in CO_2 emissions in the first decades of the twenty-first century [39]. Structural changes in the energy supply become effective only in the longer term because of the inertia caused by long periods of capital turnover. With respect to alternative energy supply technologies, the A1B scenario group assumes a "balanced" approach, in which none of these technologies

Table 1 Temperature change scenarios for 2025 and 2050

Model	Institution	Temperature change by 2025 (°C)	Temperature change by 2050 (°C)
CCCMA-31	Canadian Centre for Climate Modelling and Analysis (CCCma)	+0.95–1.01	+2.03–2.23
CNRM-CM3	Centre National de Recherches Meteorologiques, Meteo France, France	+0.96–1.02	+2.12–2.22
NCAR-PCM1	National Center for Atmospheric Research (NCAR), NSF, DOE, NASA, and NOAA	+1.02–1.05	+2.20–2.36
MIROCMED	CCSR/NIES/FRCGC, Japan	+0.65–0.74	+1.47–1.57
CSIRO-30	CSIRO, Australia	+0.52–0.64	+1.35–1.65
UKHADGEM	Hadley Centre for Climate Prediction and Research, Met Office United Kingdom	+0.99–1.15	+2.1–2.85
MMD average		+0.74–0.95	+1.83–2.12

Table 2 Precipitation change scenarios for 2025 and 2050

Model	Institution	Precipitation change by 2025 (%)	Precipitation change by 2050 (%)
CCCMA-31	Canadian Centre for Climate Modelling and Analysis (CCCma)	−5.6 to −2.1	−8.6 to −4.8
CNRM-CM3	Centre National de Recherches Meteorologiques, Meteo France, France	−6.8 to −5.5	−4.2 to −2.8
NCAR-PCM1	National Center for Atmospheric Research (NCAR), NSF, DOE, NASA, and NOAA	−1.5 to +0.8	−3.1 to +1.9
MIROCMED	CCSR/NIES/FRCGC, Japan	−1.3 to +1.7	−2.9 to +3.8
CSIRO-30	CSIRO, Australia	+1.5 to +7.5	−1.8 to +3.3
UKHADGEM	Hadley Centre for Climate Prediction and Research, Met Office United Kingdom	−4.8 to −1.9	−9.4 to −4.8
MMD average		−1.4 to +0.4	−0.3 to +1.7

gain an overwhelming advantage. This scenario group includes the A1B marker scenario developed using the AIM model [36].

In the A1B–AIM marker scenario, the global average per capita final energy demand grows from 54 GJ in 1990 to 115 GJ in 2050 and to 247 GJ in 2100 [36]. At the same time, the final energy carbon intensity declines relatively slowly until 2050 (from the current 21 tC per TJ of final energy to 16 tC per TJ), which results in a steep increase in CO_2 emissions in the first decades of the twenty-first century. After 2050, when structural changes in the energy sector take effect, carbon intensity declines rapidly to reach 7.5 tC per TJ. Emissions peak around 2050 at a level 2.7 times (16 GtC) that of 1990 and fall to around 13 GtC by 2100, which is about twice the current level. The total, cumulative 1990–2100 carbon emissions in the A1B–AIM scenario equal 1,499 GtC [39].

All the five models used in our studies predict temperature increase close to 1°C or slightly less (MMD average varies geographically from 0.74°C to 0.95°C) by 2025 and about 2°C by 2050 (MMD average varies from 1.83°C to 2.12°C). Canadian, French, NCAR, and UK model have very similar temperature change projections temperature over +2°C by 2050, whereas Japanese and Australian models predict a temperature increase about 1.5°C.

Like in many other parts of the world, precipitation change projections for Turkmenistan are less certain and scenarios by different models differ significantly, with MMD averages ranging geographically between −1.4% and 0.4% for 2025 and between −0.3 and 1.7% by 2050. There are significant differences between the models, as well as different parts of Turkmenistan. Both Canadian and UK models predict precipitation decline order of 8–9% by 2050, while Japanese and Australian models predict increase of 3.3–3.8% in some parts of the country.

There are several sources of uncertainties associated with climate change scenarios predicted by the computer models. The spatial resolution of AOGCMs is quite coarse (a horizontal resolution of between 250 and 600 km, 10–20 vertical layers in the atmosphere and sometimes as many as 30 layers in the oceans). Many physical processes, such as those related to clouds, occur at more detailed scales and cannot be adequately modeled; instead, their known properties are averaged over the larger scale in a technique known as parameterization [36]. Other uncertainties relate to the simulation of various feedback mechanisms in models concerning, for example, water vapor and warming, clouds and radiation, ocean circulation and ice and snow albedo, and land cover dynamics [20, 40, 41].

4 Land Use and Land Cover Changes

During the last few decades, Turkmenistan has experienced widespread changes in land cover and land use following the socioeconomic and institutional transformations of the region catalyzed by the USSR collapse in 1991. The decade-long drought events and steadily increasing temperature regimes in the region came on top of these institutional transformations, affecting the long-term and landscape scale vegetation responses. A few studies have been conducted to examine land cover changes in Turkmenistan and other parts of the Central Asian region using coarse-resolution (8 km at 10–15 days) NDVI (Normalized Difference Vegetation Index) data. By analyzing the Pathfinder AVHRR Land (PAL) data from 1981 to 1999, de Beurs and Henebry [42, 43] found three distinct patterns of significant difference in land surface phenology (LSP) models that linked the NDVI with accumulated growing degree days to describe the seasonal course of vegetation activity. Using similar modeling techniques, Henebry et al. [44] focused on irrigated areas in Turkmenistan and Uzbekistan in two time periods spanning the disintegration of the Soviet Union: 1985–1988 and 1995–1999. They found no significant change in land surface phenology in the irrigated areas of Karakalpakstan between periods. However, there were significant changes in the LSP patterns in the

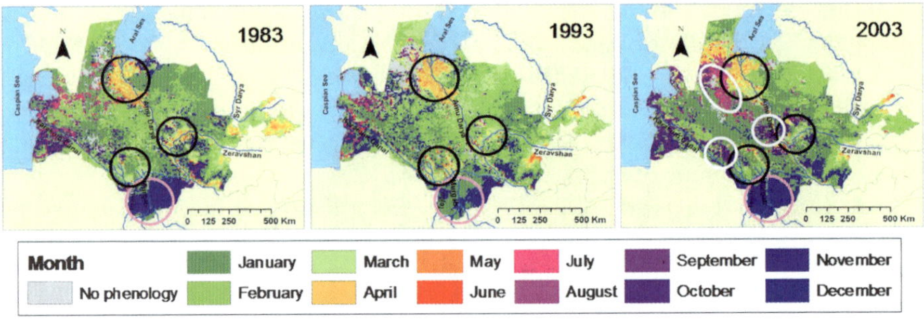

Fig. 1 Examples of imagery representing the dates for the start of the growing season derived from the NDVI time series data for 1983, 1993, and 2003 for Turkmenistan and Uzbekistan. *White circles* highlight the newly irrigated and expanded areas in Turkmenistan, *black circles* identify traditionally cotton crop areas, and *pink circles* – climate-driven fluctuation in the start of season in nonagricultural Badghyz–Karabil semidesert zone in Turkmenistan [17]

irrigated areas along the Zarafshan River in southern Uzbekistan and along the Karakum Canal in Turkmenistan.

Kariyeva [17] also found significant LSP changes following institutional changes using a different modeling approach, namely, phenological metrics extracted from the GIMMS NDVI data set [45] using the Timesat algorithm [46]. She found differences in land cover trends in Turkmenistan before and after the USSR collapse with shifts in crop preferences and expansion of agricultural areas (Fig. 1).

In the years since independence, the agricultural sector of Turkmenistan has seen significant shifts in the overall agricultural production as well as the composition of that production (Fig. 2). Estimated agriculture production rose by 189% between 1992 and 2010 [47]. The Turkmenistan economy shifted away from reliance on cotton exports and food imports toward more balanced food security portfolio. Cotton production decreased by 17% and its importance in the agricultural sector fell from 43% of production in 1992 to just 12% in 2010 (Fig. 2). Cereals production rose by almost 800%, moving from 12% of the agricultural production in 1992 to roughly 40% during the first decade of the twenty-first century; animal production rose by over 300%, rising from 21% in 1992 to 31% by 2010; and production of fruits and vegetables rose by a comparatively modest 109% but dropped its share from 24% in 1992 to 18% by 2010 (Fig. 2).

The drought of the last decade resulted in relatively low NDVI values compared to the 1990–1995 years. Increased precipitation in Central Asia during this interval is linked to warm ENSO (El Niño/La Niña-Southern Oscillation) phases [16] and the 1990–1995 ENSO event was the longest warm ENSO phase since 1882 [48]. The Southern Oscillation refers to variations in the temperature of the surface of the tropical eastern Pacific Ocean (warming and cooling known as El Niño and La Niña, respectively) and in air surface pressure in the tropical western Pacific. The two variations are coupled: the warm oceanic phase, El Niño, accompanies high air surface pressure in the western Pacific, while the cold phase, La Niña, accompanies

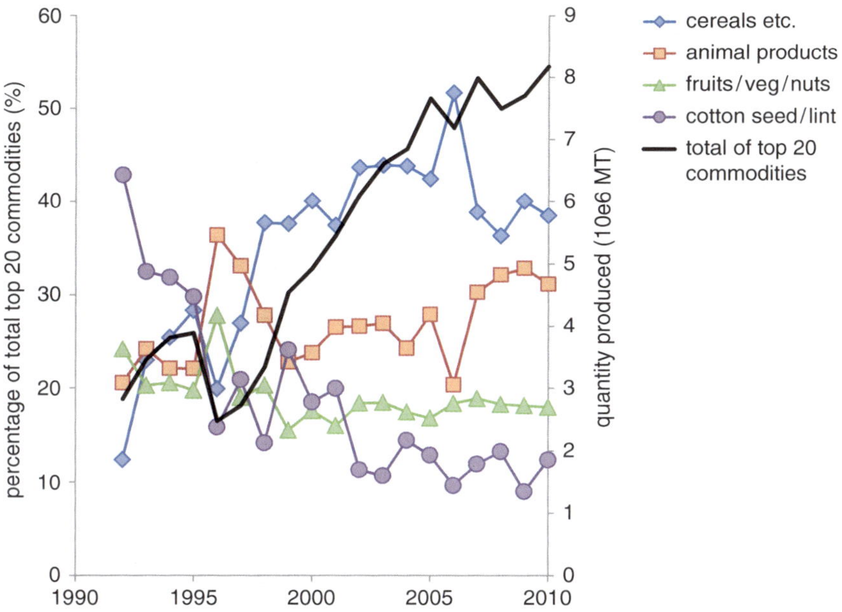

Fig. 2 FAO agricultural production data for Turkmenistan 1992–2010 [47]

low air surface pressure in the western Pacific. There is correspondence between ENSO phases and annual averaged NDVI in Central Asia [17], which tends to increase and decrease with warm and cold ENSO phases, respectively, in nonagricultural areas of Turkmenistan (Fig. 3).

Two-decade long institutional transformations have resulted in substantial changes in now independent and market-based economic priorities of the Central Asian countries: shifts in crop preferences, favoring food crop (wheat) over cash crop (cotton) production [17], increases in the agricultural labor force, particularly in Turkmenistan and Uzbekistan [49], expansion of agricultural areas, and withdrawal of more water for agriculture from the Amudarya and Syrdarya rivers, tributaries of the drying Aral Sea. The rivers are fed by seasonal snowmelt of snowpacks and glaciers of the Pamir and Tien Shan Mountains of Tajikistan and Kyrgyzstan. Intensified melting of the glaciers and snowpacks occurs under conditions of warming climate [20], causing temporary increase in river runoff, which further contributes to expansion of agricultural land use at the expense of converted natural areas. Unpredictability in runoff dynamics results from a simultaneous interplay of two dynamic processes: climate change impact coupled with transformations in socioeconomic priorities in newly independent countries. This unpredictability introduces a new concern related to changes in land use and land cover. Given a growing water demand by an increasing population in five Central Asian countries that are no longer a single institutional and administrative entity, the aforesaid concern relates directly to sustainable development of the region: the long term projected exhaustion of the water source is expected to impact the

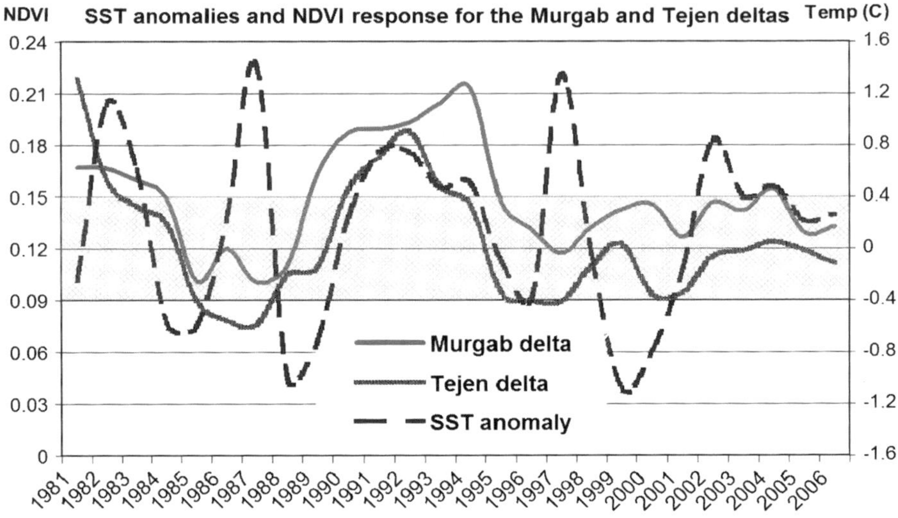

Fig. 3 NDVI time-series based phenological trajectories for the Murghab River delta zone and the Tedjen River delta zone and time series of the Sea Surface Temperatures (SST) anomalies (1981–2006) from *Niño 3.4 Region* dataset spanning 5°N to 5°S and 170° to 120°W used to define ENSO phases [17]. *Note*: ENSO phase thresholds at ±0.4°C [55]

agricultural potential of newly developed agricultural sites and coping capacities of regional ecosystems to natural and human-driven perturbations.

Kariyeva [17] explored the relationship between change patterns in land surface phenology and long-term climate variation in Central Asia. Statistical analysis included assessment of the standardized measures of linear associations between climatic and phenological variables. The phenological metrics of irrigated agricultural zones showed the lowest variability in response to precipitation regimes over time. Temperature regimes were the most noticeable climate drivers for the growing season dates in irrigated areas and explained the greatest amount of spatial variation in vegetation dynamics in. Start and peak of growing seasons were shown to have earlier timing onsets with warmer spring and summer temperatures. The duration of the growing season was shown to be longer with the increasing temperatures during spring, summer, and fall seasons. The productivity (greenness) metric was shown to have increased values with increasing rainfall and summer temperature regimes. Earlier season start and longer season length with increasing spring temperatures are occurring in almost all riparian and irrigated cropland areas of the Amudarya delta. However, the end dates of growing season are not changing for irrigated croplands of the Amudarya, which means that the end of growing season is not shifting accordingly to season start date. These observations can either be attributed to changed crop preferences that have longer growing season periods during the last three decades of agricultural production and/or decreased amount of meltwater reaching the delta. Nonagricultural areas in Turkmenistan had later season start dates with increasing temperatures with no changes in season length dates. This pattern can be explained by the fact that these dry and precipitation-driven desert ecosystems are affected by decreased trends in winter and spring precipitation. Combination of increasing temperatures and decreasing precipitation in the drylands of Turkmenistan can increase rates of potential evapotranspiration, leading to further water stress conditions for vegetation productivity. Shifting climate regimes might have direct implications for agricultural and natural land cover types in the region. Natural land cover types might experience reduction of overall vegetation productivity and irregularity in the growing season, which altogether could lead to phenological asynchrony across trophic levels caused by temporal mismatches between supply, availability, and demand.

5 Future Outlook

Arid zone of Central Asia represent an area with diverse and overlapping environmental, social, and economic stresses. The well-being and security of Turkmenistan depends on interplay of several groups of internal and external factors, such as climate variability and change, institutional changes and the subsequent regional land use changes, and internationalization of economy or globalization.

Turkmenistan is projected to become warmer and probably drier during the coming decades. Aridity is expected to increase in all republics of Central Asia, but especially in the western part of Turkmenistan. The temperature increases are predicted to be particularly high in summer and fall, but lower in winter. Especially significant decrease in precipitation is predicted in summer and fall, while a modest increase or no change in precipitation is expected in winter months. These seasonal climatic shifts are likely to have profound implications for agriculture, particularly in western Turkmenistan and Uzbekistan, where frequent droughts are likely to negatively affect cotton, cereals, and forage production, increase already extremely high water demands for irrigation, exacerbate the already existing water crisis, and accelerate human-induced desertification. The ongoing series of severe droughts of the past decade and continuous degradation of the Aral Sea and its tributaries, the Amudarya and Syrdarya, has already resulted in multiple water disputes and increased tensions among the states of the Aral Sea basin. The arid interfluve lowlands of both, the Amudarya and Syrdarya are already experiencing the effects of climate change with increased drought frequency and glacier recession [20]. Increased melting of the glaciers and snowpacks that is occurring under conditions of warming climate [20, 23, 50] will most likely cause temporary increases in water runoff over the next couple of decades and promote further expansion of unsustainable in the long-term run agricultural land use. Knowing that the aridity and water stress are likely to increase, new political and economic mechanisms are necessary to ease such tensions in future.

The ability of this western subregion of Central Asia to adapt to hotter and drier climate is limited by the already existing water stress and the regional land degradation and poor irrigation practices. Central Asia inherited many environmental problems from the Soviet times but many years after independence, the key land and water use-related problems remain the same. Deintensification of agriculture immediately after independence, documented by agricultural statistics, was significant enough to produce a signal in the temporal series of remote sensing data. Agricultural transformation had extremely high social cost, but agricultural reforms and transition to market have remained problematic across most of the region. Increasing rural poverty and unemployment, particularly among females, growing economic inequality, and shortage of adequate living conditions, medical care, and water management infrastructure have significantly increased human vulnerability of the majority of population in the region.

To cope with the multiple regional stresses in the context of multiple increasing stresses, both related and nonrelated to climate change, it is important to consider adaptation strategies that could place equal importance on environmental, social, and economic considerations. Development of such adaptation strategy involves inevitable trade-offs between environmental, economic, and sociocultural and political considerations and priorities. There is compelling evidence from around the world that development and implementation of adaptation strategies and policies are successful only when they are driven by the interests of stakeholders, groups of individuals, and communities vulnerable to the risks of climate change [51–53]. At the national and regional scale, adaptations are usually undertaken by the governments on behalf of the entire society or particular groups but, regardless the geographic scale, these decisions, policies, and projects must be driven by "place-based" initiatives and integrate the needs of various communities at multiple scales. Communities rarely face only one effect or risk of climate change at a time and the interaction of multiple vulnerabilities often can lead to amplification of risks [53, 54]. Climate change impacts are interconnected with land use changes, socio-economic changes, and many other processes that interact in the human–environmental system. Therefore, adaptations can be sustainable only if they target multiple processes and risks in the integrated manner, reaching across various aspects of human life (food security, water resources, health, quality of life, etc.) at multiple geographic and temporal scales. For example, further reduction of cotton monoculture, diversification of crops including drought-resistant varieties, and application of no-tillage techniques in agriculture could not only help to increase food security but also would decrease the use of water, improve soils through the nitrogen fixation in soil. Moreover, these practices could be useful as climate change mitigation measures to increase carbon sequestration. The renovation of the existing irrigation network and introduction of more advanced irrigation techniques, such as drip irrigation not only could significantly reduce the loss of water resources but also would improve crop productivity, reduce the soil losses due to salinization, and help to reduce the risks of water contamination and transmission of many vector-borne and water-borne diseases.

Turkmenistan signed and ratified the UN Framework Convention on Climate Change in 1995. In January 1999, Turkmenistan ratified Kyoto Protocol and published the document titled "Turkmenistan: Initial National Communication on Climate Change." Being covered under the Kyoto Protocol's Clean Development

Mechanism, Turkmenistan can trade carbon credits with the countries that fall under the Joint Implementation Mechanism [10]. The major sources of carbon emissions in Turkmenistan include the oil and gas extraction, petroleum refining, chemical industry, as well as motor transport concentrated mainly in Ashgabat, Turkmenbashi, Balkanabat, Mary, Turkmenabat, and Dashoguz. Turkmenistan has taken some steps to reduce carbon emissions, such as massive tree plantation drive throughout the country (Green Belt Project), modernization of Turkmenbashi and Seyidi refineries to conform to modern ecological standards, and removal of the cement factories away from inhabited areas. However, the widespread poverty, recent decline in the educational system, and economic dependence on cotton and hydrocarbon exports still leave Turkmenistan very vulnerable to high climatic variability, desertification, and droughts.

References

1. Hanmamedov MA, Rejepov OR (2007) On rational water use in Turkmenistan. Problems Deserts Dev 1:26–27 (in Russian)
2. Qi J, Evered KT (eds) (2008) Environmental problems of Central Asia and their economic, social and security impacts. Springer, Dordrecht
3. The World Bank Data (2012) http://www.worldbank.org. Accessed Apr 2012
4. Glantz M (1999) Creeping environmental problems and sustainable development in the Aral Sea Basin. Cambridge University Press, Cambridge
5. Saiko TA, Zonn IS (2000) Irrigation expansion and dynamics of desertification in the Circum-Aral region of Central Asia. Appl Geogr 20(4):349–367
6. Rahmatulina GG (2008) Current state of intergovernmental relationships with regards to water resources in Central Asia. In: Chufrin GI (ed) Current state and perspective of use of water resources in Central Asia. Asia Strategy Foundation for Strategic Studies of the Central Asian Region, Almaty-Bishkek-Tashkent
7. Adger WN, Agrawala S, Mirza MMQ et al (2007) Assessment of adaptation practices, options, constraints and capacity. In: Parry ML, Canziani OF, Palutikof JP et al (eds) Climate change 2007: impacts, adaptation and vulnerability. Contribution of working group II to the fourth assessment report of the IPCC. Cambridge University Press, Cambridge
8. Parry ML, Canziani OF, Palutikof JP et al (2007) Climate change 2007: impacts, adaptations and vulnerability. Contribution of working group II to the fourth assessment report of the IPCC. Cambridge University Press, Cambridge
9. Barnet J, Adger WN (2007) Climate change, human society and violent conflict. Polit Geogr 26(6):639–655
10. Lioubimtseva E (2008) Climate change in Turkmenistan. In: Philander SG (ed) Encyclopedia of Global Warming and Climate Change. Ed. Thousand Oaks, CA: SAGE, 2008. 988-89. SAGE Reference Online. Web. 11 Jul. 2012 http://www.sage-reference.com/view/globalwarming/n644.xml?rskey=vSQsa8&result=1&q=Lioubimtseva
11. Small EE, Sloan LC, Hostetler S, Giorgi F (1999) Simulating the water balance of the Aral Sea with a coupled regional climate-lake model. J Geophys Res 104(D6):6583–6602
12. Lioubimtseva E (2003) Arid environments. In: Shahgedanova M (ed) Physical geography of Northern Eurasia. Oxford University Press, Oxford
13. Orlovsky N (1994) Climate of Turkmenistan. In: Fet V, Atamuradov KI (eds) Biogeography and ecology of Turkmenistan. Kluwer Academic, Dordrecht

14. Maman S, Blumberg DG, Tsoar H et al (2011) The Central Asian ergs: a study by remote sensing and geographic information systems. Aeolian Res 3:353–366
15. Barlow M, Cullen H, Lyon B (2002) Drought in central and Southwest Asia: La Niña, the Warm Pool, and Indian Ocean precipitation. J Climate 15:697–700
16. Syed FS, Giorgi F, Pal JS, King MP (2006) Effect of remote forcings on the winter precipitation of Central Southwest Asia, part 1: observations. Theor Appl Climatol 86(1–4):147–160
17. Kariyeva J (2010) Land surface phenological responses to land use and climate variation in a changing Central Asia. Dissertation, The University of Arizona
18. Kostianoy AG, Lebedev SA, Solovyov DM (2012) Satellite monitoring of the Caspian Sea, Kara-Bogaz-Gol Bay, Sarykamysh and Altyn Asyr Lakes, and Amu Darya River. In: Zonn IS, Kostianoy AG (eds) The Turkmen Lake Altyn Asyr and water resources in Turkmenistan. Springer, Heidelberg
19. Kes AS, Mamedov ED, Khondkaryan SO et al (1993) Plains of Northern Central Asia during the late Pleistocene and Holocene: stratigraphy and palaeogeography. In: Velichko AA (ed) Evolution of landscapes and climates of the Northern Eurasia. Nauka, Moscow
20. IPCC (2007) In: Solomon S et al (eds) Climate change 2007: the physical science basis. Contribution of working group I to the fourth assessment report of the Intergovernmental Panel on Climate Change. Cambridge University Press, Cambridge
21. Schlüter M, Rüger N, Savitsky A et al (2006) Tugai: an integrated simulation tool for ecological assessment of alternative water management strategies in a degraded river delta. Environ Manage 38(4):638–653
22. Thevs N, Zerbe S, Schnittler M et al (2008) Structure, reproduction and flood-induced dynamics of riparian tugai forests at the Tarim River in Xinjiang, NW China. Forestry 81(1):45–57
23. Alamanov SK, Lelevkin VM, Podrezov OA, Podrezov AO (2006) Climate change and water problems in Central Asia. United Nations Environment Program (UNEP) and World Wildlife Fund (WWF), Moscow-Bishkek (in Russian)
24. Boomer I, Aladin N, Plotnikov I, Whatley R (2000) The palaeolimnology of the Aral Sea: a review. Quat Sci Rev 19(13):1259–1278
25. Boroffka N, Oberhänsli H, Sorrel Ph et al (2006) Archaeology and climate: settlement and lake-level changes at the Aral Sea. Geoarchaeology 21(7):721–734
26. Lioubimtseva E, Cole R, Adams JM, Kapustin G (2005) Impacts of climate and land-cover changes in arid lands of Central Asia. J Arid Environ 62(2):285–308
27. Lioubimtseva E, Cole R (2006) Uncertainties of climate change in arid environments of Central Asia. Rev Fish Sci 14(1–2):29–49
28. Micklin P (1988) Desiccation of the Aral Sea: a water management disaster in the Soviet Union. Science 241:1170–1176
29. Vinogradov AV, Mamedov ED (1991) The Aral Sea in the Holocene. In: Andrianov BV et al (eds) Aral crisis. USSR Academy of Sciences, Moscow
30. Lioubimtseva E (2004) Climate change in arid environments: revisiting the past to understand the future. Prog Phys Geogr 28(4):502–530
31. Tarasov P, Williams JW, Andreev A (2007) Satellite- and pollen-based quantitative woody cover reconstructions for northern Asia: verification and application to late-Quaternary pollen data. Earth Planet Sci Lett 264(1–2):284–298
32. Lioubimtseva E, Simon B, Faure H et al (1998) Impacts of climatic change on carbon storage in the Sahara-Gobi desert belt since the late glacial maximum. Global Planet Change 16–17:95–105
33. Sorrel P, Popescu S-M, Klotz S et al (2007) Climate variability in the Aral Sea basin (Central Asia) during the late Holocene based on vegetation changes. Quat Res 67:357–370
34. Lioubimtseva E, Henebry GM (2009) Climate and environmental change in arid Central Asia: impacts, vulnerability, and adaptations. J Arid Environ 73:963–977
35. Durdyev AM (2006) How to reduce climate change effects in Turkmenistan. Problems Deserts Dev 3:3–6 (in Russian)

36. IPCC (2001) In: Houghton JT et al (eds) Climate change: the scientific basis contribution of working group I to the third assessment report of the Intergovernmental Panel on Climate Change (IPCC). Cambridge University Press, Cambridge
37. Lioubimtseva E (2007) Possible changes in the carbon budget of arid and semi-arid Central Asia inferred from land-use/landcover analyses during 1981–2001. In: Lal R, Suleimenov M, Stewart BA, Hansen DO, Doraiswami P (eds) Climate change and terrestrial carbon sequestration in Central Asia. Taylor & Francis, London
38. Wigley TML (2008) MAGICC/SCENGEN 5.3: user manual (version 2). NCAR, Boulder
39. Nakicenovic N, Alcamo J, Davis G et al (2000) IPCC special report on emissions scenarios. IPCC special reports. Cambridge University Press, Cambridge
40. Arnell NW (2004) Climate change and global water resources: SRES emissions and socio-economic scenarios. Glob Environ Chang 14:31–52
41. Pitman AJ et al (2009) Uncertainties in climate responses to past land cover change: first results from the LUCID intercomparison study. Geophys Res Lett 36:L14814. doi:10.1029/2009GL039076
42. de Beurs KM, Henebry GM (2004) Land surface phenology, climatic variation, and institutional change: analyzing agricultural land cover change in Kazakhstan. Remote Sens Environ 89(4):497–509
43. de Beurs KM, Henebry GM (2005) A statistical framework for the analysis of long image time series. Int J Remote Sens 26(8):1551–1573
44. Henebry GM, de Beurs KM, Gitelson AA (2005) Land surface phenologies of Uzbekistan and Turkmenistan between 1982 and 1999. Arid Ecosyst 11(26–27):25–32
45. Tucker CJ, Pinzon JE, Brown ME et al (2005) An extended AVHRR 8-km NDVI dataset compatible with MODIS and SPOT vegetation NDVI data. Int J Remote Sens 26 (20):4485–4498
46. Jönsson P, Eklundh L (2004) TIMESAT – a program for analyzing time-series of satellite sensor data. Comput Geosci 30(8):833–845
47. FAOSTAT (2012) Food and Agriculture Organization Statistics. http://www.fao.faostat.org. Accessed Apr 2012
48. Trenberth KE, Hoar TJ (1996) The 1990–1995 El Niño-southern oscillation event: longest on record. Geophys Res Lett 23(1):55–60
49. Stanchin I, Lerman Z (2007) Water in Turkmenistan. Hebrew University of Jerusalem, Department of Agricultural Economics and Management, Jerusalem
50. Podrezov OA, Dikih NA, Bakirov KB (2001) Climate change and glaciers of the Tien-Shan for the last 100 years. Kirghiz-Russian Slavic Univ Bulletin 1(3):33–40 (in Russian)
51. Adger WN (2006) Vulnerability. Glob Environ Chang 16(3):268–281
52. Polsky C, Neff R, Yarnal B (2007) Building comparable global change assessments: the vulnerability scoping diagram. Glob Environ Chang 17(3–4):472–485
53. Schröter D, Polsky C, Patt AG (2005) Assessing vulnerability to the effects of global climate change: an eight step approach. Mitig Adapt Strat Glob Chang 10:573–596
54. Leichenko RM, O'Brien KL (2002) The dynamics of rural vulnerability to global change: the case of southern Africa. Mitigation Adaptation Strategies Global Clim Change 7:1–18
55. Trenberth KE (1997) The definition of El Niño. Bull Am Meteorol Soc 78(12):2771–2777

Water Resources of Turkmenistan

Igor S. Zonn

Abstract In arid conditions of Turkmenistan featuring high air temperatures and very low precipitations, water becomes the synonym of life. Economic development of the rich natural resources of the country is wholly based on water resources use. Water factor plays the most important role in the country's economy at all stages of its development. The river network is developed poorly. The water resources of the country are represented by large transborder rivers (Amu Darya, Tedjen, and Murghab), smaller rivers, springs, and karizes as well as groundwaters. The potential of natural resource use is limited by the renewable water reserves in different sources, first of all, in rivers. The main threat for the water resources of the country is a sustainable tendency to depletion and pollution of both surface and groundwaters caused by a complex of natural and anthropogenic factors, including decreasing precipitation. Steady shrinking of mountain glaciers being a source of replenishing river flow which may be intensified due to global climate warming will affect surface water runoff. The water resource issue is most acute for the north and northeast of Turkmenistan (Dashoguz Velajat and Darganatin Etrap of Lebapsky Velajat – the zone of the Aral environmental crisis).

Keywords Drainage waters, Groundwaters, Surface waters, Water resources

Contents

I.S. Zonn
Engineering Research Production Center for Water Management, Land Reclamation and Ecology "Soyuzvodproject", Baumanskaya Str. 43/1, Moscow 105005, Russia
e-mail: igorzonn@yandex.ru

I.S. Zonn and A.G. Kostianoy (eds.), *The Turkmen Lake Altyn Asyr and Water Resources in Turkmenistan*, Hdb Env Chem (2014) 28: 59–68, DOI 10.1007/698_2012_205,

1 Introduction

Turkmenistan territory extends over the southwestern part of Central Asia, near the northern border of the subtropical zone. It is characterized by the continental climate, small atmospheric precipitations, and specific hydrological cycle affected significantly by the relief.

The country has no water resources formed within its territory. As they are replenished with natural precipitations the water resources are represented by only temporary surface runoff (TSR) and subsand water lenses. The greater volume of surface water resources are formed in the mountains beyond the borders of the country and concentrate in transborder rivers that are used jointly by the Central Asian states. The surface runoff within Turkmenistan is insufficient for river formation. The rivers due to their low-water flow and water intake for irrigation get lost in the flat terrain forming closed drainless basins. This explains special acuteness of the problem of water resources being the key factor for development of agriculture, industry, and water supply.

According to estimates of scientists, the glaciation area in the Pamir–Altai Mountains where the Amu Darya River takes its origin has already shrank by 40% on the average which resulted in decreased water availability in the rivers of the country.

2 River Flow Resources

Water resources of Turkmenistan are rather modest and are represented by surface waters – large transborder rivers Amu Darya, Murghab, Tedjen, and Atrek which flow for 95% are formed beyond the country borders and small rivers flowing down from the northern slopes of the Kopetdag, springs and karizes, numerous dry ravines (over 350) of the Greater and Lesser Balkans, and Koytendag – and also by fresh groundwaters (Fig. 1). In the direction from south to north the natural river network is being gradually replaced with artificial water streams constructed for development of irrigated farming (canals, collectors, drains).

The *Amu Darya River* is one of the largest in Central Asia, belonging to the inland Aral Sea basin and playing the key role in Turkmenistan. Its flow is used by five states – Afghanistan, Kyrgyzstan, Tajikistan, Uzbekistan, and Turkmenistan. The waters of this river are divided completely among these countries. The main principles of water division with regard to the balance of interests are set forth in several interstate treaties of the Central Asian states on cooperation in joint management, utilization, and protection of international water resources (1992) to which Afghanistan has not so far acceded and the treaty between Turkmenistan and Republic of Uzbekistan (1996). Under these treaties, Turkmenistan is appropriated a quota of 22 billion m^3 of water. After confluence of Vakhsh and Pyanj, the river

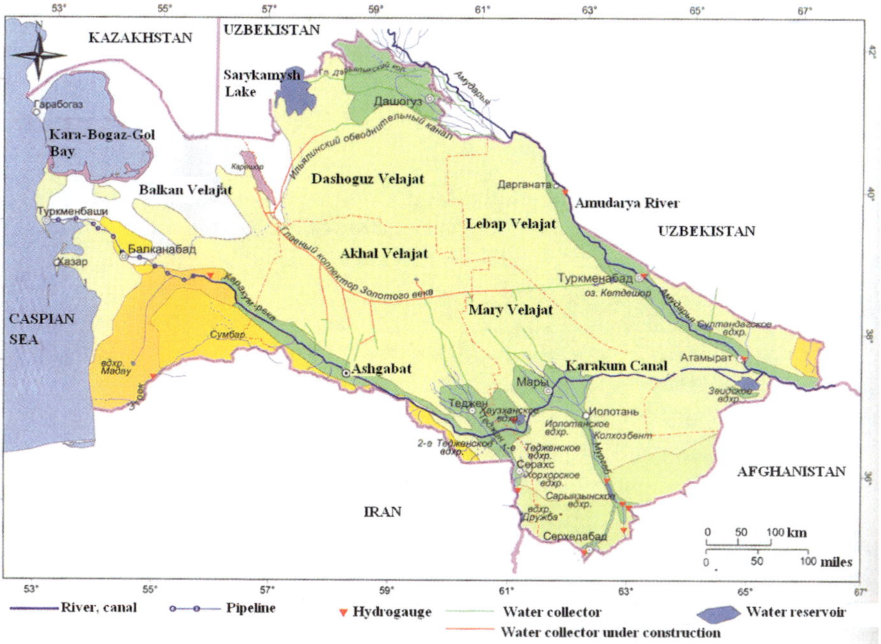

Fig. 1 Hydraulic network in Turkmenistan

gets the name of Amu Darya. Its length is 1,415 km and from the origin of the Vokhjura River – 2,620 km (744 km in Turkmenistan). The Pyanj River accounts for 60% and the Vakhsh River – for about 40% of the total flow of Amu Darya. The Amu Darya basin is clearly divided into the mountain area where the flow is formed and the flat area where the flow is disseminated. In the mountain area, Amu Darya takes numerous tributaries. Only in the first 180 km, these are the Kunduz (Afghanistan) on the 12th km from the Vakhsh and Pyanj confluence, the Kafirnigan on the 38th km on the right side, the Surkhandarya on the 137th km, and the Sherabaddarya on the 180th km. The watershed area is 230,000 km^2. In its plain part for 1,200 km, the river does not have any tributaries and its flow gets lost to filtration, evaporation, and water intake for economic needs, including irrigation. By different estimates, the Amu Darya accounts for 90–95% of the whole surface water resources of Turkmenistan.

The Amu Darya water regime depends on thawing of high-mountain snow and glaciers and on rainfalls. It is characterized by high flow in summer and low flow in winter. The total mean annual flow in the Amu Darya basin varies from 68.1 to 79.5 km^3, out of which about 17–19 km^3 a year or 24% is formed on the territory of Afghanistan. In a water-abundant year with 5% probability, the Amu Darya flow may be as large as 108.4 km^3, while in a low-water year with 95% probability, it may drop to 46.9 km^3. Turkmenistan appropriates 22 km^3 a year of the river flow. The maximum river flow is observed in the first 3 months of the vegetation period:

May, June, and July – 41 km^3. The Amu Darya flow is regulated by in-channel reservoirs – Nurek (Tajikistan) and Tyuyamuyun (Uzbekistan).

The main specific feature of the Amu Darya River basin is that the territories with favorable natural and economic conditions for irrigated farming development locate rather far from the main channel of the river and their water resources are very limited.

In the Soviet times much attention was focused on irrigation development in the Central Asian republics, mostly, to increase cotton production that was widely used to meet, primarily, the needs of the defense industry. That time, the large Karakum Canal–River that withdrew annually 10–12 km^3 of water from the Amu Darya (see Chapter "Karakum Canal – Artificial River in a Desert") and several in-system reservoirs that were vital for creation of the seasonal water reserves (unused autumn–spring flows) were built.

The second largest river in Turkmenistan is *Murghab*. The basin of this river covers the territories of two states: its upper reaches are in Afghanistan and the lower reaches in Turkmenistan. Murghab originates in the Paropamisus Mountains in Afghanistan at an altitude of 2,600 m over the sea level. It ends in the Mary Velajat in the Karakum sands to the northwest of the Mary town. The total length of the river is 852 km (in other sources – 785 km of which about 350 km are in Turkmenistan; a stretch about 30 km long makes the border between Turkmenistan and Afghanistan). The flow of the river is 1.69 km^3. The area of the Murghab basin is 47,000 km^2. The largest tributaries of the Murghab are Kashan and Kushka rivers. The rivers are fed mostly by snow melt and rainfalls. The Murghab waters are characterized by great turbidity – some 6–16 kg/m^3. In winter the river does not freeze. Its waters are used completely for irrigation. On the territory of Turkmenistan the river flow is regulated by a cascade of five in-stream and off-stream seasonal reservoirs (Table 1). It is capable to provide water for about 110,000 ha. Afghanistan and Turkmenistan have no so far an agreement on the river flow utilization.

One more major river in Turkmenistan is the *Tedjen River* (*Herirud*) which basin is shared by three states – Afghanistan, Iran, and Turkmenistan. The upper reaches of the river are in Afghanistan and Iran. The length of the river is 1,124 km; the area of its basin is 70,600 km^2. It originates in Afghanistan, in the Paropamiz Mountains at an altitude of 3,000 m. As far as the Herat oasis in Afghanistan it flows as a mountain river cutting through a narrow valley and forming rapids and waterfalls. In its lower reaches it flows over a vast valley where its waters are taken for irrigation. The exit from the Herat Valley is the border between Turkmenistan and Iran where the valley gets broader. Within Turkmenistan the river flows over a vast valley with steep banks breaking into numerous arms. In the lower reaches the waters of the Tedjen are withdrawn for irrigation of lands in the Tedjen oasis. The river ends as a network of irrigation canals in the Karakum Desert. It is fed mostly from snow melt in mountains and, to a less extent, by rainfalls and groundwaters. The flood peak is in April. In spring the water flow is 450 m^3/s, while in summer – no more than 10–15 m^3/s. The average turbidity is 4 kg/m^3. The Tedjen is a nonfreezing river. Its major tributaries are left – Tagaonshlan (Afghanistan) and Keshedrud (Iran). The river flow is 0.98 km^3. From 1950 the seasonal flow of the river within Turkmenistan

Table 1 Hydraulic characteristics of reservoirs on the Murghab River [1]

Reservoir	Year of construction	Capacity, million m^3	Features of regulation
Saryyazy	1984	660	In-stream
Kolkhozbent (Soltanbent)	1941	54.6	In-stream
Yoloten (Elotan)	1910	70.0	In-stream
Middle Hindikush	1895	15.0	Off-stream
Lower Hindikush	1895	16.0	Off-stream

Table 2 Hydraulic characteristics of reservoirs on the Tedjen River [1]

Reservoir	Year of construction	Capacity, million m^3	Features of regulation
Tedjen I	1951; 1978	190.0	In-stream
Gorgor	1960	20.0	Off-stream
Tedjen II	1960	141.0	In-stream (silted)

was regulated by three reservoirs (Table 2): Tedjen I (190 million m^3 in capacity), Tedjen II (141 million m^3), and Gorgor (20 million m^3) [2]. Until now the 1926 Iran–Turkmenistan Bilateral Agreement on the Tedjen River Flow Sharing has been in effect. In 2005 Turkmenistan together with the Islamic Republic of Iran constructed the new reservoir Dostuluk (Friendship) on the Tedjen River that permitted irrigation of over 60,000 ha of new lands.

A rather large river in the southwest of Turkmenistan is Atrek (Etrek). It flows along the border between Turkmenistan and Iran. Its watershed basin is shared by these two states. It flows into the Caspian Sea. Together with its tributaries – Sumbar and Chendir – the total length of the river is 635 km, out of which 135 km are in Turkmenistan. The watershed area is 26,700 km^2. It takes its origin in the Turkmen–Khorasan Mountains by confluence of the Sebaz and Sulyakha rivers. The feeding of this river is mixed. No ice cover is formed here. The main (right) tributary is the Sumbar River 247 km long. After getting into the Turkmenistan territory its average annual flow is 240 million m^3. In the lower reaches it often dries out due to water intake for irrigation, mostly in Iran. The Atrek waters reach the Caspian Sea only in spring during floods (March–May). The whole flow getting into Turkmenistan is fully accumulated in three off-channel reservoirs: Mamedkul with a capacity of 20 million m^3, Delilin with a capacity of 16 million m^3, and Kizilay – 5.3 million m^3.

The Atrek River flows over the territory composed of loose, mostly sandy and clay soils which determines high water turbidity – 25 kg/m^3 on the average (six times higher than turbidity of the Amu Darya), while its maximum may be over 170 kg/m^3. This is the most turbid river in Central Asia. Accumulation of great quantities of sediments led to formation in the Atrek lower reaches of a flat sandy–clay marshy delta cut by traces of old arms of the river that changed its bed more than once.

Water salinity during floods is 1 g/L, in the lower-water period – 2 to 4 g/L – and in some years to 10 g/L.

Table 3 Water resources of Turkmenistan

River/source	Hydrological post	Many years average data of water flow	
		m^3/s	$km^3/year$
Surface water			
Amu Darya	Atamyrat	697.6	22[a]
Murgab	Tagtabazar	51.8	1.631
Tejen	Aulata	27.6	0.869
Etrek	Chat	11.3	0.354
Small rivers of the Kopetdag and Koitendag mountains, large springs and karizes of the northeast slope of the Kopetdag	–	4.8	0.150
Subtotal		793.1	25.004
Groundwater	–	–	1.269
Total	26.273		

[a]Notes: Share of Turkmenistan for water intake from Amu Darya, in accordance with the agreement of five Central Asia states (1992) and the Agreement between Turkmenistan and Republic of Uzbekistan (1996)

In 1957 the former USSR and Iran concluded the Treaty on Joint Utilization of Water and Energy Resources of the Atrek Border Sections. At present its waters are utilized on the 50/50 principle by Turkmenistan and Iran.

According to many-year observations the water resources of transborder rivers, except the Amu Darya, vary significantly as they are not regulated in the cross-border areas. The whole water flow of rivers is formed beyond Turkmenistan and depends not only of natural water availability but on water intake beyond the Turkmenistan borders.

Turkmenistan built 16 reservoirs for irrigation with a total capacity of 3.7 km^3, out of which 5 on the Karakum River with a total capacity over 2.5 km^3. These reservoirs are designed to accumulate winter flow of the rivers for its later use in the vegetation period.

In the recent years measures are taken to increase the capacity of the earlier built reservoirs. Thus, the capacity of the reservoir "15 Years of Independence" (formerly Zeid Reservoir) is increased from 1.25 to 1.5 billion m^3; Saryyazyn Reservoir, from 331 to 400 million m^3; and Khauzkhan, from 700 to 800 million m^3. The increased capacity of these reservoirs allows for addressing the problems that may arise during irrigation of agricultural crops in the low-water periods of the Amu Darya.

Accordingly, the average many-year volume of surface water resources of the country (Table 3) is 25.0 km^3/year and according to other estimates – 27.1–32.8 km^3/year [3, 4]. Here Turkmenistan accounts for no more than 500–700 million m^3/year. If we add here the flow of karizes on the northeastern

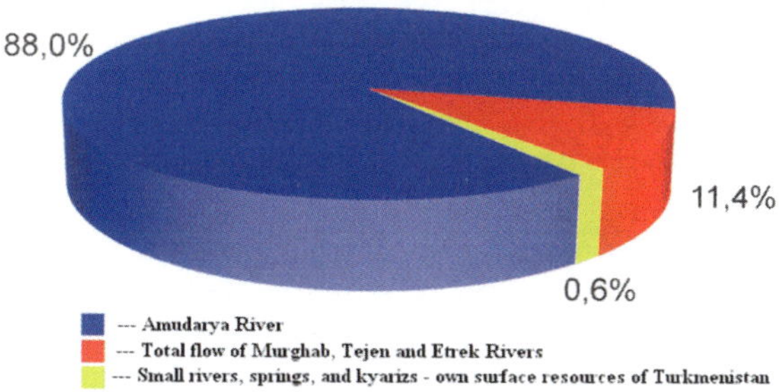

Fig. 2 Share of water resources in Turkmenistan

slope of the Kopetdag making about 350 million m^3/year, the volume of own surface water resources of Turkmenistan will be 850–1,050 million m^3/year only (Fig. 2).

3 Temporary River Flow

The TSR is made up of the waters of large subsand lenses, waters of lenses under irrigation canals and large rivers, the flow from takyrs and takyr-like watersheds, and from some watershed areas in the Karakums. They are characterized by not large sizes and significant variability in time and area.

The average many-year volume of TSR formed on the takyrs and takyr-like soils in the Karakum Desert is equal to 332 million m^3 that are used in small scale in distant range animal husbandry, mostly for irrigation of pasturelands. In the recent time TSR was somewhat reduced due to initiation of irrigation of some takyr lands.

From old times the local population has been using freshwater lenses. In some cases they dig pits to the lens level, in others – create lenses artificially: they divert rainwater into an open pit with the salt water in the bottom. A lens of freshwater collected from the takyr surface is formed over the salt water.

The total area of takyrs in Turkmenistan is about 20,000 km^2, and the area of the takyr-like watersheds is nearly the same [3]. The area of takyrs is not large, while the area of takyr-like watersheds may reach 100 and even 200 km^2. During a year the runoff from "good" takyrs located mostly in the south of the country may form a water layer of 20–25 mm thick on the average with regard to all water losses, while from "worse" takyrs – 10–15 mm thick. A takyr area of 3 km^2 may provide 40,000 tons of water. The area of freshwater surface in sub-takyr lenses varies from several square meters to 10 km^2, while the volume of freshwaters in a lens – from 1,000 to 100,000 m^3 [4].

Eight large subsand fresh groundwater lenses were found and studied in the Karakums. The greater freshwater resources occur in such large subsand lenses as Yaskhan, Chilmamedkum, Dzhillikum, Balkuin, and others. The total strategic resources of freshwaters in them are 69 km^3 (Table 4). The Yaskhan freshwater lens locates in the western part of Karakums. In economic terms it is very important

Table 4 Freshwater reserves in subsand lenses

Lenses	Contour area, km^2 (up to 1 g/L)	Static reserves, km^3
Yaskhan	2,000	10
Cherkezli	400	2
Balkuyi	650	0.45
East Zaunguz	1,000	3.4
Jilikum	2,950	8.4
Repetek (up to 3 g/L)	300	0.84
Karabil	6,765	25
Badghyz	3,000	19
Total	17,065	69.09

and interesting. It sizes 65 × 25 km, and its thickness (when filled not completely) in the center is over 70 m. The Chilmamedkum freshwaters contain both ground and interlayer artesian waters in the Apsheron and Akchagyl deposits. The East Zaunguz lens occurs in the east of the Zaunguz Karakums and is linked with the ancient buried Neogene Valley filled with sands. Its width varies from 70 to 25 km.

Only two lenses – Yaskhan (water reserves 10 km^3) and Chilmamedkum (water reserves 4 km^3) – are utilized on a commercial scale. The full-scale water intake is constructed on the Yaskhan lens providing 30,000 m^3 of water daily or 10.95 million m^3 a year. It supplies water to Balkanabat and other towns in the Balkan Velajat. The example of more than 30-year operation of the Yaskhan deposit confirms reliability of water reserve estimates, quality of groundwaters, and also methods of their operation. On the Chilmamedkum lens, the water intakes for pasture water ducts are built. They withdraw about 1 million m^3 of water a year or 2,740 m^3 a day. Therefore, the total potential water resources of Turkmenistan amount to 130.7 km^3.

In the south of Turkmenistan there are about 15 dry ravines over 10 km long. After heavy rains or active snow melt, the mudstone streams rush along them forming a channel several meters wide and to 2 m deep. As an experiment, the Zirik ravine was crossed with a dam that retained 200,000 tons of water. At the foot of the Kopetdag there are such giants as Gyaurli – 103 km long, 10 m deep with the channel width 50 m. During a year Gyaurli collected 2 million m^3 of water, on the average. But all they are operated in a pulse mode.

In some ravines (Atryk, Adzhider, Avgez, and Obochai), the water flow during short periods may be as high as 1,000 m^3/s. However, the average annual flow of mud streams does not exceed 100 million m^3. Some of this amount is lost to evaporation, while the other replenishes the fresh groundwaters.

Freshwater lenses are formed under large rivers and main irrigation canals; their total reserves are forecasted at 307 million m^3 [5].

The greatest reserves of fresh and slightly saline groundwaters are confined to the Neogene–Quaternary deposits. They occur in places of intensive feedings (piedmont

Table 5 Projected and proven groundwater stocks and their consumption by velayats, 1,000 m^3/day

Velayats	Projected resources	Proven resources (as of 1 January 2002)	Water intake
Akhal	2,600.4	1,770.6	929.7
Balkan	598.9	342.5	76.8
Dashoguz	492.8	265.4	72.2
Lebap	2,995.3	855.7	16.1
Mary	2,738.8	241.5	38.1
Total	9,426.2	3,475.7	1,132.9

plains) and in areas composed of unfixed sands (Yaskhan, Chilmamedkum, Eastern Zaunguz, Southeastern Karakums).

By 2000 over 200 deposits and areas with fresh groundwaters were investigated. For 81 of them, the State Commission for Reserves (SCR) approved groundwater supply reserves of 3,475,700 m^3/day or 1,269 million m^3/year (Table 5).

With regard to the approved reserves of fresh groundwaters, the total (summed-up) water resources of Turkmenistan are evaluated at 26.27 km^3/year. In the balance of water resources the fresh groundwaters account for less than 2%. In the nearest years the amount of extracted groundwaters may be doubled. However, further growth of production is required for performance of feasibility studies. Out of the approved groundwater resources the greatest amount of groundwaters of high quality is concentrated in the Akhalsky Velajat – at the foot of Central Kopetdag. Freshwaters in the Southeastern Karakums are confined to continental Neogene deposits. The depth of their occurrence varies from 20 to 250 m; their full thickness exceeds 80 m.

Apart from freshwaters and VPS waters, there are also secondary or anthropogenic water resources – drainage waters amounting to about 7 km^3 a year. These waters make a part of active water resources that are transformed in the course of irrigation: surface runoff–groundwaters–drainage flow.

The area of irrigated lands in the country is about 1.8 million ha and the land reclamation stock suitable for development – over 17 million ha or only 3.5% of the stock [6]. Here 1.6 million ha or 73% of irrigated lands are affected by salinity to different extent. According to 2011 estimates, the length of irrigation systems is 44,200 km, including on-farm 34,200 km. The total length of the collector–drainage network is 36,600 km, including on-farm 27,300 km [7].

The main water users in the country are irrigated farming (about 94% of the total volume of consumed water), industry (3%), municipal sector (2.7%), cattle breeding complexes and pasture irrigation (0.6%), and fishery (0.1%).

For Turkmenistan the problems of water resources, their utilization, and protection are of key priority due to the growing anthropogenic pressure on water bodies. In the future the pace of economic development and augmentation of the production potential of the respective sectors will lead to the increased water consumption. Insufficiency of water resources in the country and their condition prove that these problems may be addressed only by applying management, financing, and "know-how." The latter includes introduction of water-saving technologies and effective and wise management of irrigation water.

References

1. National Environmental Action plan by the President of Turkmenistan Saparmurat Turkmenbashi on Environment Protection in Turkmenistan, Ashgabat, 2002
2. Shtepa BG, Kudrin BA, Zonn IS, Orlova VP, Volynov AM (1975) Land reclamation in the USSR, Kolos, Moscow, 271 pp
3. Babaev AG (1983) Desert as it is. Molodaya Gvardiya, Moscow
4. Shevchenko AG (1982) Regularities in distribution and formation of fresh-water lenses in deserts and experience of their rational management. Ylym, Ashgabad
5. Babaev AG, Kolodin MV (1985) Water resources of Turkmenistan. Problems of desert development, N 4
6. Sustainable Development of Turkmenistan, RIO + 10. National overview (2002) Ashgabad
7. Water Economy of Turkmenistan (2011) Ashgabad

The Caspian Sea and Kara-Bogaz-Gol Bay

Aleksey N. Kosarev, Andrey G. Kostianoy, Igor S. Zonn, and Sergey S. Zhiltsov

Abstract The Caspian Sea and its Kara-Bogaz-Gol Bay play an important role in different branches of the economy of Turkmenistan. Turkmen shores of the Caspian Sea have a big potential as a national and international resort area which is developing quickly. The Caspian Sea water (after desalination) is an immense source of potable and technical water for the country, living in the desert conditions. This chapter describes the main geological, physical, chemical, biological, and climatic characteristics of this largest enclosed water body in the world. Special attention is given to the Kara-Bogaz-Gol Bay, which is located in the territory of Turkmenistan and during a century played a key role in the chemical industry of the Turkmen Soviet Republic and today in Turkmenistan.

Keywords Climate, Geology, Hydrology, Kara-Bogaz-Gol Bay, Meteorology, The Caspian Sea

---

A.N. Kosarev
Geographic Department, M.V. Lomonosov Moscow State University, Vorobyevy Gory, Moscow 119991, Russia

A.G. Kostianoy (✉)
P.P. Shirshov Institute of Oceanology, Russian Academy of Sciences, 36, Nakhimovsky Prospekt, Moscow 117997, Russia
e-mail: kostianoy@gmail.com

I.S. Zonn
Engineering Research Production Center for Water Management, Land Reclamation and Ecology "Soyuzvodproject", 43/1, Baumanskaya Street, Moscow 105005, Russia

S.S. Zhiltsov
The Peoples' Friendship University of Russia, 6, Miklukho-Maklaya Street, Moscow 117198, Russia

Diplomatic Academy at Ministry of Foreign Affairs of Russian Federation, 53/2, Build. 1, Ostozhenka Street, Moscow 119992, Russia

I.S. Zonn and A.G. Kostianoy (eds.), *The Turkmen Lake Altyn Asyr and Water Resources in Turkmenistan*, Hdb Env Chem (2014) 28: 69–94, DOI 10.1007/698_2013_228,

Contents

1 Introduction

The Caspian Sea is located to the west of Turkmenistan and plays a very important role in the economy of the country (Fig. 1). Turkmenistan is connected with the other Caspian Sea countries by shipping routes. Offshore, there are oil and gas fields, which are developed by national and foreign companies. The sea is reach in bioresources, and fishery is a part of the economy of the country. The Kara-Bogaz-Gol Bay during a century played a key role in the chemical industry of the Turkmen Soviet Republic and today in Turkmenistan. National tourism is developed at the coast of the warm Southern Caspian Sea, and national tourist zone "Avaza" (near Turkmenbashi town) with several modern hotels, restaurants, cafes, sandy beach, artificial canal, and park zone is progressively developed. In 2010, new very comfortable international airport in Turkmenbashi was built with a capacity of 800 passengers per hour. Finally, the Caspian Sea water (after desalination) is an immense source of potable and technical water for the country, living in the desert conditions.

The Caspian Sea is the largest enclosed water body in the world which occupies a large depression in the earth's crust [1, 2]. By the end of 2011 its sea level was located at −27.6 m abs (below the world ocean level) [3]. The area of the Caspian Sea in these conditions is more than 390,000 km^2, volume of water – about 78,000 km^3, average depth – 208 m, maximum depth – 1,025 m. From north to south the sea stretches for 1,030 km and is 300–400 km wide. The vast size and large meridional extent determine the natural diversity of the different regions of the sea. This chapter describes the main geological, physical, chemical, biological, and climatic characteristics of the Caspian Sea. Special attention is given to the Kara-Bogaz-Gol Bay.

Fig. 1 The Caspian Sea drainage basin (2007). In UNEP/GRID-Arendal Maps and Graphics Library. Retrieved on 21 April 2011 from http://maps.grida.no/go/graphic/the-caspian-sea-drainage-basin

2 The Caspian Sea

2.1 Geological History

The geological history of the Caspian Sea is linked with the history of the ancient Tethys Ocean, which in the late Mesozoic and early Paleogene occupied an area of the modern Mediterranean, Black, and Caspian seas. In the Paleogene and Neogene repeatedly occurred connection of the Caspian and Black seas to the ocean and the separation from it. During these events, the primary marine life died out partly and partly modified. In the middle Pliocene, the Caspian Sea for the first time was separated from the Black Sea part of the freshened Pontian Sea and became completely isolated water body. Brackish fauna of the Caspian type was formed in the sea, which preserved to our time. Thus, the middle Pliocene (6–7 million years ago) may be considered a time of formation of the Caspian Sea. From that time until the present day the Caspian remains an independent, isolated water body, only occasionally getting a short-term relationship with the Black Sea. In Quaternary the size of the Caspian Sea, its salinity, flora and fauna continued to change, the sea experienced multiple transgressions and regressions. However, their causes in pre-Quaternary and Quaternary were different. In the Neogene these were mainly tectonic processes and the associated deflections of the crust. In Quaternary, transgressions and regressions of the Caspian Sea are the result almost entirely of the climate change. Isolation of the Caspian Sea from the ocean was again violated only in our days – artificially, as a result of construction of navigation canals, which connected the sea with the Sea of Azov, the Black, Baltic, and White seas.

2.2 Physical and Geographical Characteristics

According to the physical and geographical conditions and bottom topography, the Caspian Sea is divided into three parts: the Northern, Middle, and Southern Caspian Sea (Fig. 2). The northern shallow (15–20 m) part of the sea is located entirely on the shelf. It is separated from the Derbent Depression of the Middle Caspian (maximum depth of 788 m) by a steep decrease of depth. Underwater Apsheron Ridge separates the Derbent Depression from the South Caspian Depression, where the greatest depths of the sea (1,025 m) are located. Notional boundary between the Northern and Middle Caspian is the line "Chechen Island – Cape Tyub-Karagan", and between the Middle and Southern Caspian – the line "Zhiloy Island – Cape Kuuli" (Fig. 2). The volume of sea water in these parts of the sea equals to 0.5%, 33.9%, and 65.6% of the total volume of the Caspian Sea, respectively.

There is a little number of islands in the Caspian Sea with a total area of about 2,000 km^2. Most of the islands is located in the Northern Caspian Sea. Here, in the western part the largest island Chechen (120 km^2) is located, as well as Tyuleny Island (68 km^2), and in the eastern part – Kulaly Island (73 km^2). Many small

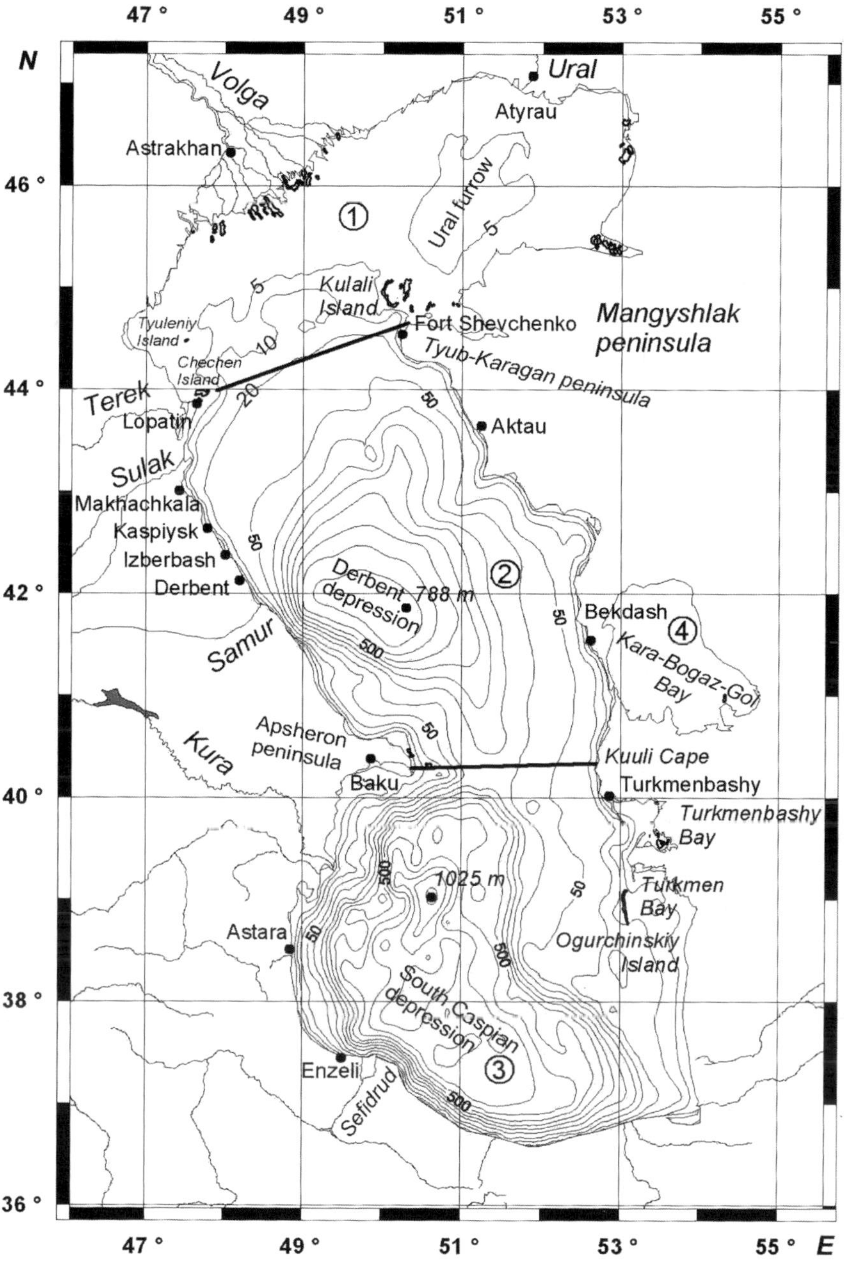

Fig. 2 Bottom topography of the Caspian Sea and its division on the Northern (1), Middle (2) and Southern (3) Caspian

islands are scattered at the Volga River mouth. In the Middle Caspian, south of Baku Bay there are several small islands of the Baku Archipelago. At the eastern coast there is Ogurchinsky Island (Fig. 2).

Approximately 130 rivers reach the Caspian Sea, only a few of them are large. Sea drainage basin covers an area of 3.5 million km^2 (Fig. 1). The ratio of the sea surface and the watershed area (1:10) explains the significant role of processes occurring in the Caspian Sea basin in the state of the sea. The greatest importance for the Caspian Sea has the Volga River basin, which is 1.4 million km^2, i.e. almost 40% of the total catchment area.

The river network in the Caspian Sea is very unevenly distributed. All the major rivers flow into the Northern Caspian Sea and to the western coast of the sea. Mean annual runoff of these rivers – the Volga, Ural, Terek, Sulak, Samur, Kura – exceeds 90% of the total flow of all the rivers in the sea. The rest of it falls on the rivers of Iran and small streams of the western coast of the sea. Eastern coast has no constant river flow into the sea. The intensive use of river water resources of the Caspian Sea, regulation of major rivers (except the Ural River), has led to a reduction of surface runoff into the sea and to its seasonal redistribution.

The coastline length of the Caspian Sea at the sea level of about 27 m below the ocean level, with the coastline of the islands, is about 7,500 km, but it varies with the sea level change. The nature of the Caspian coast is diverse. The Caspian Lowland is located at the northern boundary of the sea (Fig. 1). From Agrakhan Peninsula in the west to Buzachi Peninsula in the east the coasts are low-lying, semi-arid. Only a huge (up to 15,000 km^2) Volga River Delta is full of life. Western coast of the sea between Makhachkala and Baku cities is mountainous (Figs. 1 and 2). Caucasus Mountains stretch along the western coast, in some places come close to the coastline (Fig. 1). Narrow coastal strip between the eastern slopes of the Caucasus Mountains and the sea is a plain, cut by numerous channels of the rivers, the largest of which is Samur (Fig. 2). Especially diverse are landscapes at the western coast of the Southern Caspian. From Baku City to Cape Alyat foothills of the Caucasus Mountains stretch along the coast. Further south, the mountains give way to a wide Kura-Araks Lowland, where Kura River flows into the sea (Figs. 1 and 2). Further south Talysh Mountains reach the sea coast and the Lenkoran Lowland stretches along the coast. To the south it narrows, and Bogrovdag Ridge extends along the Caspian Sea.

Southern coast of the Caspian Sea is bordered throughout by Elburz Mountain Range (Fig. 1). This mountain range stretches along the coastline at a distance from 3–5 to 30–50 km. The average height of Elburz Mountains is 2,000 m. The highest of them is the cone-shaped, snow-capped Mount Damavand – an extinct volcano 5,604 m high. From the mountains a lot of small rivers drains to the Iranian sea coast, the most important of which are Babol and Gorgan rivers. Southern shore of the Caspian Sea is very picturesque. Warm and humid subtropical climate of the coast promotes diverse vegetation, including valuable crops (citrus, tea, rice, tobacco).

The entire eastern coast of the Caspian Sea is the desert. From Mangyshlak Peninsula to Kara-Bogaz-Gol Bay the Ustyurt Plateau borders the sea with steep ledges (chinks). South of the Kara-Bogaz-Gol Bay stands Turkmenbashi Peninsula with low mountains of Kubadag Mountain Ridge. Further south, Turkmenbashi Bay is bounded on the south by Cheleken Peninsula. The eastern coast of the Southern Caspian is low; here sands of the western part of the Karakum Desert come close to the sea.

Several large bays are allocated in the Caspian Sea (Fig. 2). Agrakhan and Kizlyar bays are located on the western coast of the Middle Caspian, Kyzylagach Bay is in the Southern Caspian. In the northeastern part of the sea, a large Mangyshlak Bay is found between the peninsulas Buzachi and Tyub-Karagan. Large Kazakh, Kara-Bogaz-Gol, and Turkmen bays intrude deep in the eastern coast of the Caspian Sea. Coastline of the Caspian Sea is undergoing significant changes due to the interannual sea level variations. Especially susceptible to this effect is the coast in the shallow northern part of the sea.

2.3 Bottom Relief and Sediments

The bottom relief of the Caspian Sea is closely associated tectonically with the geological structure of the surrounding land. The Northern Caspian is located within the ancient Russian Platform and younger Scythian–Turan Platform. The western part of the Middle Caspian and Southern Caspian are all in the Alpine folded belt. Bottom of the Northern Caspian Sea is a shallow flat plain with numerous islands and ancient riverbeds of the Volga, Ural, Terek rivers. Northern and Middle Caspian are divided by the Mangyshlak Sill. In the bottom topography it is expressed in the form of shallow area, extending from Chechen Island to Kulaly Island, then to Tyub-Karagan Peninsula. Southern edge of the sill is artificially limited by 20 m isobath (Fig. 2).

In the bottom topography of the Middle Caspian the main morphological elements: the shelf, slope, bed of the depression are clearly highlighted. Depression in the Middle Caspian is asymmetric: in the western part a shelf is narrow and slope is steep (except for the northern part), in the eastern side a shelf is wide and slope is shallow (Fig. 2). The width of the western shelf of the Middle Caspian ranges from 15 to 130 km, the width of the eastern shelf varies from 50 to 130 km. The upper part of the slope is separated from the shelf by a bend. The western slope of the middle part of the sea has a width of 20–60 km. The eastern slope is a slightly inclined plain 40–150-km wide. Underwater slope at depths of 600–700 m goes to the abyssal flat plain, called Derbent Depression (Fig. 2). It stretches along the western coast to 250 km, with a width of 40–80 km. In the depression there is a maximum depth of the Middle Caspian – 788 m.

Middle Caspian Depression is separated from Southern Caspian Depression by the Apsheron underwater sill, crossing the sea from the Absheron Peninsula to Cape Kuuli. The greatest depths over the central part of the sill are of 160–180 m.

Location of the Southern Caspian in the geosynclinal region causes significant complexity and ruggedness of its bottom relief. Western shelf has a width of

15–60 km. In the northern part there is a lot of small islands and banks mostly of mud-volcanic origin. Eastern shelf of the southern part of the sea is much wider – up to 190 km (Fig. 2). At the southern coast of the Caspian the shelf is very narrow (5–10 km) and steep. The structure of the continental slope is more complex. On the western slope we can observe a series of elevations up to 500 m with peaks which represent mud volcanoes. Eastern and western slopes in the Southern Caspian Depression are relatively steep and have a stepped character. The foot of these slopes is located at a depth of 700–800 m. The Southern Caspian Depression differs from the Derbent one, because its central part has an abyssal plain, where elevations alternate with local depressions. In the central part of the Southern Caspian Depression there is Abich Trough with the greatest depth of the sea – 1,025 m. In the southern part of the Southern Caspian Depression there is Near-Elburz Deflection, expressed as a relief depression.

The distribution of sediments in the Caspian Sea is due to the bottom topography, water dynamics, and hydrochemical conditions. Northern and Middle Caspian receive more than 90% of the liquid flow and over 75% of the solid runoff. Main enrichment by terrigenous material occurs in the river mouth areas, and its further advection is related to the currents. Bottom sediments of the Caspian Sea are represented by limestone and sandstone sediments. Coarse sediments (silt and terrigenous sands) are characteristic for the Northern Caspian. On the western shelf and slope of the Middle Caspian Sea to the depths of 30–60 m terrigenous muds dominate. In the eastern part of the Middle Caspian, with a wider shelf, in an arid climate there is an intense accumulation of carbonate sediments with biogenic component. On the eastern slope, at depths of 200–400 m there is an enrichment of sediments by silica, which is associated with upwelling of deep waters rich in silicon. The bottom of the deep Middle Caspian Depression is covered by limestone mud of gray and gray–green color.

In the Southern Caspian, at the western shelf gradual change of sands by limestone mud is observed with depth. The bottom of the Southern Caspian Depression is covered by limestone clayey silt. On the eastern slope and shelf of the Southern Caspian there are calcareous silts with high carbonate. Despite high rates of sedimentation, on the seabed there are some areas with low sediments, and even rock outcrops. They are common along the edge of the continental shelf and in some parts of the continental slope, as well as on the top of the hills, and on the Apsheron Sill [1].

2.4 Climate

Because of the large meridional extent the Caspian Sea is within several climatic zones: the northern part – in a temperate continental climate, the western coast – moderately warm climate, southwestern and southern parts of the sea – subtropical climate. For the eastern coast the desert climate is typical.

In winter in the Northern and Middle Caspian weather is conditioned by the continental polar air related to the Siberian Anticyclone, and Arctic air which spreads from the Kara and Barents seas. In the Southern Caspian southern cyclones are often observed. The weather is rainy and unstable in the west; dry in the east. In summer the spurs of the Azores Maximum impact the sea, most of all in the west and northwest. In the southeast, the sea is influenced by a large Iran–Afghan Minimum. In summer, over the Caspian Sea the weather is stable and dry [1, 4, 5].

The air temperature in winter (January–February) varies from −10°C (in the most severe winters – to −30°C) in the northeastern part of the sea to +8 to 12°C in the south. During the summer (July–August) average temperature throughout the sea is 24–26°C, and the maximum (more than 40°C) is observed on the eastern coast. Figure 3 shows mean annual air temperature at 2 m above displacement height for the Caspian Sea region for the period 2000–2012. A cold area on the west is related to the Caucasus Mountains. The highest mean air temperatures are observed in the Southern Caspian Sea and at the border between Turkmenistan and Iran.

Atmospheric precipitation in different regions of the sea falls unevenly from 100 mm in the arid eastern coast to 1,700 mm in the southwest (near Lenkoran town in Azerbaijan). In the open sea the average rainfall is 200 mm/year [1, 4]. Figure 4 shows mean monthly amount of precipitation for the Caspian Sea region for the period 2000–2010. We observe a significant decrease of precipitation from west to east along the whole Caspian Sea. The most dry are lands in Kazakhstan and Turkmenistan. Maximum of precipitation is observed in the Caucasus Mountains.

In relation with typical air mass transfer, the following general steady winds are established over the Caspian Sea: the northerly (NW, N, NE), south-easterly and the vortical. Northerly winds are observed in 40% of cases in average (in the summer – up to 50%), and almost half of them are north-westerly winds. South-easterly winds are repeatable up to 36% in winter and about 40% in spring [1, 4, 5].

Speed of the prevailing north-westerly and south-easterly winds, on average, is of 5–9 m/s (moderate winds). Strong winds (more than 10 m/s) of these directions do not exceed 4–6% of cases. The frequency of severe storms (more than 25 m/s) over a large area of the sea is very small. The annual average wind speed in the Caspian Sea is 6–7 m/s. The highest average wind speed is observed in the Middle Caspian of 6–7 m/s, and in the Apsheron Peninsula of 8–9 m/s. At the eastern coast the average wind speed is of 5–6 m/s, with a maximum near Mangyshlak Peninsula. In the Southern Caspian, where strong winds are rare, the annual average wind speed is of 3–4 m/s, the frequency of the light winds is up to 90%. The number of days with storms (wind speed more than 15 m/s) in the Southern Caspian is no more than 20–30 per year; in the Northern Caspian and at the eastern coast of the Middle Caspian there are 30–40 storms per year. Most storm activity is observed near the Apsheron Peninsula (50–60 day/year), which is explained by the orographic effect. Over the Northern Caspian, the easterly winds prevail (36–50% per year), as well as westerly and north-westerly winds. Mean wind speed here is of 5.8 m/s. Strong winds are observed very rarely [1, 4, 5].

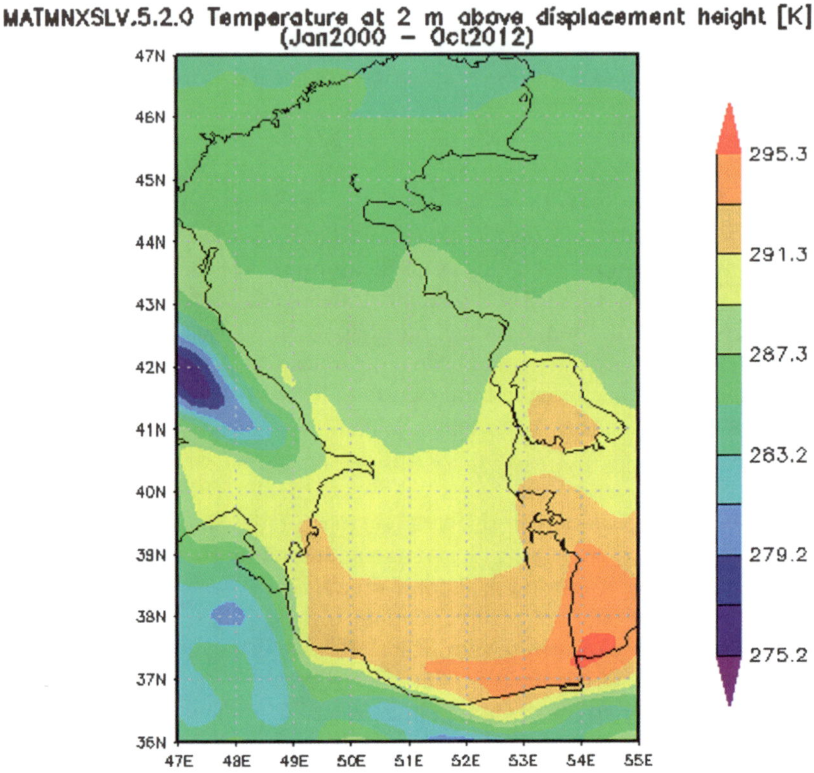

Fig. 3 Mean annual air temperature (°K) at 2 m above displacement height for the Caspian Sea region for the period 2000–2012. Note that 273.15°K equals to 0°C. The graph was constructed using MERRA Monthly Historical Data Collections, Giovanni, NASA

In accordance with the prevailing winds in the Caspian Sea, in the open sea waves are propagating from the north and northwest (32%), or from the southeast and south (32%). Waves coming from the east are very rare (about 12%), and in about 20% of cases waves are small or the sea is calm. The most frequently observed waves correspond to wind speed of 10 m/s, the most rare – to the wind speed exceeding 25 m/s. Storm waves are usually generated in winter and spring when the northerly winds blow with speed up to 20 m/s. Strong, long-lasting storms (mainly north-westerly and south-easterly) is a characteristic feature of the central part of the Middle Caspian. Because of the orographic effect, there is a shift of storm activity to the western coast of the Middle Caspian. Along the eastern coast, wave field is generally twice weaker than along the western one [1, 4, 5].

During northerly storms the highest waves are observed in the region of the Apsheron Archipelago. At the epicenter of highest waves at Neftdashlary banks wave height may reach up to 8 m, and in extreme storms – even 9–10 m. But most often, in the area there are waves up to 2 m high. At the Turkmen coast, the

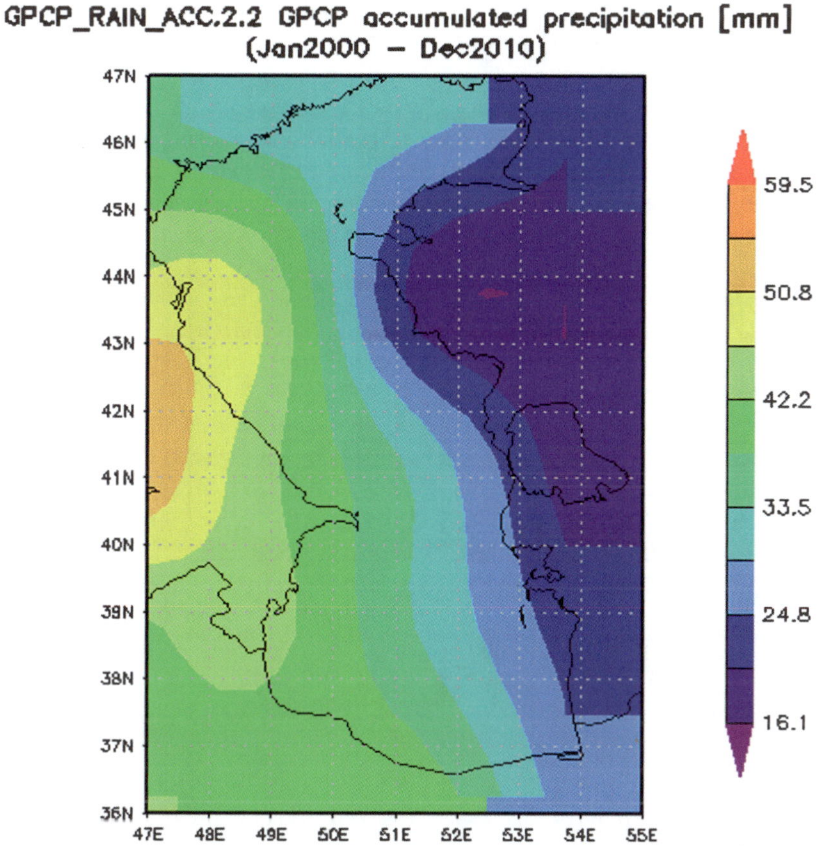

Fig. 4 Mean monthly amount of accumulated precipitation (mm) for the Caspian Sea region for the period 2000–2010. The graph was constructed using Monthly Global Precipitation (GPCP) and TRMM Online Visualization and Analysis System (TOVAS), Giovanni, NASA

moderate and strong north-westerly winds (5–15 m/s) cause waves up to 1 m high, and the stormy winds – up to 2–3 m [1, 4, 5].

At the south-easterly winds greatest waves are generated in the north of the Middle Caspian, in the region of Makhachkala and Derbent towns, and Mangyshlak Peninsula, where the wave height can reach 6–7 m, as in the open sea area. Easterly winds, even the most powerful may cause waves not higher than 2–3 m [1, 4].

In the shallow northern Caspian waves directed to the northwest, east and southeast dominate (70%). The greatest possible wave height increases from north to south with increasing depth. With winds of 15–20 m/s wave height increases from 0.5 m near the Volga Delta to 4 m on the border with the Middle Caspian. The most quiet time in the Northern Caspian is summer, when calm is often installed in the vast area.

The Caspian Sea partially freezes every year. In the Middle Caspian ice cover is small, and in mild winters it does not occur. The Northern Caspian freezes every year, and ice covers much of it. The average southern boundary of the ice cover goes between Northern and Southern Caspian between Chechen Island and Kulaly Island, and farther to Cape Tyub-Karagan. In mild winters ice formation begins in the northeastern part of the sea from the middle of November and the end of the month, the ice extends along the northern coast. In January, the whole Northern Caspian is covered by ice. In December, in the Middle Caspian ice appears in the shallow bays and inlets of the eastern coast, and in January – in Makhachkala region on the western coast. From the second half of February an intensive destruction of ice begins. First of all ice is totally melting along the coast of the Middle Caspian, then in some areas of the Northern Caspian, which becomes totally ice free in late March – early April [4].

In very severe winters, ice formation occurs much earlier than usual, and the time of melting is delayed by 2–3 weeks. In such winters large masses of floating ice have drifted along the western coast as far as to the Apsheron Peninsula. On the eastern coast, the Turkmenbashi Bay can freeze, as it was observed in severe winters of 2007/2008 and 2011/2012. In warm winters, ice formation may be delayed by almost a month, and ice destruction starts much earlier than usual. There were years when the Volga River at Astrakhan did not freeze. The average thickness of the ice in the Northern Caspian Sea varies between 20 and 30 cm offshore and 50–60 cm in the northeastern part. Height of ice hummocks may reach 1–1.5 m [1, 4].

2.5 Water Budget

For the Caspian Sea the incoming part of the water balance is surface runoff, precipitation and groundwater flow, which are balanced by the evaporation from the sea surface and flow of sea water to the Kara-Bogaz-Gol Bay, where it evaporates. For the water balance of the Caspian Sea river runoff and evaporation are the most important, the ratio of which mainly determines the interannual changes in the volume of sea water and the sea level. Because of the different calculation methods and the lack of observational data the assessment of the water balance components should be accepted with certain assumptions.

River runoff is the main incoming part of the water balance, which gives up to 80% of water coming to the sea. It is very variable, in the past century its average annual value was about 300 km^3, but it changed from 332 km^3 in 1900–1929 to 240 km^3 in 1970–1977. Volga River brings about 80% of the total river runoff or about 240 km^3/year in average. Ural, Terek, Sulak, Samur, and Kura rivers bring about 15% of the river flow into the Caspian Sea. Small rivers, including those of the Iranian coast, give about 5% only. Noticeable variation of the river flow to the Caspian Sea may cause changes up to 50 km^3 in the water budget and the corresponding sea level fluctuations. It should be noted that anthropogenic withdrawal of river water is about 40 km^3/year, of which 25 km^3 belong to the Volga. Without these losses the Caspian Sea level in the 1955–1990s would be 1.6 m higher than that was observed.

The volume of rainfall is much smaller than the river runoff and evaporation, and it contributes only 18–25% to the incoming part of the water balance. Long-term average volume of precipitation for 1900–2000 is 75 km^3. The role of groundwater flow is negligible; indirect estimates show an average value of 4 km^3.

Evaporation from the sea surface is the main expendable component of the water balance. Interannual changes in the evaporation are much smaller than those of the river runoff. In the twentieth century, its value in different periods of time varied from 349 to 397 km^3/year. To these values we have to add a flow of sea water to the Kara-Bogaz-Gol Bay, which is conditioned by the difference in the sea and bay levels, as well as by morphometry of the Kara-Bogaz-Gol Strait connecting them. In the last century, due to various reasons, the flux to the Kara-Bogaz-Gol Bay varied from 20 to 25 km^3/year to almost 0 km^3. The ratio of the incoming and outcoming components of the water balance determines the sea volume change and the sea level.

2.6 Sea Level

Long-term changes in the Caspian Sea have an impact on all key processes in the sea, as well as on the economic activity on its coast. The main factors causing the long-term variations in the Caspian Sea level are geological (change in the volume of the depression) and climate (variations of the sea water balance), although their contribution to the sea level variability is not equal [1]. Tectonic movements have played a key role in the initial stages of the Caspian depression formation. But already in the Holocene (more than 10,000 years ago), the main cause of large-scale changes in the Caspian Sea level is the change of climatic conditions in the basin of the sea. Over the last 2,000 years sea level fluctuations were up to 7 m. Analysis of instrumental measurements of the sea level, which began in 1830, showed that from the beginning of the twentieth century until 1929, the Caspian Sea level was close to the mark of −26.2 m abs (Fig. 5). But then the sea level began to decline sharply, and by 1956 it has dropped by almost 2 m. This was caused by a severe drought in the basin of the Volga River and the reduction of its runoff. In the 1950s, precipitation over the Caspian Sea drainage basin has increased, but in these years a lot of riverine water was taken for the economy needs. Therefore, in the 1950s and 1960s, the Caspian Sea level has not increased, but only stabilized. In 1970s it was observed a new decline in the sea level caused by a decrease of the Volga runoff and increased evaporation from the sea area. In 1977 the sea level dropped to a mark of −29 m abs – the lowest in the last 400–500 years. The total value of the sea level decrease in the twentieth century was 3 m from which 1.5 m was caused by anthropogenic reduction of the river runoff. The fall of the sea level has reduced its water area by about 40,000 km^2, mainly due to shallowing of Northern Caspian [1, 5].

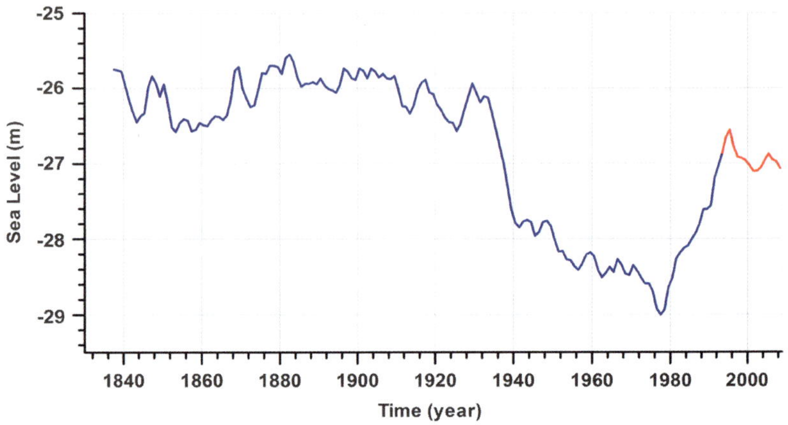

Fig. 5 The Caspian Sea level (m) variations from 1836 to 2009. *Blue line* shows instrumental records and *red line* shows satellite altimetry measurements

Since 1978 it has been observed a rapid rise of the sea level till 1995, when it reached the level of −26.6 m abs (Fig. 5). This increase was caused by changes in the water balance on the drainage basin of the Volga high flow, which connection with the sea level variability is reliably established. Since that time we observe a general decrease of the sea level, which by the end of 2011 was located close to −27.6 m abs [3]. Forecast of the future changes in the Caspian Sea level is of great scientific and practical interest, but this is an extremely difficult task. It is connected with the need for climate prediction in the vast region covering the Caspian Sea drainage basin. This explains the fact that the minimum of 1977 (and the followed sharp rise of the sea level) and a maximum of 1995 (and a followed decrease of the sea level) have not been predicted by scientists. Therefore, only probable estimates of the Caspian Sea level changes for the coming decades could be done. According to them, with relatively stable physical and climatic conditions for the Caspian Sea drainage basin, the fluctuations of the sea level are most likely in the range of about 2 m, i.e. from −26 to −28 m abs [1, 6]. We have to note that such fluctuations of the Caspian Sea level is a natural phenomenon that reflects the "breath" of the water body. This should be taken into account when considering both scientific analysis and planning of economic activities in the coastal zone of the sea.

2.7 Water Circulation

Circulation of the Caspian Sea is influenced by the wind, especially in the upper layers, and the unevenness of the density field in the water column [7, 8]. Coastlines and bottom topography have a significant impact on the field of currents. In the shallow Northern Caspian weak, unstable currents, governed by the wind and the

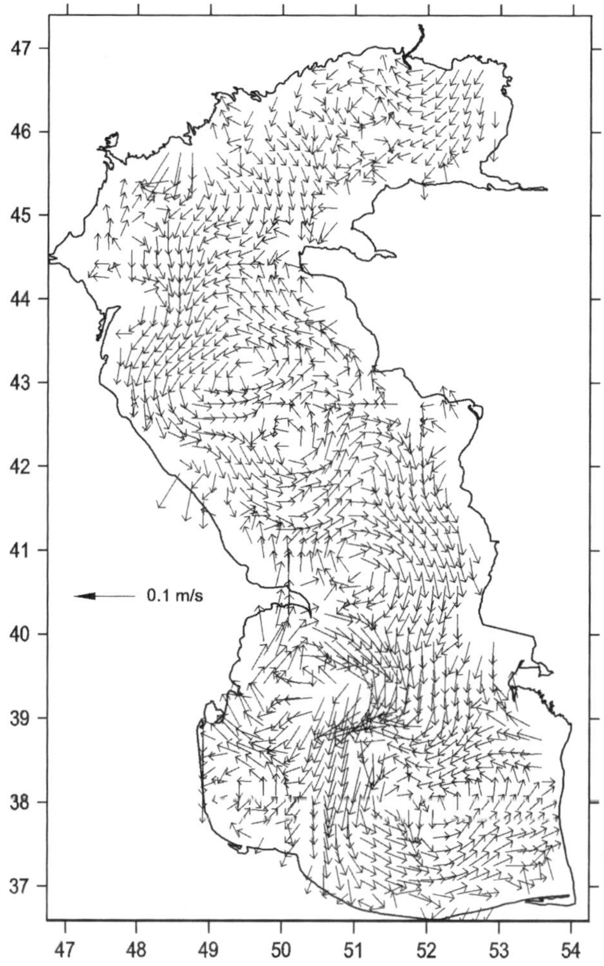

Fig. 6 Mean annual climatic field of the Caspian Sea upper layer currents according to numerical model [7]

river flows, are observed. A significant part of the Volga riverine water is advected to the south along the western coast of the Middle Caspian. In the deep-sea basin circulation is rather complicated and has a vortex character. In the Middle Caspian two gyres can be detected: the cyclonic one in the northwestern part and anticyclonic one in the southeastern part (Fig. 6). In the Southern Caspian also there is a "vortex dipole", but of opposite direction: the anticyclone in the northwestern part and the cyclone – in the southeastern part. In different seasons the location, size, and velocity of these gyres are changing. The gyres are clearly manifested in the upper 100 m layer, where average velocity is of 5–30 cm/s. Strong northerly and southerly winds can cause a short-term intensification of currents up to 50–60 cm/s, and sometimes even higher.

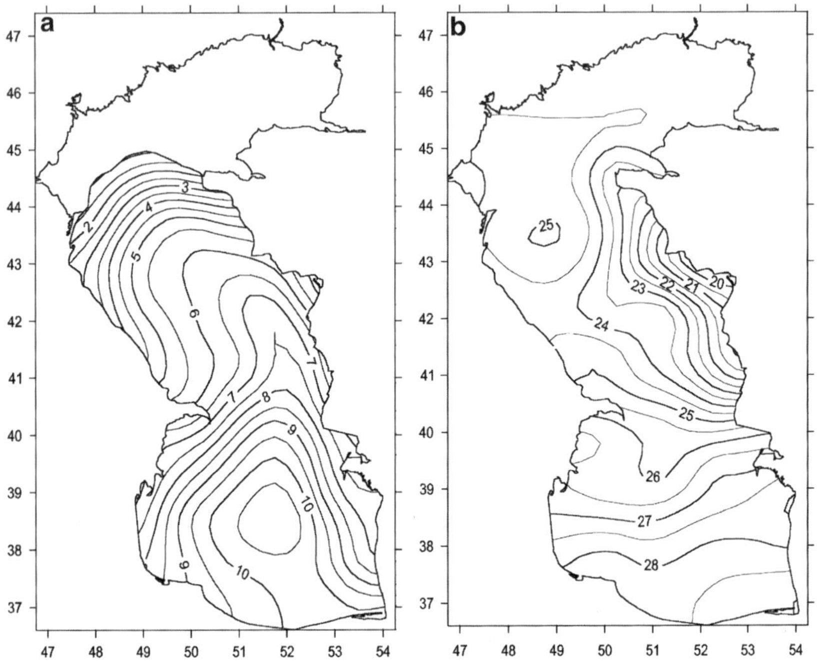

Fig. 7 The Caspian Sea surface temperature (°C) in February (**a**) and August (**b**). The February field is confined in the north by the floating ice edge

2.8 *Temperature and Salinity*

Sea surface temperature (SST) in the winter (January–February) changes from 0°C to 0.5°C in the north to 10–11°C in the south (Fig. 7a) [1, 8–10]. The Northern Caspian Sea freezes from November to March, although ice cover is unstable. Due to different severity of winters, formation and melting of ice start in different months, and ice cover area varies considerably.

In the summer the water temperature throughout the sea is more uniform and varies from 23°C to 28°C (Fig. 7b). However, in July and August along the eastern coast of the Middle Caspian the SST drops to 12–17°C in the zone of seasonal coastal upwelling.

In winter, the vertical distribution of water temperature is uniform, thanks to the development of processes of density (convective) mixing. In summer, at depths of 20–30 m a sharp jump in temperature (thermocline) is formed, separating the upper warm layer from the rest of the water column with lower temperatures. In autumn, with the beginning of the cooling the thermocline is destroyed. In the bottom layers of the deep valleys the temperature in the Middle Caspian is of 4.5–5.5°C, in the Southern Caspian – 5.8–6.5°C [1, 8–10].

Large volume of the river runoff in the closed Caspian Sea causes low salinity of its waters (12.7–12.9‰), which is almost three times lower than the average salinity

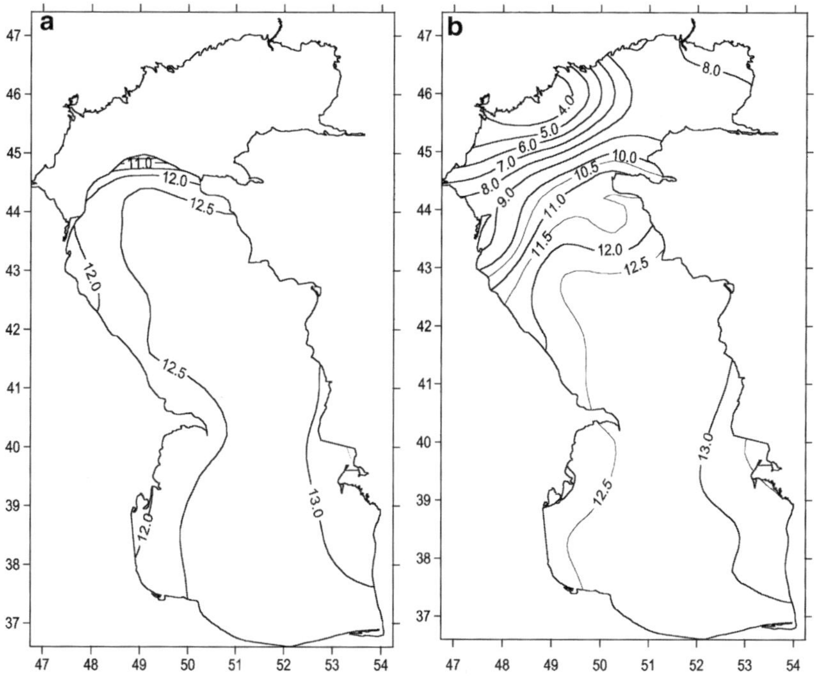

Fig. 8 The Caspian Sea surface salinity (‰) in February (**a**) and August (**b**). The February field is confined in the north by the floating ice edge

of the World Ocean (35‰). The salinity is the most variable in the Northern Caspian Sea, where it increases from 0.1‰ to 0.2‰ near the Volga and Ural deltas to 10–12‰ at the border with the Middle Caspian (Fig. 8). Thus the greatest horizontal salinity gradients are observed in the frontal zone between the riverine and sea waters. In the Middle and Southern Caspian there are small differences in salinity, and its values are within the limits 12.5–13.4‰, rising from the northwest to southeast. Distribution of salinity is very homogeneous in the water column, where it increases from the surface to the bottom only by 0.2–0.5‰ [1, 8–10].

Thanks to the uniform salinity water column of the Caspian Sea is well mixed. In winter, vertical circulation in the Middle Caspian reaches a depth of 150–200 m, in the Southern Caspian – 80–100 m, and in severe winters – much deeper. The deepest layers of the sea are ventilated by downwelling of cold, dense water formed in the northern and eastern shelf areas. Dissolved oxygen in the upper layers of the sea in winter is of 7–10 ml/l, and in the summer – 5–6 ml/l. In the bottom layers, it is equal to 2–3.5 ml/l in the Middle Caspian and 1.5–2.5 ml/l in the Southern Caspian [1, 8–10].

Since the late 1970s, a serious change in the vertical hydrological structure of the Caspian Sea has occurred. In those years, the climate processes over the Caspian Sea Basin have changed, accompanied by an increase in river runoff and, as a consequence, increase in the sea level. This was resulted in a decrease in surface salinity and an increase in water column stratification. Ventilation of the deep basins of the Caspian Sea became less intensive that is unfavorable for the ecological state of the water body [1, 8–10].

2.9 Fauna and Fishery

Fauna of the Caspian Sea is various in origins [1]. Dominated by species of the Caspian original complex (approx. 65%) – the descendants of marine Tertiary fauna have changed during the evolution of the sea. These include the sturgeon, herring, sprat, gobies, zebra mussel, and kardium clams, most of the crustaceans. About 25% belong to the freshwater species that have invaded the Caspian Sea during periods of desalination, such as fish – carp, perch. At the end of the Ice Age some arctic invertebrates, fish (salmon, white salmon) and the seal penetrated to the Caspian Sea, which still breeds on the ice of the Northern Caspian. Finally, the fourth set are Mediterranean species. Most of them have gotten into the sea spontaneously, especially after 1950s, when the waterway was set between the Caspian and Azov seas through the Volga-Don canal. In addition, in 1930s–1940s mullet and two species of invertebrates – Nereis and Abra ovata have been deliberately introduced into the Caspian Sea. They successfully acclimatized and became the main food of sturgeon. However, introduction of the new species has not always a positive effect. At the end of the 1990s, the jellyfish Mnemiopsis Leidyi penetrated from the Black Sea to the Caspian Sea, and actively spread there. Mnemiopsis eats plankton and livelihoods of sprat – the main object of the Caspian fishery. As a result, the population of sprat has greatly reduced. Measures to combat the spread of Mnemiopsis are under development [1].

Phytoplankton creates the basis of the primary production of organic matter, which includes approximately 450 species (diatoms, blue-green, green, and other algae). Phytobenthos includes more than 350 species. Zooplankton has more than 300 species. Zoobenthos includes about 400 species of invertebrates (mollusks, crustaceans, worms) [1].

The total number of species and subspecies of fish found in the Caspian Sea and the river deltas is more than 120. They are divided into four groups: (1) marine fish, which the entire life cycle takes place in the sea (sprats, some herring, mullet); (2) marine fish living in the sea, but migrating for the reproduction in the river for long distances (sturgeon, beluga, stellate sturgeon, salmon, herring, etc.); (3) fish, living in the brackish areas of the sea, and migrating for the reproduction to the deltaic reservoirs (roach, bream, carp, perch); and (4) river fish, living in the freshwater estuarine waters of the rivers [1, 2].

The Caspian Sea is a traditional region for fishery. Herrings and some other fish made a basis of the catches in the past. Sturgeon fishing was of great importance, as well as production of smoked sturgeon and caviar. At the beginning of the last century, sturgeon annual production was about 40,000 tons, which was up to 80% of the world sturgeon catch. So far, the largest population of sturgeon remains in the sea, although their number was strongly reduced, mainly due to marine pollution and illegal fishing. To restore the sturgeon population the comprehensive measures, agreed by all the littoral states, are required [1, 2].

The Caspian Sea is famous for an abundance of waterfowl. Some of them arrive here from the north for the wintering; the others are flying from the south for nesting. The former include geese, ducks, swans, gulls, loons; the second – the eagles, etc. In accordance with the Ramsar Convention four coastal regions of the Caspian Sea – the Volga Delta, Kyzylagach, Turkmenbashi and North Cheleken bays are recognized as sites of international rank [2].

3 Kara-Bogaz-Gol Bay

3.1 Physical and Geographical Characteristics

On the eastern coast of the Caspian Sea the Kara-Bogaz-Gol Bay (in Turkmen "black mouth bay") juts in the land (Figs. 1 and 2). According to the ancient legend, the Bay absorbed the Caspian Sea water and even ships, which risked going to this huge lagoon. The well-known Russian traveler G.S. Karelin was the first who explained the riddle of Kara-Bogaz-Gol Bay. In 1836 he had the courage to enter the Bay and found that the Caspian water does not disappear there, but very intensively evaporates under the dry and hot climate of the surrounding desert. In 1847, the Russian Navy Lieutenant I.M. Zherebtsov on the steam corvette "Volga" sailed around the shores of the Bay for the first time, made their description, and draw a geographical map. In 1897, the Russian Ministry of Commerce and Industry sent to the Kara-Bogaz-Gol Bay an expedition on the ship "Krasnovodsk" led by hydrologist I.B. Spindler to explore the salt resources of the Bay, and the causes of fish death in it. It was then compiled the first geographic description of the Bay [1, 2, 11].

Kara-Bogaz-Gol Bay is a shallow depression in relief with a flat bottom and variable coastline. This is the largest lagoon in the Caspian Sea of about 18,000 km^2, separated from the sea by two sand spits (Fig. 9). Between them there is the Kara-Bogaz-Gol Strait 7–9-km long, 120–800-m wide, and 3–6-m deep (Fig. 10) [1, 2, 11]. The Bay is located on the territory of Turkmenistan.

Morphometric characteristics of the Kara-Bogaz-Gol Bay vary significantly depending on the level of the sea and water in the Bay. Because of the difference in the levels of the sea and the Bay, the Caspian Sea water flows at a rate of 50–100 cm/s through the Kara-Bogaz-Gol Strait into the Bay, where it evaporates up to 800–1,000 mm/year. Thus, taking in mind that the average annual amount of precipitation in the area is not more than 110 mm, the Kara-Bogaz-Gol Bay represents a huge natural evaporator of sea water. Due to such a large evaporation, the Bay is filled by a brine ("rapa") – salt sea water with a salinity of 270–300‰. This is a concentrated solution of different salts (sodium chloride, magnesium and potassium, magnesium sulfate, and a small amount of rare earth elements) [1, 2, 11].

Fig. 9 Satellite view of the Kara-Bogaz-Gol Bay on 11 June 2003. Image courtesy Jeff Schmaltz, MODIS Land Rapid Response Team at NASA GSFC (http://earthobservatory.nasa.gov/IOTD/view.php?id=3550)

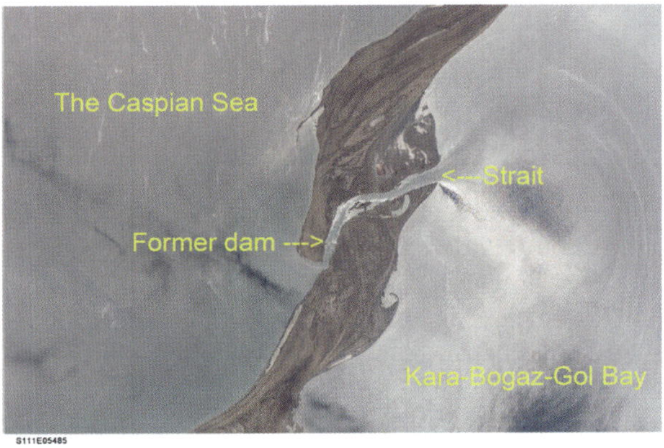

Fig. 10 High resolution photo of the Kara-Bogaz-Gol Strait made from Space Shuttle on 17 June 2002. Astronaut photograph STS111-E-5485 was provided by the Earth Sciences and Image Analysis Laboratory at Johnson Space Center (http://earthobservatory.nasa.gov/IOTD/view.php?id=2611)

Physiography of the Kara-Bogaz-Gol Bay and its natural regime, which depend on the level of the Caspian Sea and flow to the Bay, have significant interannual changes. In the first decades of the twentieth century the level difference between the sea and the Bay was small (about 0.5 m), and between them there was a hydraulic connection. The level of the Kara-Bogaz-Gol Bay (relative to World Ocean) was close to −26.5 m. Seasonal changes in evaporation led to the Bay level variability of about 2 m. The Bay surface was about 18,000 km^2, the volume of water – 130 km^3 and prevailing depths of 8–10 m. The Caspian Sea water inflow into the Bay reached 18–25 km^3/year, and the Bay yearly received about 330–380 million tons of salt. But the role of the Bay in the desalination of Caspian Sea remained very small – 0.2–0.3‰ for the last 100 years [1, 2, 11].

3.2 *Chemical Industry at the Bay*

The huge reserves of minerals and steady water–salt balance between the sea and the Bay became the natural basis for industrial development of salt resources of the Kara-Bogaz-Gol Bay. The total salt reserves in the Bay are estimated as billions of tons. This is the largest accumulation of sodium sulfate in Eurasia, and the only deposit in the world, where permanent natural crystallization of mineral salts is the source for their production on an industrial scale. The Bay has rich reserves of mirabilite (Glauber's salt) – sodium sulfate decahydrate $Na_2SO_4 \cdot 10H_2O$. Sodium sulfate is a valuable chemical feedstock used in the chemical, pulp and paper, textiles, glass industry, as well as in agriculture and medicine. Hundreds of the Soviet Union factories received the chemical products of the Kara-Bogaz-Gol Bay [11].

In the first decades of the twentieth century, production of chemical raw materials was based on the natural processes. The upper layers of the Bay contain sodium chloride, NaCl, magnesium chloride $MgCl_2$, and sodium sulfate Na_2SO_4. Every year in late November, when the temperature of water decreased to 5.5–6°C, water became saturated by sodium sulfate, and mirabilite was released in the form of colorless crystals. They precipitated on the bottom of the Bay, and winter storms brought them on the shore. By mid-March, when water temperature exceeded 6°C, minerals start to dissolve in the waters of the lagoon. From November to March, the locals collected crystals of mirabilite on the shores. In summer, in dry climate crystals lose water and turned into a dry mineral thenardite [11].

In 1926 Trust "Turkmensol" established a production of sodium sulfate on the southern shore of the Bay near Cape Umchal. In 1929, a single Trust "Karabogazsulfate" was created for commercial production of salts. Production was based on a new "pool" method when surface brine was pumped to the nearby dry hollows and due to natural evaporation mirabilite was deposited on the bottom. The brine without sulfate was dumped to the other pool, and mirabilite was collected by special machines and sent to the customers [1, 11].

Beginning from 1930, due to a sharp drop in the Caspian Sea level, the natural behavior of the Kara-Bogaz-Gol Bay has changed significantly. In 1939, the flow of sea water in the Bay has dropped to 6 km^3, water and salt balance has been disturbed. In August of that year, as a result of saturation of the brine of the bay, crystallization of edible salt began. The increase in salt concentration in the brine, and changes in its chemical composition make commercial production of salt from surface brines unprofitable. It was replaced by a new method of extraction of raw materials from the buried interstratal brines, occurring between the salt layers under the bottom of the Bay, where the salt composition was constant. Since 1954, Trust "Karabogazsulfate" worked only on the buried brines. In addition to sodium sulfate, magnesium chloride (bischofite) and magnesium sulfate (epsomite) were mined. In the long term, it was planned a complex processing of brines for a production of a wide range of chemicals including bromine, boron, magnesium oxide, and some rare earth elements [11].

3.3 Period of Desiccation

As the Caspian Sea level was decreasing, the Bay area declined, depth decreased, the length and depth of the Strait increased. In the 1960s, solid rocks came to the surface and the erosion of the bottom of the Strait almost ceased. The difference in levels between the sea and the Bay has increased, and at the mouth of the Strait the world's only sea waterfall 3.7 m (in 1970) high was formed. Maximum flow velocity in the waterfall exceeded 2.5 m/s. The Strait with blue sea water flowing among the yellow sands, a waterfall, a unique fauna of birds and animals created a unique natural complex on the deserted eastern coast of the Caspian Sea.

By the end of the 1970s, the Caspian Sea water flow into the lagoon dropped to 5–7 km^3, water level in the lagoon decreased to −32 m, the lagoon area was reduced to 10,000 km^2, the volume of water – to 20–22 km^3, and the salinity of the brine has risen to 270–300‰. In order to slow down the fall of the Caspian Sea level, in March 1980, the Strait was closed by a sand dam which prevented the flow of sea water into the Bay (Fig. 10). However, by this time the level of the Caspian Sea has already raised by 0.5 m and continued to increase rapidly until 1995 [11].

After isolation of the Bay from the sea it started to dry rapidly. By the end of 1983, the Bay area was about 1,000 km^2, the volume –0.2 km^3, and depth –0.2 to 0.3 m. By mid-1984 the Bay almost completely dried up and turned into a dry salt lake. Isolation from the sea caused significant chemical changes in the composition of salts, degradation of chemical raw materials, and complexity of its processing technology [1, 11].

3.4 Period of Revival

With the rapid rise of the Caspian Sea level, to preserve the unique salt deposit it was decided to restore the sea water flow to the Kara-Bogaz-Gol Bay. In September 1984, a flow of Caspian water into the Bay was renewed in the volume of 1.5–1.6 km^3. However, this did not lead to the restoration of hydrological and hydrochemical conditions in the Bay. In June 1992, the dam has been destroyed, and the natural flow of sea water into the Bay was resumed. The contour of the lagoon gradually returned to its previous shape. At a high level of the Caspian Sea in 1993–1995 the flow into the Bay reached 37–52 km^3, well above the previously observed values. By mid-1996, after a complete filling of the Bay, water flow

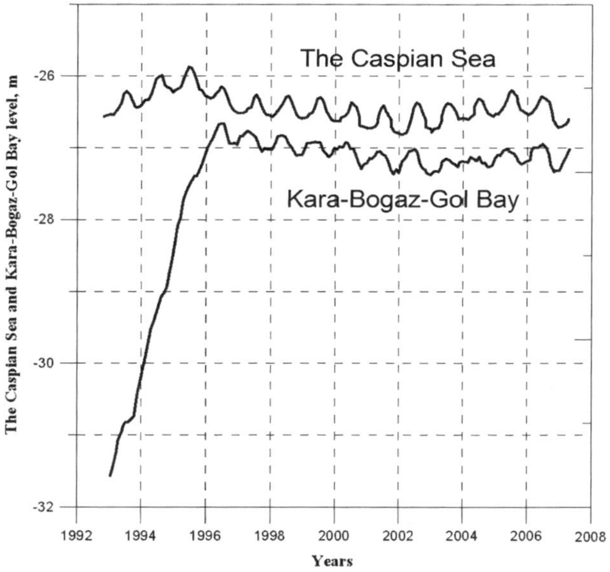

Fig. 11 Interannual and seasonal variability of the Caspian Sea (the *upper line*) in Kara-Bogaz-Gol Bay (*lower line*) level shown by the altimetric data of the *TOPEX/Poseidon* and *Jason-1* satellites over the period from January 1993 to April 2007 according to the project "Surface water monitoring by satellite altimetry," LEGOS, France (http://www.legos.obs-mip.fr/soa/hydrologie/hydroweb/)

through the Strait began to be determined only by evaporation from the Bay surface, and the flow decreased to 17 km^3 in 1999. The difference in levels of the sea and the Bay decreased from 6.9 m in 1992 to 0.2–0.6 m in 1996. Between the water level in the sea and the lagoon a direct relationship was established: the increase in the sea level led to a rise of the Bay level and vice versa (Fig. 11) [1, 11].

The process of filling the Bay and establishment of a new hydraulic regime with the Caspian Sea is well traced by satellite altimetry data (TOPEX/Poseidon, Jason-1 and others) with high spatial (5–6 km) and time (5–10 days) resolution [1, 3, 5, 11]. It was recorded that up to mid-1996 an active filling of the Bay at a rate of 170 cm/year was observed (Fig. 11). Then the elevation of the Bay stopped and only seasonal changes became evident, which were in a good agreement with the seasonal variations of the Caspian Sea level (Fig. 11). Currently fluctuations of the Kara-Bogaz-Gol Bay occur at elevations of about −28 m.

Significant volume of sea water flow into the Bay in 1992–1995 led to slow the rise of the Caspian Sea level by more than 30 cm. Today, the Kara-Bogaz-Gol Bay has completely restored its shape in the former boundaries, and its water area has returned to 18,000 km^2. There is an opportunity to revive the unique natural landscape and the environment of the Kara-Bogaz-Gol Bay, as well as to continue using the rich natural resources of the Bay.

4 Conclusions

The Caspian Sea thermohaline structure responds clearly to changes in large-scale external factors, such as heat and water fluxes through the sea surface and river runoff. Changes in the Caspian Sea hydrological parameters inevitably reflect on functioning of marine ecosystems. The analysis of the long-term (1956–2000 period) alterations in the Caspian Sea hydrology and external thermohydrodynamic factors enabled to recognize two main states of this system [1, 8, 10]. During the first one reduced river runoff coupled with relatively high winter severity results in strengthened ventilation, which involves the whole water column. This has an effect of water salinity increase and vertical salinity distribution homogenizing. This is accompanied by vertical expanding of seasonal thermocline, decrease of temperature gradients, and cooling of deep waters.

The other sea state corresponds to the opposite combination of the external factors – increased river runoff and milder winters. This leads to a stable vertical salinity stratification, which impedes vertical exchanges of heat and salt and distorts regular ventilation of near-bottom layers. In this case the lower boundary of the actively ventilated zone is located at 200–300 m depth. In winter, intermediate layers are intensively fed by colder surface waters; it results in a drop of water temperatures within the lower part of the thermocline. The thermocline itself becomes thinner, while its gradients become higher.

The two states lasted different periods of the Caspian history. While the first state was observed during 10 years from late 1960s to late 1970s, the followed second state persisted through 1995 (i.e., during 17 years). The duration of these states is likely to be controlled by macro-circulation processes such as North-Atlantic Oscillation and others [8].

The latter of two states appears to be less favorable for the Caspian ecosystem, because the increase of vertical water density stratification is normally accompanied by a drop of dissolved oxygen concentrations in deep waters along with weakening of supply of the surface euphotic layer with nutrients. The decrease in thickness of the upper quasi-uniform layer and sharpening of summer thermocline contribute to concentration of pollutants in surface waters. This poses a serious hazard to the sensitive Caspian Sea biota. Analysis of rapid and dramatic changes in the Caspian hydrological regime highlights the need for complex environmental monitoring and adequate environmental protection system.

The Kara-Bogaz-Gol Bay has completely restored its level and borders within their former outlines and occupies an area of about 18,000 km^2. This shows the possibilities for revival of the unique natural landscape and environment of the Kara-Bogaz-Gol region, and for the protection and use of the richest natural resources of the Bay. Once again, the history of Kara-Bogaz-Gol Bay has been instructive for decision-makers and scientists and showed that without multidisciplinary and comprehensive research of the ecological, economic and social aftereffects one should not alter natural equilibriums reached over thousands of years, and the human intervention into the complicated processes proceeding in the environment may lead to catastrophic results.

The Caspian region, including Turkmenistan, is rich in hydrocarbon resources (oil and gas), the development of which is actively conducted in many parts of the coast and the sea area. Oil pollution is a serious threat to its environment [3, 12]. Development and implementation of an effective environmental policy and integrated monitoring system is the most pressing contemporary problems of the Caspian Sea.

Further sea level, wind and waves, SST, suspended matter and chlorophyll concentration, and oil pollution monitoring at various points of the Caspian Sea and Kara-Bogaz-Gol Bay with the use of satellite remote sensing and other observations should allow us to follow the ecological state and future changes which are extremely important for designing, constructing, and operating industrial installations and infrastructure in the sea and on its coasts, and first of all for providing ecological security for economic activities in the Caspian Sea region [3, 12]. In 2011 we proposed to organize in Ashkhabad the National Center for Satellite Monitoring and Regional Climate Change in Turkmenistan with the support of European Space Agency (ESA) and Committee on Space Research (COSPAR) [12]. This center could effectively monitor the ecological state, oil pollution, and sea level of the Caspian Sea waters of Turkmenistan, especially in such sensitive areas as the National tourist zone "Avaza", as well as in the areas of the oil and gas production offshore, and in the areas of new offshore projects, such as the Trans-Caspian gas pipeline.

For those readers who are interested in the Caspian Sea region, its geography, nature, ecology, resources, history, ethnography, and economy we recommend the Encyclopedias of the Caspian Sea we published in Russian edition in 2004 [13] and second Russian edition in 2013 [14], as well as in English edition published in Springer in 2010 [2]. These encyclopedias also include a chronology of historical events having relation to the Caspian Sea development and study for recent 300 years, from the time of Peter the Great to the present.

References

1. Kostianoy AG, Kosarev AN (eds) (2005) The Caspian Sea environment. The handbook of environmental chemistry. Vol. 5: water pollution, Part 5P. Springer, Berlin, 271 pp
2. Zonn IS, Kostianoy AG, Kosarev AN, Glantz M (2010) The Caspian Sea Encyclopedia. Springer, Berlin, 527 pp
3. Kostianoy AG, Lebedev SA, Solovyov DM (2013) Satellite monitoring of the Caspian Sea, Kara-Bogaz-Gol Bay, Sarykamysh and Altyn Asyr Lakes, and Amu Darya River. In: Zonn IS, Kostianoy AG (eds) The Turkmen Lake Altyn Asyr and Water Resources in Turkmenistan. Springer, Berlin
4. The Caspian Sea (1992) Hydrometeorology and hydrochemistry of the seas. Project "Seas," Vol. VI, Issue 1, Hydrometeorological conditions. Gidrometeoizdat, St. Petersburg, 359 pp
5. Lebedev SA, Kostianoy AG (2005) Satellite altimetry of the Caspian Sea. Sea, Moscow, 366 pp (in Russian)
6. Kosarev AN, Nikonova RYe (2010) Caspian water balance and current sea-level changes. In: The Caspian region: environmental consequences of the climate change, Moscow State University, Moscow, pp 253–256

7. Trukhchev D, Kosarev A, Ivanova D, Tuzhilkin V (1995) Numerical analysis of the general circulation in the Caspian Sea. C R Acad Bulg Sci N11–N12:31–34
8. Kosarev AN, Tuzhilkin VS, Kostianoy AG (2004) Main features of the Caspian Sea hydrology. In: Nihoul JCJ, Zavialov PO, Micklin PhP (eds) Dying and dead seas. Climatic versus anthropic causes. NATO ARW/ASI Series, Kluwer Academic, Dordrecht, pp 159–184
9. Kosarev AN, Yablonskaya EA (1994) The Caspian Sea. Academic Publishing, The Hague, 259 pp
10. Kosarev AN, Tuzhilkin VS (1995) The Caspian Sea climatic thermohaline fields. State Oceanographic Institute, Moscow, 96 pp (in Russian)
11. Kosarev AN, Kostianoy AG, Zonn IS (2009) Kara-Bogaz-Gol Bay: physical and chemical evolution. Aquat Geochem 15(1–2):223–236. doi:10.1007/s10498-008-9054-z
12. Kostianoy AG, Lebedev SA, Zonn IS, Lavrova OYu, Solovyov DM (2011) Satellite monitoring of Turkmenistan. Signal, Moscow, 16 pp (in Russian and English)
13. Zonn IS (2004) The Caspian Sea Encyclopedia. Mezhdunarodnye otnosheniya (International Relations), Moscow (in Russian)
14. Zonn IS, Kostianoy AG, Kosarev AN, Zhiltsov SS (2013) The Caspian Sea. Encyclopedia, 2nd edn. Vostochnaya kniga (Eastern book), Moscow (in Russian)

Karakum Canal: Artificial River in a Desert

Igor S. Zonn

Abstract The idea to use the water of the Amudarya River for irrigation of the Karakum Desert was shaped in the eighteenth century and it was partially realized during the tsarist time in Russia. But only in the 1950s the Karakum Canal, the world's major hydraulic engineering project, was designed and constructed. After Turkmenistan became an independent state this canal was renamed Karakum River. The artificial Karakum River ("outflow" from the Amudarya by analogy with inflow) connected Amudarya, Murghab, and Tedzhen rivers into a single water system making the basis for economic development of the country. This artificial river permitted to extend the irrigated lands for growing cotton, fodder crops, vegetables, and melon crops; to create fishery farms; to water desert pastures and, accordingly, stimulate development of distant-range grazing of cattle; to develop shipping and use the waters of this river in industry and power engineering.

Keywords Irrigation, Karakum Canal, Water resources

Contents

I.S. Zonn (✉)
Engineering Research Production Center for Water Management, Land Reclamation and Ecology "Soyuzvodproject", Baumanskaya Str. 43/1, Moscow 105005, Russia
e-mail: igorzonn@yandex.ru

I.S. Zonn and A.G. Kostianoy (eds.), *The Turkmen Lake Altyn Asyr and Water Resources in Turkmenistan*, Hdb Env Chem (2014) 28: 95–106, DOI 10.1007/698_2012_194,

1 Introduction

Casting glance on the map of Turkmenistan you can easily see the disproportion in the geographic combination of the main natural factors: water and land are separated by the vast expanses of the Karakum sands. Water is represented by a powerful flow of the Amudarya River going over the northern boundary of the desert; large tracts of fertile lands occurring as sports along the mountain range of Kopetdag make the southern margins of the desert. Therefore, since old times the nature itself had suggested the giant task for this region – to repair the disturbed harmony, to connect water and land to form symbiosis that will provide for maximum use of productive forces, agriculture improvement and general development of economics in the republic.

This task can be realized by implementation of a large-scale project: to withdraw the required amount of water from the Amudarya, to direct it via the canal in the desert to the lower reaches of the Murghab River, further on to the lower reaches of the Tedzhen River and to the water-deficient areas in West Turkmenistan up to the Caspian Sea.

2 History of the Karakum Canal Project

It is said that in the ancient times the Turkmen people did not ask "How much land did you have?" but they asked "How much water did you have?" Almost 90% of the Turkmen territory is occupied by deserts where life depends on availability of water. The ancient archives saved many plans on irrigation development of the Karakums, the world's largest desert. There are data referring to the Neolithic Time (late fourth – early third millennia B.C.) saying that the people used mountain springs in the piedmont areas of the Kopetdag bordering on the Karakum Desert for cultivation of cereals. In the first half of the first millennium B.C. the Murghab, Tedzhen, and Atrekriver valleys were developed for agricultural purposes. In the third century B.C., the Parthian Empire was founded here which territory expanded as far as the piedmonts of the Kopetdag Mountains. The rulers of Parthia were titled "King of Kings." In the ancient times it was the largest state after Rome.[1] The economy of Parthia was based on agriculture.

Herodotus, Polybios, and Pliny (third to first centuries B.C.) wrote in their treatises about extensive application of land irrigation in Southern Turkmenistan based on construction of large canals and irrigation facilities.

[1] Nisa was the capital of Parthia and its spiritual center and, at the same time, the core of the Parthian Empire. It occupied two uplands close to the present-day settlement of Batir 18 km from Ashgabat which was, probably, a satellite of Nisa.

The wars for conquering the territories of Southern Turkmenistan occurred till the late nineteenth century in which the irrigation canals were destroyed periodically. In 1713 Khodzha Nepe, the head of one of the clans of Mangyshlak, traveled from the Caspian Sea to Petersburg to meet Tsar Peter the Great. During this meeting he told Tsar that in the ancient times the Amudarya River ran into the Caspian Sea. In the past the presently barren plains were blossoming oases. Khodzha Nepe asked the Russian Tsar to help Turkmen people to return Amudarya into its former channel. That time Russia investigated the possibility to construct a waterway from Russia to India. Thus, in 1714–1717 Peter the Great organized two expeditions to this area, but both of them ended in failure – the Tsar's treasury had not enough money for such ventures.

By the end of the nineteenth century the communal-tribal land ownership prevailed in this region. During this time about 66 thou ha of lands in the south of Turkmenistan were under irrigation, and 47 thou ha were irrigated with waters of the Murghab River. Maintenance of irrigation systems was conducted by water users. In 1881–1884 Southern Turkmenistan was the part of Russia as the Trans-Caspian Region. This deed stirred keen interest on business circles of Russia to new lands. Cotton fever embraced various strata of the Russian society. Even Tsar Alexander III owned 25 thou desyatins (1 desyatina = 1.45 ha) in the Murghab River valley for cotton growing. In 1887 after occupation by the Russians of the Merve oasis with its fertile lands the greater part of these lands were given to private ownership of the son of Alexander III – Nikolai II, the last Russian Emperor. In 1895 some works on rehabilitation of the irrigation systems applying European hydraulic engineering and even subsurface drainage were conducted, but on a rather modest scale. From that time on "the Murghab sovereign (Tsar's) estate appeared invariably in all official documents as the most significant achievement of the Tsar's administration in irrigation development in Russian Turkestan."

The idea on transfer of the Amudarya waters across the Karakum Desert to the west into the water-deficient regions of the Trans-Caspian Region was shaped on the basis of the results of geographic reconnaissance studies and after construction in 1896 of the first railroad here that connected Krasnovodsk (present-day Turkmenbashi) with Chardzhou. Famous geologist, later on Academician V.A. Obruchev conducted surveys along the railroad route. And he suggested an idea that the chain of solonchak depressions (shors) originating near the Amudarya on the border with Afghanistan and stretching northwest into the desert represented the remnants of the ancient Amudarya channel (10–20 thou years ago) that went into the Caspian Sea in the geological past (the Aral had not yet existed that time). About 10–20 thousand years ago these channel-type depressions were known as Kelif's Uzboi. Later on it was found that the Kelif's Uzboi was a relict channel of the Bakh River in Afghanistan. Discovery of the Kelif's ancient river channel attracted the attention of the Russian engineers and businessmen.

By late nineteenth – early twentieth centuries the interest to cotton growing for development of the Russian textile industry had grown enormously. And the southern regions of Central Asia if they are ensured adequate water supply had bright perspectives in these respects.

Water resources in the south of Turkmenistan are formed by the flow of two small rivers – Murghab and Tedzhen and it should be mentioned here that their flows were formed mostly beyond the border of Turkmenistan. The average annual flow delivered to the territory of Turkmenistan with the Murghab River is 1.7 bill m^3, and with that of Tedzhen – 0.9 bill m^3. In view of the unfavorable distribution of the flow of these rivers that are principally snow and rain fed it was necessary to construct water reservoirs. The solution of this problem is very complicated here due to silting processes, which is affirmed by more than century-old experience of water reservoir maintenance.

The problem of land irrigation in the valleys of the Murghan and Tedzhen rivers with water from the Amudarya River supplied via the canal going across the Southeastern Karakums was first raised by engineer-economist G.P. Sazonov who elaborated his proposal based on the literature data. In autumn 1906 this proposal was discussed at a special meeting of scientists of Russia.

In 1907 the scheme of water diversion from the Amudarya to the west via the Kelif's Uzboi shors was suggested by engineer M.N. Ermolaev who developed it on the basis of the results of field studies. He presented the Project on the Amudarya Flow Diversion to the Merve and Tedzhen Oases for Irrigation of 516,000 Desyatins of Lands in the East of the Trans-Caspian Region. In 1915 irrigation engineer F.P. Morgunenkov suggested preliminary considerations on the use of the Amudarya water for irrigation of lands near the Murghab River and also remote areas. However, the advanced ideas of the Russian engineer required performance of significant volumes of works in extremely difficult conditions, such material and money input that for many years they were only the dream and they could be accomplished only in distant future.

Even during V.I. Lenin's Plan GOELRO it was claimed that many hopes were with connected with land reclamation and power industry. And later on these hopes proved too great. Already in the 1920s the main reliance was made not on the free initiative labor of the people, but on the forced labor of GULAG prisoners, cheap and mass. The nature transformation plan started with construction of the Belomor Canal. This plan also included construction of the Dnieper-Danube, Volga-Urals, Aral-Caspian and Great Turkmenian canals and many others.

Water supply and irrigation of the Karakum Desert became the concern of the State which was confirmed in February 1925 at the First All-Turkestan Congress of Soviets where the question on the Karakum Canal construction was raised and the resolution on the water and land reform was adopted. Soon after the Congress the results of the long work of the complex expedition to the Southeastern Karakums headed by F.P. Morgunenkov were received. He had proposed to use the natural slope of the Kelif's Uzboi to direct water by gravity from the Amudarya to the Murghab oasis. In 1927 the pilot water release was conducted into the Kelif's Uzboi from Basaga-Kerki Canal constructed in the Amudarya River valley. The water flowed by gravity in this canal to the southeastern Karakums for 100 km. Further on this experience was repeated many times. It showed that the salt cover of the Kelif's shors rapidly disappeared under the Amudarya silt (the Amudarya carried four times more of suspended solid than the Nile; 1 m^3 of water contains 4 kg of such solids) and did not cause saltation of inflowed water, and seepage losses reduced gradually due to silting of sandy soils.

Thus, to a certain extent, already during that time the doubts expressed by some prominent engineers were refuted. Thus, Tsinzerling [1] called the project of the Karakum Canal construction a risky venture in technical terms, and D.D. Bukinich considered that this project, in general, was "a priori doomed to failure."

Large-scale surveys on various alternatives of the Karakum Canal route were developed beginning from 1940 after adoption of the Resolution of the Soviet of People's Commissars of the USSR and the Central Committee of the Communist Party "On Measures for further development of agriculture, in particular, as concerned long-stapled cotton cultivation in the Turkmenian SSR." In spite of military time, there was compiled a project plan for construction of the Karakum Canal according to the so-called "southern altlernative" under the leadership of the Ashgabad's engineer I.V. Boltenkov. The last alternative in contrast to the "northern alternative" (suggested formerly by F.P. Morgunenkov) envisaged construction of the canal through the eastern part of the Kelif's Uzboi having almost latitudinal direction. Further on, when the Kelif's shors moved to the northwest, the canal route preserving the latitudinal orientation approached the Murghab at sufficiently high elevations. This gave the opportunity to use gravity irrigation at maximum area of the Murghab delta plain lying on the elevations below the canal. When using the Northern Alternative, utilizing almost the whole Kelif's Uzboi, commanded irrigation area in respect to the canal is considerably reduced, and the cost of construction increased. Besides, there also increased inevitability of large water losses at the tremendous idle run through the Karakums and leaving the Murghab-Tedzhen lands with scanty water "ration." In favor of the Southern Alternative was also the fact that for the route from Kerki to Mary and Tedzhen there was already prepared the report of 37 volumes mentioned above. The Southern Alternative made it possible to involve local population in construction and land development in the old-developed and densely populated regions, to rest upon the railroad, the Amudarya and the cities from Chardzhou to Krasnovodsk with their manufacturing industry [2].

The Second World War had suspended the way of studies and construction works. After the end of the war I.V. Stalin initiated adoption of the Resolution of the Government of the USSR "On Construction of the Main Turkmenian Canal." This Resolution stipulated that the canal route should go from Takhia-Tash to the coast of the Krasnovodsk Bay withdrawing from the Amudarya 350–400 m^3/s of water with a possibility to increase this figure to 600 m^3/s without water diversion into the Caspian Sea.

The construction of the Main Turkmenian Canal (MTC) was started in 1950. However, in connection with discovery of considerable freshwater resources in the Yaskhan freshwater lens (which water was quite sufficient to satisfy the needs of Western Turkmenistan), the construction of MTC was ceased in 1953 [3].

3 Description of the Karakum Canal

The construction of the Karakum Canal named after N. Niyazov was resumed in 1954 following the Southern Alternative (until 1990 this canal was named after V.I. Lenin and from this time on and until 2007 it bore the name of First President of Turkmenistan S. Niyazov). It was the largest hydraulic structure of the irrigation and land reclamation importance in the deserts of the world. At present this canal branches off from the left bank of the Amudarya at Basaga settlement where it takes water and runs as far as Kazanjyk (now Bereket), the town in the northwestern foothill plain framing the slopes of the Kopetdag. Its length is 1,380 km, the head structures take out annually about 13.5 km^3 of water from the Amudarya River. After completion of construction when the canal would bound the western edges of the Kopetdag and come to the coast of the Caspian Sea its length would make to 1,400 km and the annual flow 18 km^3.

The eastern and southern regions of Turkmenistan covering one-third of the whole territory of the state are involved in the sphere of irrigating measures carried out on the canal's runoff. The region of Lebap, Murghab and Tedzhen oases, vast territory of the Kopetdag plain and the industrial regions of West Turkmenistan are supplied with the Amudarya water by means of the canal irrigation system.

In 2010 in the canal area over 900 thou ha were irrigated, almost four times more than before its construction. In the future, by means of the canal there will be irrigated up to 1 mln ha of fertile lands.

Now the width of the canal in its upper reaches is, on average, not less than 100–120 m. With lake-type expansions at certain sites it reaches 1–1.5 km. The canal is navigable from the Amudarya to Mary. By its sizes, hydrological indices, peculiarities and the scale of processes taking place in it the canal is inferior to many rivers of Central Asia delivering four times more water than all irrigation sources in the south of Turkmenistan [4]. It is quite rightfully that the Karakum Canal is called the man-made river, Karakum-Darya or Karakum River [5]. The designers used the natural slope of the land surface from 250 m on the Amudarya to minus 28 m near the Caspian Sea thanks to which the water moves in the canal by gravity from east to west.

The enormous route through which the Amudarya water now flows was not made at once. Canal construction was carried out by stages providing irrigation of certain areas independently of readiness of its subsequent sites. The canal was built without lining and had the river intake which reduced its cost and accelerated its construction. Apart from this the indicated structure of the water tract makes it possible to change the section of the ready channel according to water intake increase from stage to stage, but not to construct it from the very beginning for all water discharge required in the far future. Therefore, with each subsequent stage of construction there simultaneously expands the active part of the headworks structure till the beginning of the construction site.

The Karakum Canal was planned as a year-round functioning waterway. For this purpose, for the first time in the world practice of large irrigation canals construction, the internal runoff regulators, i.e. reservoirs at the end of the canal sections for all construction phases were stipulated by the plan. Water reservoirs envisaged by the canal construction make it possible under conditions of continuous action to accumulate water reserves during autumn–winter period which stipulated the increase of the summer flow for irrigation by 30–35% without increase of the discharge capacity of the water main [6].

Stage I. Amudarya-Murghab Section. The first stage of the canal from the Amudarya to Murghab has the length of 397 km, out of which about 300 km pass in the desert sands. The unlined canal construction began in 1954 and already in 1959 water year arrived to the Murghab delta. Headwater discharge was 130 m^3/s, water intake of the annual Amudarya flow – 3.5 km^3. The canal provided 88 thou ha of lands with irrigation in the Murghab oasis and together with irrigated lands in the Murghab basin – about 170 thou ha. There could be made such comparisons: the first stage of the canal with earthwork volume of 129 mln m^3 and indicated above sizes was constructed within 5 years. The Panama Canal with the volume of earthwork of 212 mln m^3 and length of 81.6 km was constructed during 34 years, and the Suez Canal 173 km long with earthwork volume of 75 mln m^3 – during 11 years.

When designing and constructing the first stage of the canal there had to be solved a number of cardinal problems and almost insuperable difficulties overcome because the world practice had no such experiment of water delivery to long distances through the sand desert. The main problem concerned the seepage losses from the canal. That time it was possible to make a forecast of seepage change in interaction with forming regime of groundwater. It appeared that with large-scale of initial losses (2.4 m^3/s per 1 km of the canal) the volumes of seepage will be gradually reduced due to bed colmatation and bedrock saturation under the canal. However, if you first excavate the canal and then supply water into it then with water intake of 100 m^3/s water will reach the Murghab not less than within the year after the end of construction and in minimum quantity.

This fact obliged instead of “dry” method of construction to elaborate never formerly used method of “behind water delivery.” Its essence consists in the fact that water is admitted into the narrow and short (up to 10 km) pilot cut, excavated by bulldozers and excavators, by the use of which the canal bed is enlarged to the design section by means of dredgers. It made possible to solve simultaneously several problems: to accelerate the reduction of seepage losses by ground saturation after pilot cut drive, provide for all communications by water tract (instead of land, impassable), supply builders in the desert with water, establish floating dwellings and auxiliary premises on the barges instead of surface ones, etc.

Stage II. Murghab-Tedzhen Section. The second stage of construction proceeded from 138 km from the Murghab to Tedzhen, out of which 70 km run across a sandy desert. Simultaneously, there was extended the prepared site to discharge 198 m^3/s with annual water intake from the Amudarya of 4.7 km^3. To accumulate the free autumn-water flow there was constructed the Khauskhan water reservoir 460 mln m^3 in capacity (design full storage is 1.5 km^3). Construction of the second stage began in 1960 and within unprecedented short term – 7 months, the Amudarya water was supplied to the Tedzhen River and the length of the canal reached 553 km. Irrigation area increased by 72 thou ha.

Stage III. Tedzhen-Geoktepe Section. The third phase of the canal construction started in 1961 by building of the pilot cut from Tedzhen to Ashgabad. Already on the 12th of May 1962 the residents of Turkmenistan's capital met the Amudarya water. Its capital, suburban rural economy and the zones of recreations were provided with water, and there appeared the Kurtly (Western) water reservoir of 48 mln m^3 in capacity, (Eastern) and Sport water reservoir of 6.3 mln m^3. Headwater discharge increased to 317 m^3/s with corresponding reconstruction of already built part of the canal, the length of which made up 793 km. Simultaneously the canal was elongated by 44 km from Ashgabad to Geoktepe and completed by it tail Kopetdag water reservoir of 190 mln m^3 of useful capacity (now 500 mln m^3); the volume of the Khaukhan water reservoir increased to 875 mln m^3. The canal irrigated 50 thou ha of virgin lands at the area near the Kopetdag plain and provided stable irrigation of 20 thou ha, formerly irrigated by local flow. Construction of the third stage was completed in 1975. The maximum head discharge reached 400 m^3/s and the total irrigation area reached 514 thou ha (Fig. 1).

Stage IV. Geoktepe-Karandjik Section. In 1971 with the pilot cut Geoktepe-Kazandjik the construction of the fourth stage began. Simultaneously, there was expanded the active part of the canal for head flow rate of 580 m^3/s (annual flow – 13.5 km^3). The canal reached Karandjik in 1981 and the canal length was 1,150 km.

Stage V. Karandjik-Atrek Section. In 1992 there began the construction of the fifth stage of the canal into the direction of the Atrek River for irrigation of subtropical virgin lands in Southwestern Turkmenistan. Four thousand years ago 160 thou ha were irrigated here or nearly 1 ha per each inhabitant. Construction of the canal's southwestern branch going as far as Kyzylatrek with its terminal reservoir (250 mln m^3 in capacity) will permit to irrigate about 30–35 thou ha.

The Ashgabad-Erbent water pipeline of more than 100 km long was constructed for supplying water to the cattle breeding pastures and for irrigating the arid lands of the Central Karakums. The 250-km long water pipeline stretched from Gazanjik to Nibetdag (now Balkhanabat). Turkmenbashi and Cheleken are supplying water to these large industrial centers.

At present the Zeid headwater reservoir of 3.5 km^3 in capacity is being built substituting the Kelif's Lakes which would be completely silted. In 2002 its capacity reached 1.2 km^3. Instead of the river intake the canal will receive water from the headworks built on the Amudarya River nearby the Kyzylayak settlement. It is intended for the regulation of suspended and coarse sediments of the Amudarya at the canal head and for spawning of herbivorous fish.

In general, five reservoirs with a capacity over 2.5 km^3 are built on the Karakum River. They are designed to accumulate the winter river flow for its subsequent use in the vegetation period (Fig. 2).

Fig. 1 Scheme of irrigation of lands in the region of V.I.Lenin Karakum Canal (till 1966)

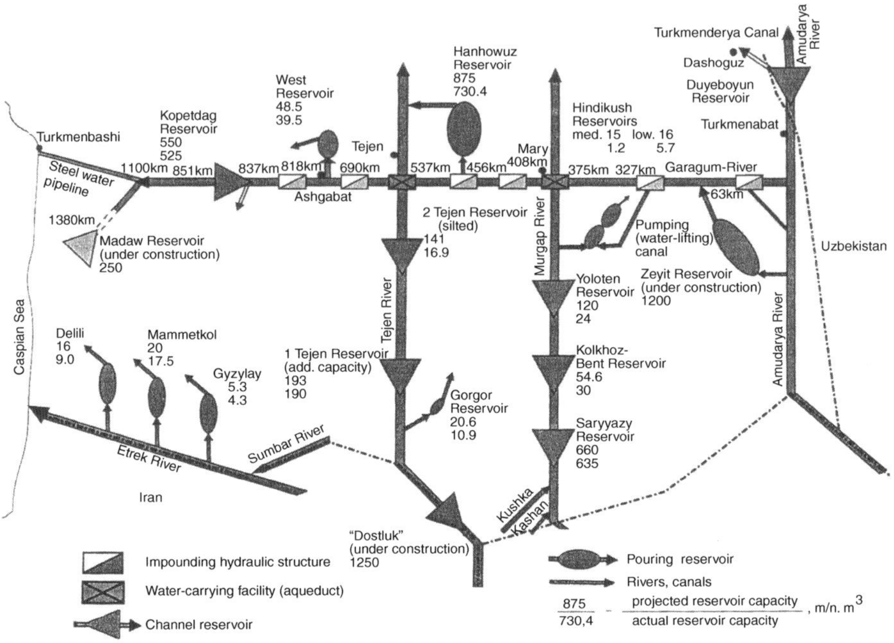

Fig. 2 Complex irrigation system of the "Karakum River" [7]

4 Importance of the Karakum Canal

When the canal was dug out and water came to the regions called "Dzhanakhyr" that could be translated as "the end of life," the people mounted donkeys and crossed hundred kilometers to look at the canal. Many of them were carrying away water in phials being afraid that other people would not believe them.

The importance of the Karakum Canal for the national economy of Turkmenistan can be hardly exaggerated. From 1958 to 1990 the cotton production in the canal zone increased 4.5 times. In 1990, out of 1,457 thou tons of raw cotton harvested in Turkmenistan the half was grown up in the canal zone, including all fine-fiber cotton varieties. The arrival of "big water" intensified cattle breeding, water supply of pastures, growth of the oil and gas industry and helped to resolve the problem of the population water supply. The canal caused reconstruction of all industrial production, stipulated formation of the new agro-industrial complex the key element of which is a large irrigation construction (Fig. 3).

The canal attracted thousands of waterfowl that chose to stay here for wintering, flight rest and feeding having changed the migration routes along the canal. The number of wild boars increased and deer – decreased because the canal traversed their migration routes. The Karakum Canal became a part of the general program of desert development.

Fig. 3 Editors of the book (Igor Zonn – *left*, Andrey Kostianoy – *right*) at the Karakum Canal near Ashkhabad in November 2011

One of the problems is overgrowing of the canal with reeds that narrows the waterways and aggravates its passability. So far in order to clear it there were used dredgers or the reeds were cut out. Now of wide use are the vegetable-food fish, such as Amur whitefish that eats only reeds, but also returns food to the people by its meat, including famous “Tolstolobik” locally called “Muksun”, “Topuga” that is known abroad as silver carp. Wonderful balyks could be prepared of this fish. The Amur whitefish or grassy carp eats up to 2 kg of plants a day per each kilo of its weight and its weight may reach 25–30 kg.

In addition to real achievements that made possible construction of the Karakum Canal the extremely unfavorable ecological situation was formed around it. Withdrawing about 12 km^3 of water per year from the Amudarya, the canal, according to calculations of some Turkmenian specialists, loses to 3 km^3 to seepage and evaporation, while according to estimates of the Uzbek specialists – 5 to 7 km^3. These water losses were due to seepage. Because of water seepage in the area of the Karakum Canal at a distance to 10 km and over there formed inundated and saline lands. Many ancient channels and depressions are filled with seepage water and when you are flying over the canal route there is an impression that down in the desert the lake area appeared – “Karakum Venice.” It should be noted that as a result of excessive irrigations the large land areas in the oases were taken out of use due to salinization problems. For 4 years after water supply to Ashgabad the groundwater level at 50 m from the canal rose to 10–11 m at within 1–2 km distance. Abrupt uplift of groundwater on the northern fringe of Ashgabad demanded construction of over 100 wells for vertical drainage to lower groundwater level [8].

All these negative consequences of the canal construction were unavoidable because during its design as an advanced project and with the purpose to shock the world by the scale of nature reformation the problems of consequences and impact on the environment were not considered and even raised. It is already during the period of glasnost and perestroika the people began to speak about the canal that it became the cause of ruin of the Aral and some called it the "State crime." Today the zone of canal influence covers about 300 thou km^2 with population of over 2.5 million.

Construction of the Karakum Canal being the wide-scale flow transfer has changed radically the territorial redistribution of water resources across Turkmenistan. It helped to eliminate the century-old imbalance between the extensive areas of fertile lands locating in one part of the country and the water resources locating in the other part. Thus, about 80% of arable lands concentrate in the south and southeast of Turkmenistan, while water resources – in the east of the country.

The Karakum River is important and needed. It permits to address simultaneously several burning economic issues related to development of irrigation, agriculture and forestry, power engineering, industry, transport, and urban construction. This canal supplies water to all major industrial centers, such as Ashgabad, Turkmenbashi, Mary, and Balkhanabat.

References

1. Tsinzerling VV (1927) Irrigation in the Amudarya Basin. Publishing House of the Water Management Board of Central Asia, Moscow, USSR (in Russian)
2. Zhmuida VB (1951) Main Turkmen Canal. Geografgiz, Moscow (in Russian)
3. Grinberg LM (1965) The Karakum Canal. Ashgabad, Ylym (in Russian)
4. Orlovsky NS (1999) Creeping environmental changes in the Karakum Canal zone of impact. In: Glantz MH (ed) Creeping environmental problems and sustainable development in the Aral Sea Basin. Cambridge University Press, Cambridge
5. Brummell P (2005) Turkmenistan. The Bradt travel guide. The Globe Pequot Press, Connecticut, USA
6. Yanshin AL, Melua AI (1991) Lessons of environmental errors. Mysl, Moscow (in Russian)
7. National Environmental Action plan by the President of Turkmenistan Saparmurat Turkmenbashi on Environment Protection in Turkmenistan, Ashgabat (2002), Mega, Turkey
8. Sustainable Development of Turkmenistan. RIO+10. National Report. 2002, Ashgabad (in Russian)

Sarykamysh Lake: Collector of Drainage Water – The Past, the Present, and the Future

Leah Orlovsky, Offir Matsrafi, Nikolai Orlovsky, and Michael Kouznetsov

Abstract Sarykamysh is one of about 2,500 artificial lakes-collectors of drainage water in Central Asia. The Lake is located in a natural depression in the northwestern part of Turkmenistan, it receives irrigation surpluses and soil washing drainage water from Dashoguz and Khoresm oases. The area of the Lake has grown from 12 km^2 in 1962 to 3,955 km^2 in 2006. In terms of volume the change is from 0.6 km^3 to 68.56 km^3, respectively. Currently, the national plan is to create a new lake-accumulator in the Karashor depression – the Golden Age Lake. Nowadays, less water is being discharged into the Lake, and in the future its area/level will decrease significantly. With average annual evaporation rates of 1.2–1.4 m/year, the drying process is expected to be rapid. The study attempts to model the possible scenarios in the development of the Lake following a change of inflow. This research deals with the retrospective study of the parameters of the lake in the past 40 years using GIS and remote sensing methods in order to suggest a forecast of these parameters. The forecasted parameters will enable the mitigation of the negative regional impacts of the Lake's changes. A three-dimensional model of the Sarykamysh depression was built using the 1940s topographic maps. Topex/Poseidon altimeter data, early Corona satellite images, and time-series of the Landsat satellite images were applied on Digital Elevation Model (DEM) together with ground measurements of the parameters of the Lake and meteorological data. The model was calibrated and validated, and the water balance of the Lake was calculated, enabling us to suggest with higher accuracy, an optimal future inflow.

Keywords Aral, Central Asia, Forecast, Model, Remote sensing, Turkmenistan, Water balance

L. Orlovsky (✉), O. Matsrafi, N. Orlovsky and M. Kouznetsov
The Jacob Blaustein Institutes for Desert Research, Ben-Gurion Universiry of the Negev, Sede Boqer Campus, 84990 Midreshet Ben-Gurion, Israel
e-mail: orlovsky@bgu.ac.il

I.S. Zonn and A.G. Kostianoy (eds.), *The Turkmen Lake Altyn Asyr and Water Resources in Turkmenistan*, Hdb Env Chem (2014) 28: 107–140, DOI 10.1007/698_2012_191,

Contents

1 Introduction

The Aral Sea Basin is one of the most ancient areas of irrigated agriculture – favorable climate and natural soil fertility contribute to the development of farm production here. The total irrigated area in the Aral Sea Basin reaches up to 8 million ha [1]. Irrational use of the land and water resources during last 3–4 decades resulted in the intensive salinization of the irrigated lands. At present more than half of all irrigated lands in the region are salinized at a different rate. In order to keep the soil fertility and/or to improve its ameliorative state the construction of the collector and drainage network started in early 1960s. Since that time about 161.8 thousand km of collectors and drainage canals and more than 6,000 drainage wells have been built [2]. In the Aral Sea Basin this network produces drainage water flow within the volume of 32.71 km^3 annually. Part of it (20.29 km^3/year) returns to the rivers, and 12.42 km^3 drain out of the irrigated area and flow to the desert depressions [3]. As a result, in the periphery of irrigated area 2,341 water bodies of new type – collectors of drainage flow within the total area of 7,066 km^2 have been formed [4]. Sarykamysh Lake is the largest one among them occupying more than 50% of the total area of such water bodies. Its volume takes even bigger portion of the total volume of newly forming lakes of Central Asia. Water of Sarykamysh Lake is highly salinized and contains biogenic matters, admixtures of pesticides, defoliants, and fertilizers.

Formation of such a vast water body in the arid region has had an important and practical significance. Nevertheless, the problem of formation and development of the water bodies – accumulators of drainage water is not adequately explored. The researchers regarded the Sarykamysh Lake as a natural object [5–7] with quickly developing fishery [8, 9]. It should be noted that before early 1960s such approach was justified, since the first discharge of the drainage water into Sarykamysh depression took place in 1961.

Hydrological and hydrochemical regime and water quality in the Lake are studied rather poorly, in spite of several publications dealing with above-mentioned problems [10–12], as well as the problem of interaction of the Sarykamysh Lake and surrounding environment [13, 14]. The future of Sarykamysh Lake depends on the plans of Turkmen government to create the "Lake of the Golden Age" in Karashor depression within the distance of approximately 80 km from Sarykamysh in order to collect the drainage water from the Turkmen irrigated areas and Khoresm oasis of Uzbekistan. According to this plan the inflow into Sarykamysh Lake will be reduced significantly – to 0.7–1.1 km^3 [15].

This paper presents an attempt to reconstruct dynamics of water and salt regime of the Sarykamysh Lake from the beginning of its infill by drainage water and to give the forecast of its future after construction of the Turkmen Lake.

2 Study Area

2.1 Geographic Features

Sarykamysh is a natural depression located about 200 km southwest to the Aral Sea, south to the Ustyurt Plateau (Uzbekistan), and north to the Karakum Desert (Turkmenistan). The depression is located to the west of the Dashoguz agricultural massif – the most densely populated area in Turkmenistan. The depression stretches approximately 150 km from the north to the south, and by 90 km from the east to the west. The deepest point in the depression rests at altitude of −46 m ASL.

The Lake borders west Ustyurt cliffs where the altitude differences are 50–80 m. Northwest of the lake is another depression – Assake-Audan, which stretches northwestward and is surrounded by the Ustyurt plateau. North of the Sarykamysh there is another small depression, which (like the previous one) is connected to the Sarykamysh by steep slopes and not by cliffs. Further on the north and northeast, the borders are still part of the Ustyurt cliffs. In the east the Lake borders with the ancient delta of the Amu-Darya, with the area adjacent to the Lake being flat lowland, with an altitude of 58 m ASL, and a moderate slope to the northwest.

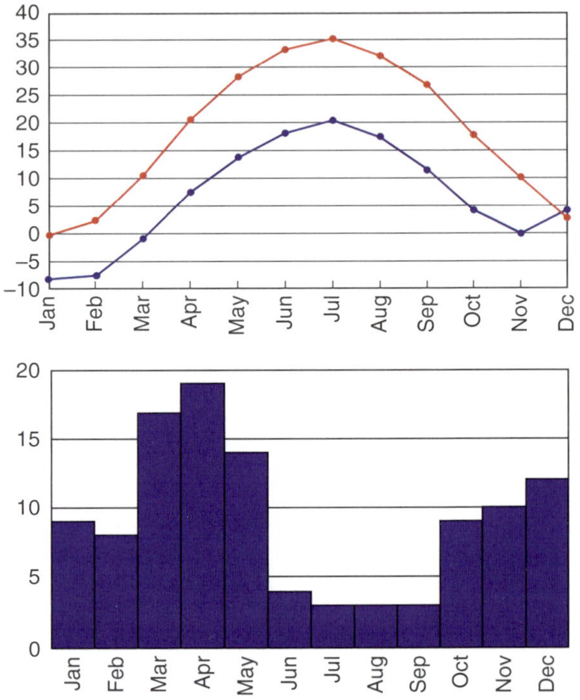

Fig. 1 Climatic parameters at Kunya-Urgench meteorological station. Average summer (*red line*) and winter (*blue line*) temperature distribution (*upper*) and precipitation (*lower*) (after www.allmetsat.com)

2.2 Regional Climate Characteristics

The climate is defined as extremely continental. The region is characterized by very hot and dry summers, combined with relatively cold winters with very little snow. Approximately 190–210 days a year are above 10°C. The air is very dry and creates enhanced evaporation conditions. The average annual precipitation in the area is 99.7–110 mm (measured at the Shakhsenem and Kunya-Urgench meteorological stations, respectively, for 1953–2006 years). The temperature in Kunya-Urgench (located 120 km northeastern to Sarykamysh and 70 km northern to Shakhsenem) can range between a maximum of 35°C (Fig. 1), while minimum temperatures can drop down to −9°C. The depression bed consists mainly of 30–50 m clay formations, making it almost impermeable. Average infiltration rates reported by Kes' [16] regarding the pre-flooding era, ranged between 0.3 and 0.6 m/day. The groundwater salinity stands on 40–60 g/L.

2.3 Geological History of the Sarykamysh

The formation of the depression is related to the Neogene (Upper Tertiary System). As a result of the alpine collision, the foundation of the Turanian plate collided into the sub-plates of the Ustyurt and the Trans-Ungus faulted and lifted. The parent material consists of marls, limestones, clays, and sands of Paleogene and Miocene, which lay in the shallow depths and are sometimes exposed. These formations are evidence to the presence of the Sarmatian Sea, which covered this area during the Tertiary period.

After the formation of the graben at the end of the Pliocene, the high sandy-pebble banks were formed, as well as dense sandstones and conglomerates containing the Apsheron fauna. These banks survived the forces of erosion in the Sarykamysh and in the Assake-Audan Lakes at the heights of −5, 0, 40, 50, 75, and 80 m ASL. This phenomenon is evidence of different periods of marine cover, whereby the sandy belts imply stable periods of marine cover, whilst the sedimentations in-between imply periods where the bodies of water experienced rapid drying.

In the late Pleistocene period, for approximately 1–2 million years, the Amu-Darya River flowed through the Central Karakum. Within this period the Sarykamysh was dry with typical desert landscapes. The sandy ridges that stretched in meridianal and sub-meridianal directions were formed as a result of the Aeolian processing of Sarmatian, Akchagyl, and Apsheron deposits. Ridges were represented by the whitish and yellow Oolite sands of local origin.

In the late Khvalynian age (about 16,000 years ago) the Sarykamysh's history connected with the dynamics of the Amu-Darya River's downstream [16, 17]. This process evolved in three main stages:

1. The Amu-Darya filled the deep Khoresm depression creating a vast lake.
2. Alluvial deltaic sediments filled the lake and turned it into a wide marsh area (water logging plain). Water made its ways westward and "found" the Sarykamysh depression.
3. Discharge into the Sarykamysh brought the water level up to 58 m ASL. The water levels covered the Sarykamysh, the Asake-Audan, and the northern depression. The surplus water flowed southward via the Neogene tectonic trough and formed the Uzboi course. The flow through Uzboi to the Caspian Sea existed from the X to the II millennia B.C.E. (Fig. 2).

According to several scientists the climax of the Holocene flood period was around the fifth or sixth millennia. As a result of the major climax the whole area of the Sarykamysh and the Aral Sea was covered. Currently this area is referred to as "The Great Aral Sea" [17, 46]. Figure 3 indicates suggested stages of the Paleo-hydrological system of Western-Central Asia. It is important to note that the existence of the great Aral Sea at this phase is not accepted by all the scientific community who deal with this issue.

Along with the shift of the Amu-Darya's course northward, the outlet to the Sarykamysh also migrated. Alluvial sediments forced the water to migrate northward from the Kanga-Darya to a new channel, the Daudan. The same process forced the water northward again toward the Daryalyk, which is the current inlet to the Sarykamysh. Since the water flows through the deepest channel, the Daryalyk is also the deepest of all three channels mentioned (to 60 m).

Fig. 2 Terraces in the southern shore of the Sarykamysh Lake. Photography: March, 2007

The Uzboi absorbed the surplus water (we do not know how far) up until the second or the third millennia B.C.E. In these times, accumulated sediments ceased the Uzboi flow and forced the Sarykamysh water to seek a way out northward, i.e. into the Aral Sea, forming the Akche-Darya channel. This last event significantly reduced the flow into the Sarykamysh. Together with the geomorphologic events described, the evolution of agricultural systems unfolded.

There is still evidence that suggests the Sarykamysh Lake was present in the first millennia B.C.E. Apparently the disconnection of the Amu-Darya from the Sarykamysh took place somewhere in the end of the second millennia. In this phase the Sarykamysh absorbed only significant floods water.

2.4 Human Presence Impact

The beginning of the artificial irrigation era downstream from the Amu-Darya is related to the end of the second millennia B.C.E, when the main part of the Amu-Darya flow went to the Sarykamysh. In the middle of the first millennia B.C.E. when the Amu-Darya's delta to the Aral Sea started to form, the man-made irrigation system began to evolve as well.

Part of the new irrigation canals led water from the Amu-Darya to the Sarykamysh. Deltaic and agricultural sediments went through processes of erosion and transportation on the western slope of the Sarykamysh (barkhan dunes, hillocks, and longitudinal ridges of sand).

Around 1,500 years ago, agricultural systems were destroyed, due to the invasion of nomad tribes into the area, which put an end to the slavery system. This process was followed by the destruction of the irrigation system, and with it all the

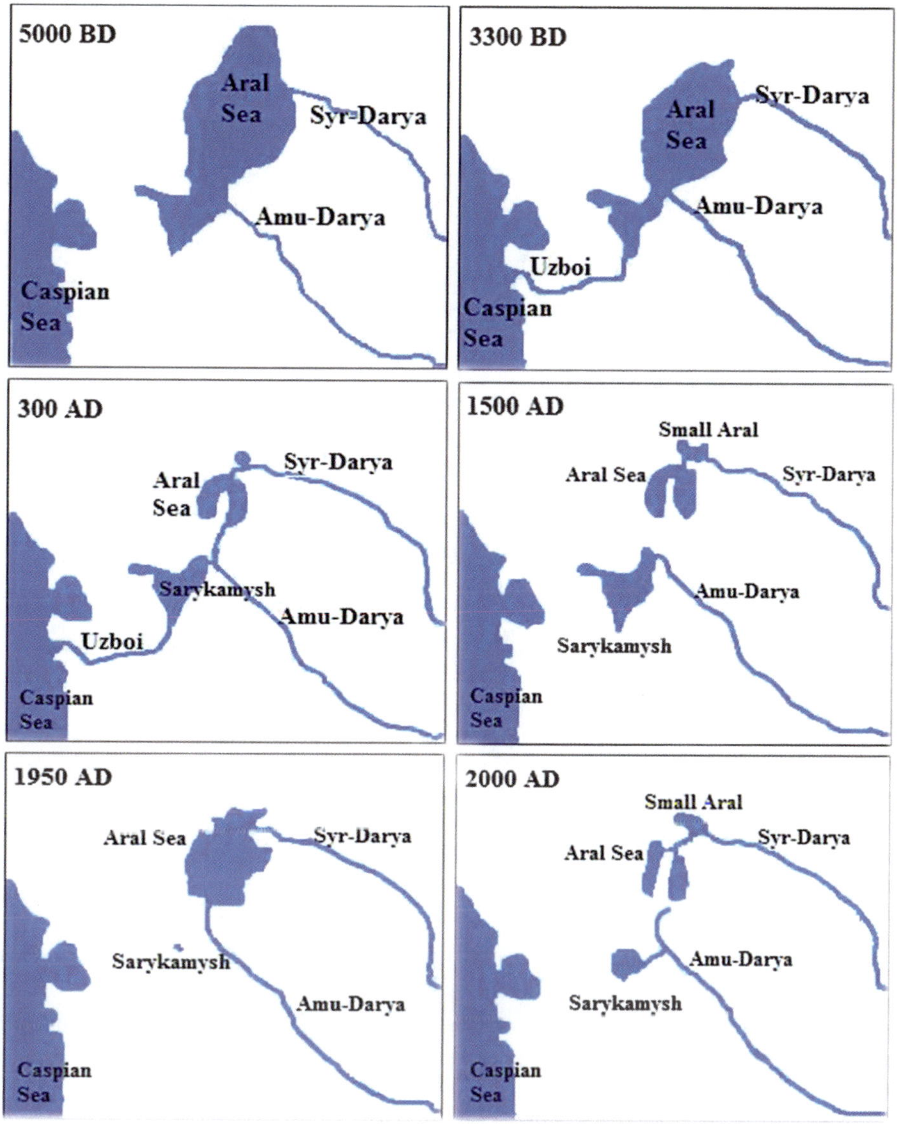

Fig. 3 Development of hydrographical systems of Central Asia (after IFAS–Aral Sea home page)

systems that diverted water toward the Aral Sea. The waters breached the basis of the Daudan and the Daryalyk. This resulted in the refilling of Sarykamysh up to the level of +52 m ASL.

The Sarykamysh is mentioned in Muslim and Persian scripts (seventh to eighth centuries), as an important and famous freshwater body in the area. Within these scripts the "Igdik-Kal'at" is referred to as an important fortress on the eastern bank of the Sarykamysh.

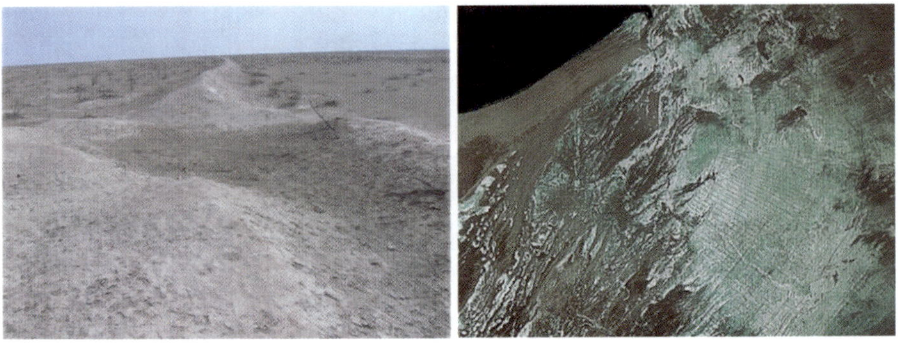

Fig. 4 Remnants of batteries in an ancient agricultural area south to the Sarykamysh Lake. The batteries apparently used to collect runoff for irrigation purposes. This set of batteries can be seen by satellite imagery (*right*). Photography (*left*): O. Matsrafi (March 2007). Source (*right*): www.Google.com

The development of the feudal agrarian system signaled the revival of the cultivation and reconstruction system. Again, the drainage was to the Aral Sea, i.e. the flow to the Sarykamysh has ceased once more.

Remnants of ancient agricultural settlements can be found around the Sarykamysh Lake. One area is so clear that it can be observed by satellite imagery about 5–10 km from the shore (Fig. 4). This settlement can be affiliated with population climax near and around the Sarykamysh during the fourth to sixth centuries or the fourteenth to sixteenth centuries [18]. Another option is that the ancient settlement was inhabited by Turkmens who lived in the neighborhood of Sarykamysh, namely the Adakly-Hyzyr tribe and created a complicated system of artificial irrigation [19].

The cultivation in the Amu-Darya's delta continued until the Mongolian invasion in the thirteenth century. The dams and irrigation canals were demolished, which created another flow to the Sarykamysh depression. Again, the water filled up the depression to the level of +50 m ASL. After a short period of rehabilitation, destruction took place at the end of the fourteenth century by Timur. In this phase the Sarykamysh was flooded, together with the Assake-Audan, which even caused surpluses to the Uzboi. A gradual reconstruction of the irrigation system led to a situation where the main part of the Amu-Darya's discharge flowed to the Aral Sea, and the Sarykamysh water level decreased down to +10–15 m ASL.

Aladin et al. [20] highlights the description by Jenkinson, who arrived in 1548 from Russia through the Ustyurt Plateau, and camped on the shore of the Sarykamysh, "the lake from which the Uzboi originates." The water he found was "fresh and sweet." Jenkinson's description teaches us that in the sixteenth century the lake's water level rose again and stabilized around +30 m ASL. At that time the western slopes of the depression were cultivated more intensively than previous times. Later the water level slowly dropped till around +10 m ASL.

The presence of the mollusk *Cardium edule* in the sediments of that period implies that the water's salinity increased and the lake became saline. The water level continued to drop until the seventeenth century, when the Sarykamysh almost completely dried up. In this phase a few heavily salinated ponds located at the deepest spots of the depression were found. For a short period, the few floods occurring in the eighteenth and nineteenth centuries disturbed the Aeolian and geochemical processes that took place during the dry period.

In 1881 an expedition mapped the area and defined two longitudinal saline lakes and between them, a natural canal. The total area of the lake was measured: 148 km^2 and the level measured was −37 ASL, and was no deeper than 5–6 m. The expedition reported salinity rates of 40–47 g/L.

Around 1914–1917 the lake was reported to be completely dry with a typical arid landscape (it is possible that the expedition did not reach the actual water body due to the size of the depression). At the end of the 1920s a lake was found in the depression. This was fed by saline groundwater (springs) from the western slopes of the depression (eastern slopes of the Ustyurt). The dehydration of the lake led to a sharp decrease of the groundwater table, resulting in all the springs going dry except for one: "Gurluk-Su" ("Saline-Water").

2.5 Landscapes and Geomorphology of Sarykamysh Depression

The landscape is well defined with the topographic variety: the water, the soil, alluvial deposits, and flora. Most of the formations are identical to the base of the Ustyurt and are sometimes exposed on the surface. In the southern, "Chirli" area ("the well"), the limestone formation "sinks" under the Quarterian deposits. Terraces found on the banks of the lake provide evidence for different levels. The terraces are built from clayey sand and in some cases with rounded pebbles that imply intensive costal-marine activity [21].

There are five large clusters of erosion and sedimentation from the Pleistocene and Holocene. In the upper part of the eastern coast five large channel entries are noticed: Kichkne-Darya, Kuruja-Uzak, Kanja-Darya, Daudan, and Daryalyk. The last two also transfer water from the CDW system. Irrigation canals can be found occasionally on the alluvial fans. These canals are evidence of an ancient agricultural activity.

The products of the Aeolian activity are connected with the dry periods. Parts of the Aeolian products result from local transport. Others are the results of Aeolian import processes from the surroundings into the depression. The imported matter tends to be white in color and finer in texture comparing to the internal transfer sands. The Aeolian Products are mainly sandy ridges from all sizes with inter ridges (honeycombs, barkhans, and sandy plates).

The position of the Ustyurt cliffs sets the trajectory of the winds and consequently the position of the dunes. In the south the dunes face the Karakum Desert. Most sandy ridges are 3–5 m high and form a mass of dunes, stretching from north–north–east to south and south–south–west of the depression toward the Uzboi corridor, where the high sandy ridges were covered by the ancient Sarykamysh Lake. The common type of dunes are 1–2 m high and are typical mainly to the bottom of the depression, while higher dunes and sandy ridges are more common in the outer circle of the sandy area. The honeycomb ridges are formed especially in the areas were the dominant winds create a whirlwind where the ridges meet uplands, especially in the center of the depression. These honeycombs are found mainly in the northern parts and in the southwestern parts of the depression toward Uzboi corridor. Barkhan dunes are found especially between the honeycomb structures. Strong winds in the hilly areas are accelerated due to the presence of these topographic features. In addition, the area receives very low amounts of precipitation, which do not contribute to vegetation development.

Solonchaks and saline lakes were an important part of the depression landscape. Conditions of extreme evaporation encourage an increase in capillar flows of water, up to the ground surface. Furthermore, the high evaporation pressure salts remain on the soil surface after the evaporation. Solonchaks are found in the lower spots of the depression because of the natural tendency of water to accumulate by gravity.

According to all the conditions described so far the Sarykamysh depression can be classified into five areas:

1. Central part – solonchaks and sands, which can be defined mainly by isolines of altitude 0 m ASL. Five deep depressions that were covered in the Pliocene. The soil is characterized by salty clay.
2. Caplarkyr – the high ridge that stretches into the depression from north–east to south–west. This ridge has a relative height of 40–50 m above the surroundings, and divides the central part into two parts northwestern and southeastern.
3. The slopes – with evidence of remains from previous floods (the sandy-pebbly terraces) all slope facing inward to the center of the depression. The slopes are divided according to different characteristics such as slope angle, morphology, and sediments composition:

   (a) The slopes of the Ustyurt in the north and in the west (from the piedmont of the Ustyurt toward the center of the depression) form a 5–6 km strip. The surface consists mainly from loamy-clay, a formation of 15–20 m thick, which thins toward the center of the depression down to several meters. Besides a few channels that cross the strip from the Ustyurt to the center of the depression, this section is pretty flat. The northeastern part of the strip is around 3 km wide and becomes wider toward the southwest area.

   (b) The northwestern area of the strip meets the Assake-Audan and the northern depression. Eastern slopes strip: generally flat area but highly scared by rills. As previously described, this is an area of deltaic sediments (ancient Amu-Darya) which are the main component covering the soil. The strip width is 15–30 km. The slope becomes steeper and deeper, from 0 m isoline upward, and rills can be found 3–10 m deep. In some places the rills are as deep as a 40–50 m. Here as well, the dominant wind trajectory is from northeast to southwest. Large sand bodies are found: Barkhans can be found between

Kichkane-Darya and the Daryalyk. In the southwestern part of the delta several local depressions are covered by saline water (4–13%), and there is higher salinity in the center of the depressions. The salt types are mainly chlorides and sulfates, where the chlorides concentration decreases and sulfates concentration increases with distance. Gypsum is also found on the edges 0–40 m. Humus concentrations are found, increasing in concentration with depth.

(c) The southern plain is stretched southward in a long fin into the Uzboi. Here the flattest slopes of the Sarykamysh depression are found. Within this area, the lithographic base is identical to the Ustyurt plateau. In the southern side of this area limestone exposures are found, which are separated by the lake's sediments layers. Northward the plain is covered by a layer of gravel and pebbles combined with takyrs and solonchaks. Many sandy ridges are found, some of which stretch up to 20 km long. The height of the Aeolian products decreases northward until approximately 3 m around the center of the Sarykamysh depression. This situation implies that the longer the area is exposed, the greater the Aeolian accumulation will be. Figure 5 presents a description of the landscape, by dividing the depression to eight landscape units separated by age and geomorphology.

2.6 Soil, Flora and Fauna

The presence of arid-land vegetation is connected not only to geomorphological history and other habitat conditions but also to lithography formation, soil types, salinity, groundwater level and its salinity. All parameters mentioned form a complicated spatial soil pattern (Fig. 6), which varies significantly from one point to another in the depression. Hence, the vegetation appears in patches.

The central part of the depression is dominated by solonchaks, where only halophytic vegetation can thrive in this environment. In most of the flooded area (1991) the vegetation is covered. The vegetation population is represented by species of *Halocnemum*, *Salsola*, *Suaeda*, and *Tamarix* [21].

Another common type found in the depression is *Haloxylon*, of two species: *Haloxylon persicum* is found more on the thick Aeolian deposits, while *Haloxylon aphyllum* is found on heavier and saltier soils. Another noticeable species in this habitat is the *Carex physodes*. The vegetation is quite sparse, 40–50 individuals per hectare, coupled with subshrubs, the vegetation cover does not exceed 20–40%. Overgrazing is affiliated as one of the underlying causes of the low vegetation cover. The *Haloxylon*'s ability to thrive is thanks to the close groundwater, and to the subsurface slope from the Amu-Darya toward the Sarykamysh.

Thin pebbly soils and relatively high groundwater are found in the Caplarkir, and in some areas in the south part of the depression that supports vegetation to the height of 1.5 m.

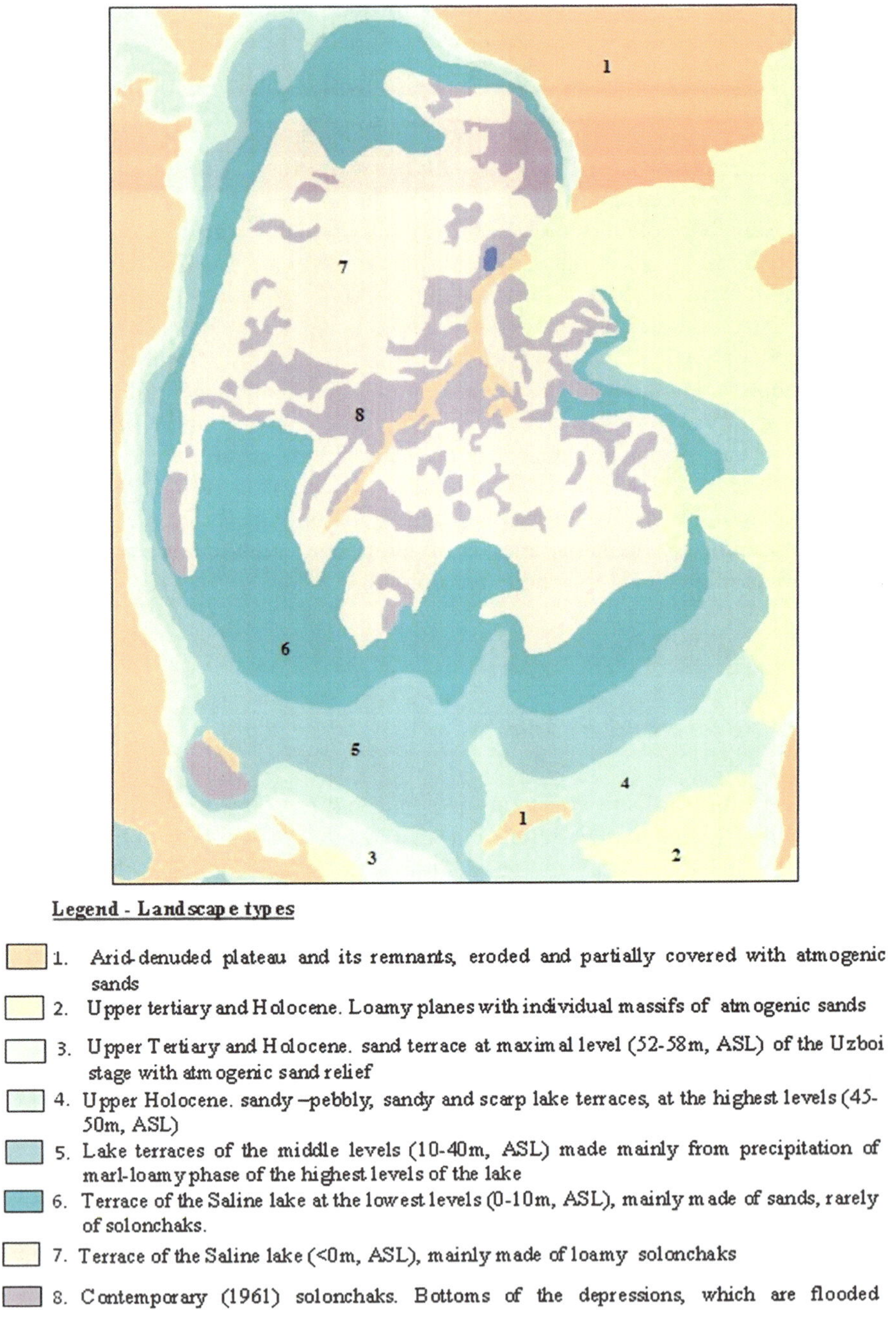

Fig. 5 Landscape map of Sarykamysh depression (after Tolstov and Kes' [5])

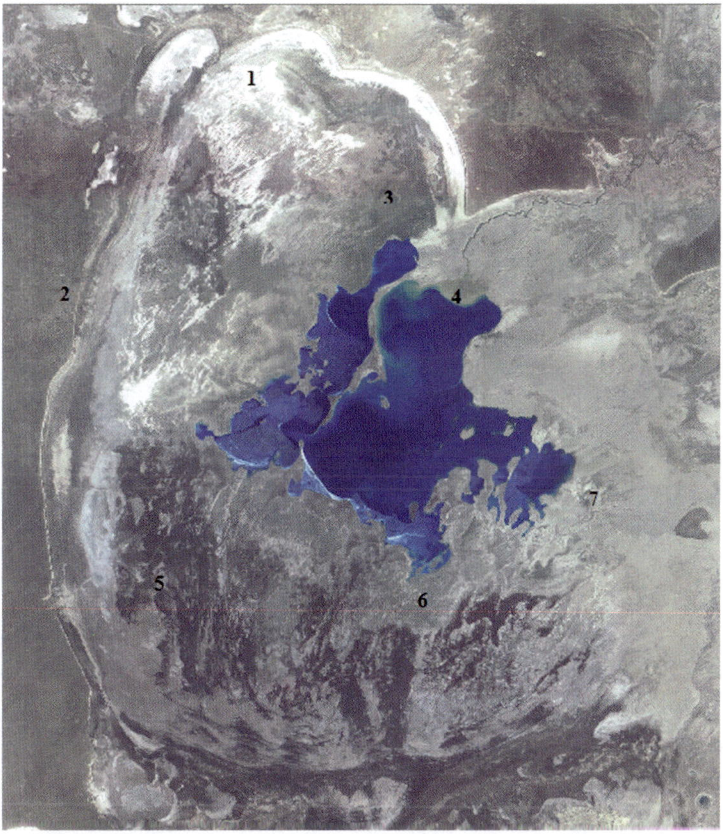

Fig. 6 Soil and landscapes types in the Sarykamysh depression as can be identified in Landsat MSS image (1973): (*1*) sands; (*2*) gypsum plateau; (*3*) sands and takyr-like soils; (*4*) alluvial sediments; (*5*) sands and solonchaks; (*6*) solonchaks; (*7*) takyrs. Satellite image view was obtained from Earth-Explorer, USGS

On gray-brown soils, which are closer to the depression's slopes, *Artemisia terrae albae* and *Salsola arbuscula* can be found. In areas with clay deposits *Artemisia diffusa* and *Salsola orientalis* can be found. *Anabasis salsa* occurs in moderately salinized soils. On the Ustyurt slopes the *Haloxylon aphyllum* stands out in its density over the area (10–50 m between individuals). *Salsola arbuscula* and *Atraphaxis spinosa* can be found in small "sinks" on the slopes of the Ustyurt. In this area the vegetative population is quite poor. Significant ephemerae presence is represented by: *Bromus tectorum*, *Alyssum aureum* and *Salsolae* semi-shrubs.

The southern plain is characterized by a large variety of soils that form a mosaic of patches between the base rock exposures. These patches are inhabited by *Artemisia kemrudica*, *Salsola gemmascens*, and *Salsola rigida* with ephemerae and annual *Salsolae*. The vegetation cover in this part of the depression does not exceed 40%. Wherever there is a thick sandy cover, the *Salsola arbuscula* and the *Haloxylon aphyllum* are dominant. With them the vegetation population is also represented by *Reaumuria fruticosa*, *Artemisia terrae-albae*, and *Alyssum aureum*. Ephemerae are represented by *Ferula assafoetida* and *Carex physodes*. The old-tugai residual-humus soils coincides with the ancient coastal sandy terraces, where *Haloxylon aphyllum* can be found mixed with *Haloxylon* growth, and with *Carex physodes,* which can also be found in the lower layer. The puffy solonchaks are bare apart from their edges where *Halocnemum strobilaceum* and *Kalidium caspicum* can be found. The sandy hills are inhabited by *Carex physodes*. In the northern area of the upper Uzboi Corridor, in the inter dunar area, biogenic crust is found which is created mainly by *Tortula desertorum*.

Generally, before and after the flooding the fauna in the Sarykamysh is very rich, with a large variety of species: from herbivores through all the food chain up to wolves and other carnivores. Previous to the flooding the fauna was represented by a long list of arid-land species. The list included birds from the Ustyurt, while the solonchaks area and the central lake "Gurluk-Kul" were inhabited by hydrophilic birds and invertebrates. The remoteness of the Sarykamysh and the very few connecting routs enabled the establishment of endemic species, such as of antelope (*Saiga tatarica*), gazelle (*Gazella subgutturosa*), and wild sheep (*Ovis ammon*). Since the area has become accessible to motorcycles, the individual's numbers have significantly decreased.

The birds population includes species such as Houbara Bustard (*Chlamydotis undulata*), Stone Curlew (*Burhinus oedicnemus*), Crested Lark (*Galerida cristata*), Finsch's Wheatear (*Oenanthe finschii*), Red-headed Bunting (*Emberiza bruniceps*), and the Eurasian Eagle Owl (*Bubo bubo*). The last species nests mainly on the Ustyurt cliffs but can be found in some places in the depression.

The reptile population includes species such as: Steppe Ribbon Racer (*Psamophis lineoletus*), Naked-toed Geckos (*Gymnodactylus russowi*), and Russian tortoises (*Testudo horsfieldi*). Among the invertebrates, rough woodlouse (Porcellio scaber) and snails (*Radula tridentata*) can be found.

The center of the depression had some uniqueness in terms of the species that inhabit it. Among these species the sand cat (*Felis margarita*) and wild cat (*Felis silvestris ocreata*) could be found together with rodents such as the Five-toed Pygmy Jerboa (*Cardiocranius paradoxus*), the European ground squirrel (*Spermophilus citellus*), and the marbled polecat (*Vormela peregusna*).

The poorest list of fauna is of the solonchaks: Plovers (*Charadrius*), and of the birds, the Common Pratincole (*Glareola pratincola*), and of the reptiles, the Reticulated toad-headed agama (*Phrynocephalus*).

2.7 Sarykamysh: Contemporary State

The modern anthropogenic stage of the Sarykamysh Lake started at the end of 1950s – beginning of 1960s – with active involvement of virgin lands to agricultural production. Until 1955 the Sarykamysh consisted of five small lakes (Fig. 7a) with

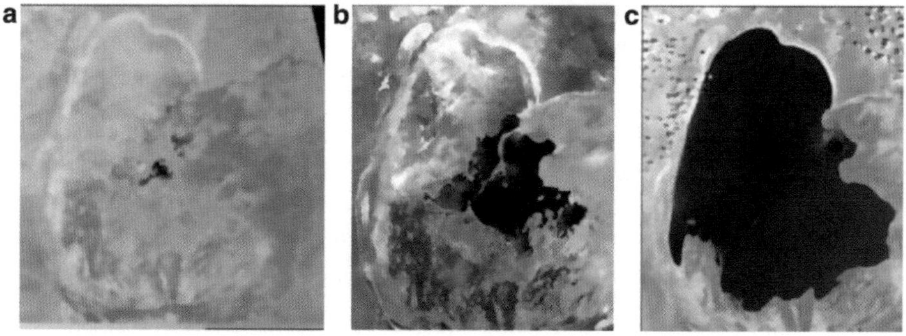

Fig. 7 Sarykamysh evolution as seen by satellite imagery from different years: (**a**) Corona (1962), (**b**) Landsat MSS (1973), (**c**) Landsat TM (2000). Source: Earth-Explorer, USGS

salinity that varied between 75 and 300 g/L. Although drainage water was dispatched with the Daryalyk channel toward the depression, no water reached the water body of the Sarykamysh (on-surface) before 1962, probably due to percolation and evaporation. Today, the Sarykamysh (Fig. 7c) is a vast water body within the area of 3,955 km^2 (2006) that receives inflow of 5.8–7.48 km^3/year.

Pavlovskaya [22] reported a lake's salinity of 12–14 g/L enabling a marine ecosystem to thrive, thanks to the natural supporting conditions and forced by natural pressure resulting from the Aral Sea desiccation.

Fishery in the Sarykamysh started in 1966, with the climax amount recorded being 2,500 tons in 1981–1985 [23]. According to Pavlovskaya [22] the Sarykamysh marine environment is optimal for a significant number of species. Thirteen species were identified as a potential base for commercial fishery. The Lake also carries problems, which are related to its sources. The water being drained from the cultivated areas contain 3% of the defoliants (used mainly over cotton), 2–3% of the pesticides, and 10–15% nutrients. The following three facts make the accumulation process more problematic since: (1) Sarykamysh is a terminal lake. Hence, all the chemicals entering do not leave; (2) the contaminated water carries the chemicals from a large number of fields; (3) an excessive usage of chemicals (20–54 kg per hectare), in the whole Amu-Darya basin.

Obviously the presence of these chemicals damages the fisheries potential. Finally the chemicals are accumulated in the water body, or may sink on its bed, or are absorbed in the biotic system. Many fish develop abnormalities, mainly in their reproduction systems, and suffer many diseases as a result of exposure to these chemicals.

There are very few studies available regarding the modern Sarykamysh Lake, or studies that focus on attempts to model the lake and its behavior, in spite of the fact that the Sarykamysh behavior is expected to cause a significant regional impact.

Alimokhamedov et al. [24] analyzed images of glaciers and lakes (including Sarykamysh) and achieved a satisfying estimation of the physical parameters for local market needs.

Between 1973 and 1985, Nuriddinov [25] monitored changes in the Sarykamysh shore line using Soviet remote sensing images, and defined remote sensing as the preferable method of monitoring lakes of this scale. The author reported that areas identified as being waterlogged on earlier images were identified as flooded by the lake on later images. Therefore, it was concluded that there is a possible subsurface flow involved in the lake's growth. This conclusion correlates with the description offered by Tolstov and Kes (1960) of sand dunes apparent in a large part of the depression.

Kikichev et al. [26] estimated evaporation from the Sarykamysh during 9 months using radioisotopic analysis. The study analyzes the distribution of natural tritium in the Lake's feeding sources, the Lake itself and the atmosphere above it. One of the advantages of this method is the achievement of definite values of estimation regardless of the Lake's size parameters.

Kes' [16] published a summary of the studies, as well as field surveys reports, and in situ data; all of which were conducted prior to the current flood. A significant part of the knowledge is based on field surveys published in a set of expedition reports regarding the Khoresm oasis by Tolstov and Kes' [5]. These surveys are highly important since they are the only scientific description of the dry Sarykamysh depression. In her description, Kes' includes a comprehensive analysis of the geology and geomorphology of the depression, climate, biotic system, and human history. Kes' also brings a mathematical analysis of the depression.

Pavlovskaya [22] studied the potential of irrigation systems and fishery according to the current state (early-mid 1990s), and the changes that irrigation water bodies and fishery have gone through. The study reveals a disturbing level of pollution and salinity within the drainage water systems where fishery takes place. The author concludes that any potential for fisheries will be terminated, unless a general drainage water system rehabilitation program is implemented.

Nezlin et al. [27] estimated the Amu-Darya flow on the basis of global atmospheric precipitation data. In their study they followed Sarykamysh water level changes (among other water bodies) using T/P data and found a correlation between precipitation patterns and trends in the lake's water levels.

This study suggests a model, which will cover the missing patch of Sarykamysh behavior under different scenarios.

3 Materials

3.1 Maps

A set of topographic maps, compiled in the 1940s (prior to the modern flood) were collected (in scale of 1:200,000). The topographic maps enabled basic delineation of the study area according to the identification of the depression and its borders. A soil type map (in scale of 1:600,000) was also collected enabling us to follow the descriptions of the geomorphologic processes in the depression.

Table 1 The list of satellite images used for the retroactive analysis of the Sarykamysh Lake

Image ID	Acquisition date	Satellite platform	Sensor	Spatial resolution (m′)
DS09034A038MC043	18 May 1962	Corona	Film camera B/W	104 m
DS09058A008MC040	29 Aug 1963	Corona	Film camera B/W	96 m
DS09065A024MC042	15 June 1964	Corona	Film camera B/W	133 m
(LM)1174031007306090	01 March 1973	Landsat	MSS	79
LM2174031007513190	11 May 1975	Landsat 2	MSS	79
LM2174031007724690	3 Sep 1977	Landsat 2	MSS	79
LM3174031008016890	16 June 1980	Landsat 3	MSS	79
LT5162031008613910	19 May 1986	Landsat 5	TM	30
(LT)4162031008920310	22 July 1989	Landsat	TM	30
(LM)5162031000024210	29 July 2000	Landsat	TM	30
LT5162031000619410	13 July 2006	Landsat 5	TM	30

Source: Earth-Explorer-USGS

Table 2 Landsat sensors and their bands

	Sensor			
Range	MSS		TM	
Blue			Band 1	0.45–0.52
Green	Band 1	0.5–0.6	Band 2	0.53–0.61
Red	Band 2	0.6–0.7	Band 3	0.63–0.69
IR	Band 3	0.7–0.8	Band 4	0.75–0.90
IR	Band 4	0.8–1.1		
Mid IR			Band 5	1.55–1.75
Short-wave IR			Band 7	2.09–2.35
Thermal IR			Band 6	10.4–12.5

Source: Earth-Explorer-USGS

3.2 Satellite Images

The only data sources available to perform a temporal survey on the development of the lake over the past 40 years were images taken over the Lake from airborne or spaceborne platforms. The earliest images available for the Sarykmysh depression area are analog images taken by Corona Satellite (declassified in 1996). Three Corona images of the Sarykamysh area were collected (1962, 1963, and 1964), which were acquired shortly after the Daryalyk has reached the Sarykamysh depression in 1961, and reflect the early stages of the development of the lake. Tables 1 and 2 show the images collected and their basic parameters.

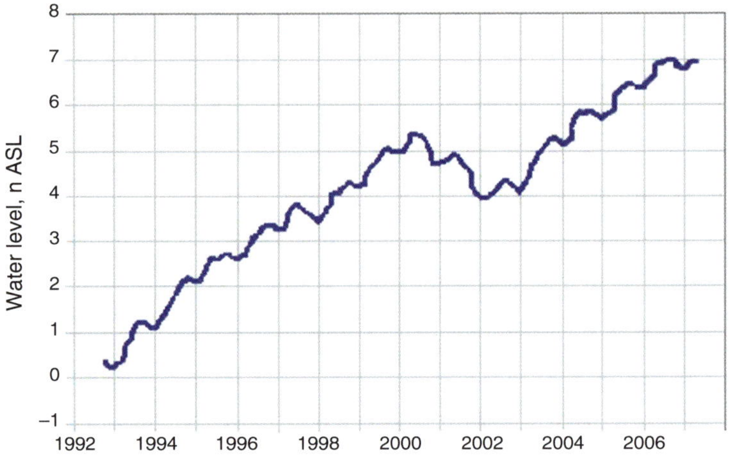

Fig. 8 Altimetry data of Lake Sarykamysh (after LEGOS, online database)

Table 3 Summary of satellite altimetry general characteristics

Satellites	Operation period	Orbital cycle	Accuracy	Minimum target area and width	
ERS2	1995–2002	35 days	>9 cm rms	>100 km^2	>500 m
ENVISAT	>2002	35 days	>9 cm rms	>100 km^2	>500 m
T/P	1992–2005	10 days	>3 cm rms	>100 km^2	>500 m
Jason-1	>2002	10 days	>3 cm rms	>100 km^2	>500 m

Source: Cretaux and Birkett [28]

3.3 Altimeters

The collected images allowed following the development of the lake at least twice in each decade since the early 1960s. There is a missing step in the sequence during the 1990s, since no images of the Lake could be found among commercial imagery suppliers. In order to complete the data sequence we used spaceborne altimeters data collected between 1992 and 2005. Spaceborne altimeters are active radar systems designed to measure altitude of ground or sea surfaces above a certain level. The measured range (R) of the altimeter from the target is calculated based on the time difference (t) between a signal and its returning echo, and the light velocity (C) as can be described by Eq. (1):

$$R = t/2 \times C \tag{1}$$

Altimetry data of the Sarykamysh (Fig. 8) were collected at the LEGOS web database. The altimetry data are an updated average of the four altimeters (Table 3).

Fig. 9 Collecting water samples from Daryalik CDW. Photography: L. Orlovsky (March 2007)

3.4 Local Data

The following local archive data had been collected with the help of the Turkmen National Institute of Deserts, Flora and Fauna, and with the Courtesy of the Turkmen National Meteorology Agency: (1) precipitation for 1953–2006 (measured at Shakhsenem station); (2) water level of Sarykamysh Lake for1970–2000; (3) surface area and volume (1962–2002); (4) CDW inflow to the Sarykamysh (1970–2002), (5) lake's salinity, salt stocks and inflow's salinity (1970–2002).

3.5 Water Samples

Water samples were collected from two areas in the Lake (one close to shore and the other in deeper water). In addition, samples were collected from the two main tributaries of the feeding CDW: the Daryalyk (Fig. 9) and the Ozerny. It is important to note that samples were collected in March 2007, prior to the irrigation and seeding seasons, during the soil washing season, when water used to wash the soil is drained into the CDW. Samples were analyzed for inorganic content and total dissolved solids (TDS) content at the Zukerberg Institution for Water Research of the Bluastein Institution for Desert Research, Ben-Gurion University of the Negev, Israel.

4 Methods

4.1 Topographic Structure Analysis

In order to predict the lake's behavior a retrospective analysis of the development of the lake over the past four decades was performed. The topographic maps were scanned and rectified. The topographic isolines were digitized and a Digital Elevation Model (DEM) was produced (Fig. 10). The plane area was calculated and volume (under the defined plane) for every height in the DEM range.

In order to estimate the accuracy of the DEM, the area and volume of the lake were used as a function of height (level) together with parallel functions based on the local measurements. In addition, nonlinear fit curves were extracted for each of the functions to enable expression of the lake parameters, based on local height measurements or forecasted ones.

4.2 Morphological Development of the Lake

The area of the Lake was extracted by supervised classification (Fig. 11) and was applied onto the DEM to extract the water level and volume of the lake at the time of image acquisition.

As seen in Table 1 there is an 11 years gap between 1989 and 2000. In order to fill up this gap the altimetric data covering most of the temporal gap have been used. Altimetry data were applied onto the DEM to derive the area and volume of the lake and fill in the gap in the images sequence.

4.3 Water Balance

The morphological changes of the lake along the years are a function of the lake's water balance. Asmar and Ergzinger [29] presented a water balance (after [30]), which they used in the analysis of the behavior of the Dead Sea. Based on their study a water balance was defined for Sarykamysh case (Eq. 2):

$$\frac{\Delta V(h)}{\Delta t} = Q_t + S(h)_t\,(P_t - E_t) + \mathrm{GW}_t \tag{2}$$

in which ΔV is the total additional volume added in time Δt (Δt will always be considered as 1 since our calculation is annual); Q represents the total inflow to the lake during time t; S represents the lake's area at time t; P and E are the total precipitation and evaporation at time t, respectively; GW represents the total inflow volume added (or reduced) by ground water. The volume and area can be described

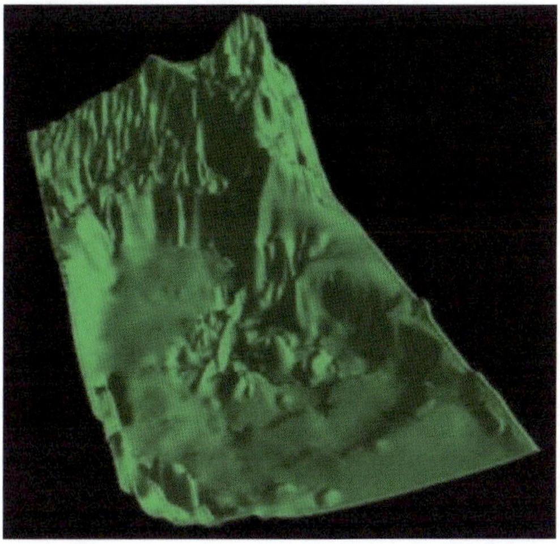

Fig. 10 A 3D view with vertical exaggerations of the Sarykamysh depression produced from topographic maps

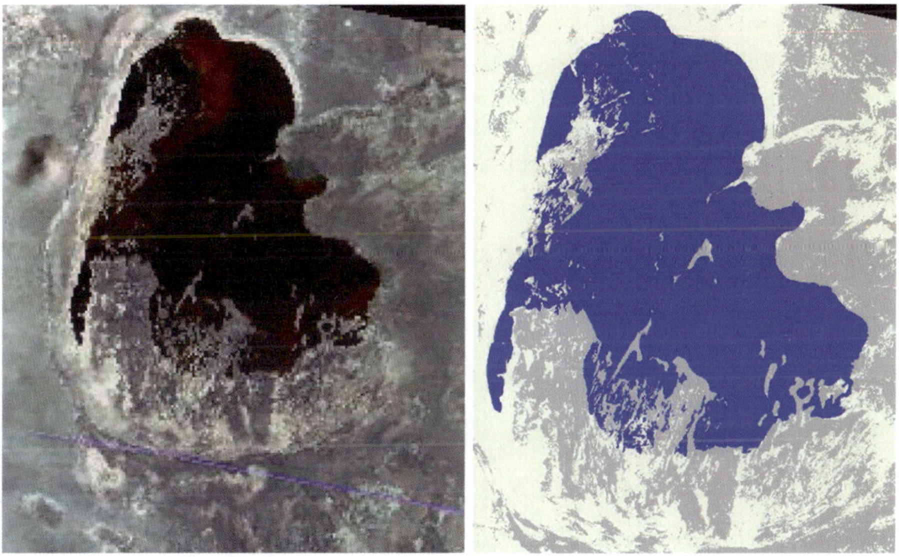

Fig. 11 Sarykamysh Lake in the 1980 MSS imagery (*left*), and supervised classification product (*right*). The flooded parts can be easily identified

as functions of the water level; inflow and precipitation data were used from the local measurements data set, mentioned previously. According to Kes' [16] the contribution of groundwater is negligible; therefore, this part of the water balance was ignored. The only element missing in the balance is evaporation. Applying all available elements into Eq. (1) enables extracting evaporation values for each year. Comparing the calculated evaporation rates with those of other authors enables evaluating the water balance and DEM.

In order to correctly compute the evaporation in the forecast one should take into account the changes in water mass that will affect salinity, which, in turn will affect the evaporation rates [31, 32]. In order to correctly calculate the salinity it is necessary to identify the elements constituting the salt balance. Glazovsky [33] and Benduhn and Renard [34] have presented a salt balance in the following way (Eq. 3):

$$\frac{\Delta SV}{\Delta t} = (S_r\, Q_r) + \left(S_{gw}\, Q_{gw}\right) + A_L\left(S_p\, P_t - S_E\, E_t\right) - e_t \tag{3}$$

where S is the water salinity; V is the lake's volume; ΔSV represents the change of total salt in the lake during time t; S_r and Q_r represent the river's (or canals') salinity and their total volume, respectively; S_{gw} and Q_{gw} represent groundwater salinity and its total volume, respectively; A represents the lake's area in time t. This variable is used as a factor to estimate the salt exchange between the lake and atmosphere together with the following parameters: total precipitation during time t (P_t) and precipitation salinity (S_p), and total evaporation (E_t) and its salinity (S_E). e_t represents the gain and loses of the total salts, which are not included in the calculated addition by inflow. For convenience this value will be referred as "residuals."

Glazovsky [33] found that atmospheric salt absorption through precipitation and loss of salt as aerosols to the atmosphere are two processes that balance each other, and therefore both can be neglected. As was mentioned previously the contribution of groundwater to the lake is negligible. Consequently, the salt balance can be expressed in a simplified way (Eq. 4):

$$\frac{\Delta SV}{\Delta t} = (S_r\, Q_r) \pm e_t \tag{4}$$

Amer [32] defined the relation between evaporation from saline water and evaporation from fresh water as a function of salinity (Eq. 5):

$$\frac{E_{sal}}{E_{fw}} = 1.0 - 0.22S \tag{5}$$

where, E_{sal} is the evaporation rate from saline water; E_{fw} is the evaporation rate from freshwater under the same conditions, S is the salinity in ds/m units. Using this expression we can calculate evaporation rates, calibrate them to freshwater evaporation values, and compare them with multi-annual averages, as measured by Orlovsky [35]. The expected evaporation values in the forecast can be calibrated for every year using the salinity values.

The forecast included applying precipitation and evaporation values together with inflow according to the scenario, and the parameters of the lake. This is defined by the functions based on local measurements or based on our DEM. Precipitation and evaporation values can be fed differently according to climate change expectations.

The last part of the salt balance is the residuals estimation. As explained, residuals will include the following: salt precipitation on the lake's bed, additional salt inflow from surrounding streams that wash in salts during floods, and salts washed from the atmosphere by precipitation, being lost by the spray into the atmosphere. The total salt stock (TSS) difference can be represented as (Eq. 6):

$$\mathrm{TSS}_t - \mathrm{TSS}_{t-1} = (S_\mathrm{r}\ Q_\mathrm{r})_t \tag{6}$$

where TSS_t represents the total salt stock accumulated in the lake (and on its bed) until time t, and TSS_{t-1} refers to the previous year; $(S_\mathrm{r}Q_\mathrm{r})_t$ represents the total inflow of salt in year t. The TSS can be defined as (Eq. 7):

$$\mathrm{TSS}_t = SV_t + \sum_{i=1}^{t} e \tag{7}$$

where $\sum_{i=1}^{t} e$ represents a sum of residuals: all salts precipitated till year t, and all gains and losses during year t. SV_t represents the total salt dissolved in the lake in time t. Using the expression of TSS (Eq. 7) in the TSS difference expression (Eq. 6) gives us (Eq. 8):

$$\left(SV_\mathrm{t} + \sum_{i=1}^{t} e\right) - \left(SV_{t-1} + \sum_{i=1}^{t-1} e\right) = (S_\mathrm{r}\ Q_\mathrm{r})_t \tag{8}$$

Therefore (Eq. 9):

$$e_t = \Delta SV_t - (S_\mathrm{r}\ Q_\mathrm{r})_t \tag{9}$$

Values of positive salts residuals will be found in years when more salt was gained than lost (mainly to precipitation) and vice versa – negative values will reflect years in which more salt was lost than gained. A multi-year average (1972–2000) of the residuals stands on 1.5% of the inflow salts per each year. Since the relative part of the residuals is very small, and due to the complexity of quantifying all the residuals components, we chose to refer to it as a negligible element in our salt balance.

5 Results and Discussion

5.1 Water Analysis

TDS analysis of the collected water samples showed that the lake water is brackish with a total salinity of 11.4 g/L, both in the costal water and in the deeper area. The TDS values in the CDWs were significantly smaller, with 2.8 g/L in the Daryalyk and 3.1 g/L in the Ozerny. Pavlovskaya [22] reported that water samples taken from the Sarykamysh contained toxic chemicals. Table 4 presents the concentration

Table 4 Representative toxic chemicals in the Sarykamysh Lake

Chemicals	Concentration found in the Sarykamysh (mg/L)	Maximal concentration recommended by WHO (mg/L)
Hexachlorine	2.5	0.0006
DDT	6.4	0.001
Organochlorine	0.0021–0.0059	0.00003

Source: Pavlovskaya [22], WHO [36]

Table 5 Summary of water sample analysis and their recommended values by WHO and US-EPA (all units in mg/L)

	Sarykamysh		CDW		MCL	
Place	Deep	Shore	Daryalik	Ozeorniy	WHO[a]	US-EPA[b]
pH	7.25	6.39	7.39	7.06		6.5–8.5
EC (mS)	17.4	17.7	4.59	5.13		
TDS	11,453	11,445	2,857	3,173	1,000	500
Cl	4,080	4,038	728	912	250	250
SO_4	3,710	3,716	996	1,028	250	500
Br	0	0	15.2	0	0.01	
NO_2	0	0	0	0		1
NO_3	3.16	3.20	3.70	7.75	50	10
CO_3	0	0	0	0		500
HCO_3	148	152	282	266		
Na	2,268	2,260	470	610		200
K	35.0	37.0	7.83	10.05	10	20
Ca	656	674	212	200		250
Mg	553	565	143	139		200
PO_4	0	0	0	0		0
$CaCO_3$	121	125	231	218		200–500

[a]Source: WHO [36].
[b]Source: US-EPA [37].

values of important chemicals and their maximal recommended concentration in drinking water, according to the World Health Organization (WHO).

Water samples from the Sarykamysh, the Daryalyk, and the Ozerny were analyzed for their major solids contents. Table 5 presents a summarized comparison of the samples with values recommended by the WHO and the US Environment Protection Agency (EPA). The results suggest that the water samples content significantly exceeds the maximal contaminant level (MCL). Since the lake accumulates the water with all their substance, the ions measurements hold higher values in the lake than in the CDWs. Obviously, both the water in the CDWs and in the lake is undrinkable, but it is important to note that the MCL values presented are relevant for drinking water only. The water can still be used for agriculture, which permits higher levels of contamination. Three components in the list stand out in

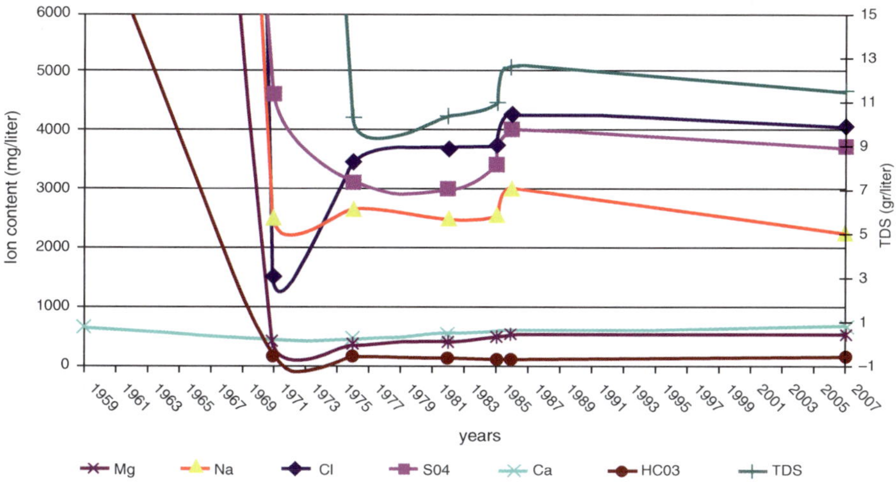

Fig. 12 Ionic content in the Sarykamysh in1959–2007

their extreme value: chloride (Cl^-), sulfate (SO_4), and sodium (Na^+). Since these ions are highly toxic [38], their extreme values can explain the cessation of fishery in the lake and in other irrigation water bodies [22].

Water samples were collected during March 2007, when water in the CDW is most likely soil washing water. Contamination values are expected to be higher during the irrigation seasons, due to the presence of fertilizers and pesticide residues.

A comparison of the ionic content reported in the Sarykamysh since 1959 (Fig. 12) indicates the existence of a few trends. There is a sharp decrease in ions content between 1959 and 1971. This decrease can be explained by the fact that drainage water, which has significantly lower salinity values, reached the Sarykamysh only in 1961 and diluted the lake's water. Between 1985 and 1986 there is a small decrease in the concentration of three ions: Na^+, SO_4^+, and Cl^- as well as in the TDS values. A retrospective analysis of the inflow and inflow's salinity of that period did not show any corresponding trends. If there was no change in the sampling and measuring methods, it is possible that the additional ionic content came from solonchaks that were covered by the lake's water at this period. Since we were not able to obtain data of water analysis between 1986 and 2007, the lack of any major changes in these ions concentrations should be considered with care.

Although there are more than a few definitions of salinity boundaries for brackish water, the Sarykamysh water is almost at the salinity level of the Caspian Sea (TDS: 12,666 mg/L) and higher than the salinity of the Aral Sea (TDS: 10,500 mg/L), as measured in 1950.

5.2 Retroactive Analysis

In order to examine the model's reliability the DEM-derived area and volume values as a function of the lake's water level (height plotted) were plotted, enabling to compare these plots to those of local measurement dataset (Fig. 13).

As seen in Fig. 13, the resemblance between the volume series is easily noticeable. However, in the area series there is a gap of up to 900 km^2 in the mid values between the local measurements and the model derivations (22% of the lake's current area). This gap is related to the water level measurements observed during the early 1970s to the mid-1980s. The gap between the two area datasets decreases, until around the height of −2 m ASL and onward the datasets are almost identical. The differences between the two datasets correspond with heights of −9 to −12 m ASL. In our DEM we can identify a topographic "shelf" or a "shoulder" between these heights, although this shelf is undetectable in the local measurements dataset. The high resemblance between the two datasets, which can be seen in the other ranges, suggests that the observed gap is not related to calibration. It is highly likely that this is a result of data loss due to differences between the sampling and measurements methods used in this study and those used in the topographic surveys of the maps (Fig. 14).

The general deviations between the two datasets can be explained in the following ways:

- Measuring methods: while computations in the present study are based on the DEM, local datasets are a collection of in situ measurements with an interpolation between them. These measurements involved field surveys in a very low spatial resolution and thus a lot of the spatial information was lost.
- During this study a wide variety of data were collected. In some of the cases not all the details regarding this data are clear. For example, the assumption was that the vertical datum used for our in situ measurements was in Krasnovodsk at the shore of the Caspian Sea (Turkmenbashi nowadays). However, at a certain stage the Soviet Union adopted a unified cartographic system, and nowadays the Baltic system is used as a vertical datum. Due to technical difficulties of finding this kind of details, it was impossible to perform a perfect calibration to the in situ data or to the topographic data. Similar difficulties occurred in the collaboration of the DEM data (Gauss-Kruger system, Krasovsky's Spheroid and Datum which is probably Pulkovo-1942) with the Landsat Data, because some spatial information would have been lost in the transfer from one system to another if one doesn't have all the cartographic details.
- A wide set of in situ measurements which runs over 30–40 years was most likely collected in more than one method, since knowledge and experience evolve. In many cases a dataset sequence is broken due to a method change, a formulae or a measuring tool.

In order to simplify the use of the database, a nonlinear regression to the DEM data has been derived, i.e., volume as a function of height (fourth order courier regression,

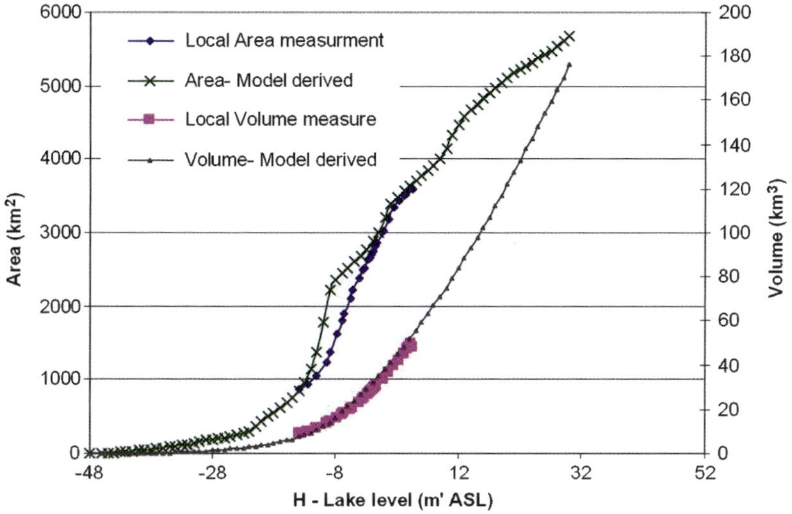

Fig. 13 Lake's area and volume as a function of water level (height), local measurements data vs. DEM produced data. It is possible to see that there is resemblance between volume series. However, a gap can be noticed between two series of area

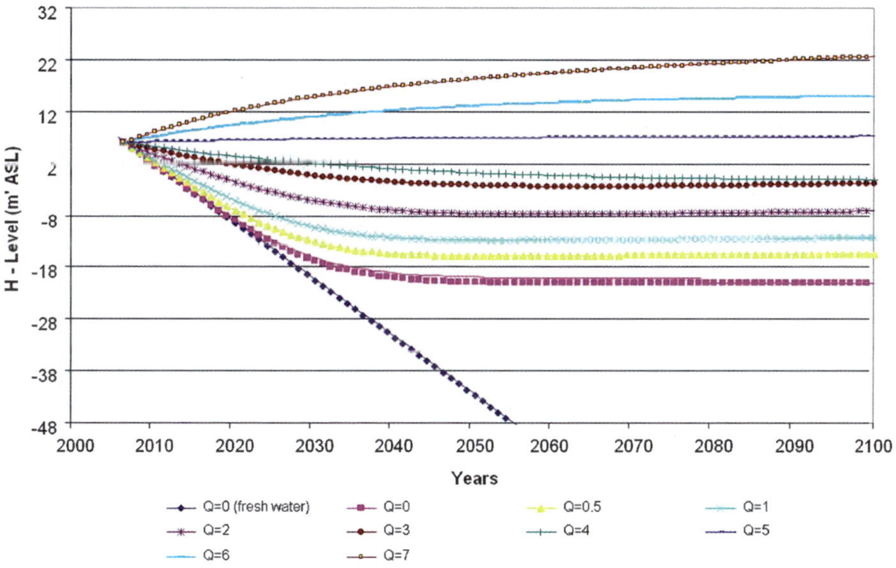

Fig. 14 The Sarykamysh Lake water level forecast till 2100, according to different inflow scenarios

Table 6 Multi annual evaporation rates as calculated with different water level database in different area and volume functions (local dataset or DEM based) and satellite imagery classification

H source	S(h)-Local dataset regression	V(h)-Local dataset regression	S(h)-DEM-regression	V(h)-DEM-regression	S- Sat. classification (interpolation)	V(h)-DEM-regression
	Area and volume derivation method					
Local gauges	Saline water: 1.428 Fresh water: 1.48					
Satellite classification derived			Saline water: 1.162 Fresh water: 1.253		Saline water: 1.190 Fresh water: 1.282	
Altimeters			Saline water: 1.276 Fresh water: 1.332		Saline water: 1.242 Fresh water: 1.297	

$R^2 = 0.9999$) and area as a function of height (ninth order polynomial, $R^2 = 0.998$). The same regressions were produced for the local measurements dataset: Area (described by third polynomial order achieved $R^2 = 0.9981$) and volume (described by exponential function $R^2 = 0.9972$). The values of the functions were used in the water balance equation (Eq. 2), presented earlier, to calculate evaporation values. Table 6 presents multi-annual evaporation values from the lake, according to the function source and the height variable source. The evaporation values based on local height measurements and with local data derived functions are very close to the evaporation values presented by Orlovsky [35]. Orlovsky explained that evaporation from the Sarykamysh must be smaller than the values he presented, because his evaporation measurements were done with a 20 m^2 water tank, and there is an opposite relationship between the volume of the water body and its heat exchange. Kikichev et al. [26] presented their calculations regarding evaporation values using isotopes analysis: from May 1986 to September 1987 they calculated an evaporation of 0.95 m. Relative calculations for this period, which were based on the DEM, showed evaporation loss of 0.93 m. The evaporation rates of the Aral Sea ranged between 0.97 and 1.05 m/year [34]. A comparison of evaporation values between the results obtained from Orlovsky [35], Kikichev et al. [26] and Benduhn and Renard [34], and the results of calculations of water balance in the current study reveal that the evaporation rates based on the local datasets are probably too high. This statement is consistent with the gap observed between the model-derived area calculation and the local area datasets (see Fig. 13). It is possible that the area values found in the local datasets are lower than the actual values, and hence the calculated evaporation values came out too high. Higher values of evaporation can explain lower growth of the dimensions of the lake. The fact that the evaporation rate based on local measurements is too high suggests that the evaporation rates based on presented in this paper datasets (DEM, classification, and altimeters) are closer to the true values of evaporation from the lake. Therefore, it is possible to refer to presented datasets as more reliable forecasts.

5.3 Forecast

Based on all the elements presented so far (water balance, salinity, and evaporation processes), a future water balance was calculated with iterations to predict the behavior of the lake until 2100. The iterations were repeated with different scenarios of inflow, inflow's salinity, annual evaporation, and total precipitation (Fig. 14), while all other parameters were kept stable along these iterations. The results of the water balance forecast suggest that the factors influencing the lake's final level are annual inflow volume, total annual evaporation, total precipitation, and inflow's salinity, in a decreasing order. Furthermore, it can be seen that in all scenarios relative stability is achieved at different levels. In all scenarios where inflow was lower than 3 km^3/year, there is a sharp decrease in the water level in the first 30–40 years, which is followed by relatively stable levels. Stable water level is achieved because the evaporation's total volume balances the annual inflow. The stability is defined as relative, since while the water gain and loss is almost balanced there is still constant addition of salt that increases the salinity values and reduces the total evaporation values, thus creating moderate fluctuations of the water level. The forecast for the lake's behavior was examined according to several aspects, i.e., no inflow scenario, lake's expansion scenario, borderline across the lake, Lake's viability and climate change.

5.3.1 No Inflow Scenario

If the inflow to the lake will cease completely, the expected salinity will reach 268 g/L around the year of 2065. The value is expected to be relatively stable (very moderate increase) since the loss of water by evaporation will be limited by the salinity. The edition of salts by inflow to the system will create an almost negligible change (around 0.2 g/L per a year).

Running the model with no salinity considerations at all can suggest that some of the natural inflow creates the original 5 small lakes. The information was collected from a variety of sources, regarding the water lake level prior to the current flooding. Kes' (1960, [16]) quotes a field survey in which the water level rests in −33 m ASL. It is possible to achieve such a level of stability by running the model in an inflow scenario of 0.17 km^3/year. The fact that the lake was observed when there was no artificial inflow leads to the inevitable conclusion that the water source is natural, i.e. precipitation or ground water. The lake's volume calculation, as observed in 1962 (when the inflow first reached the lake), is 0.6 km^3. Taking into account an average annual precipitation (100 mm) it was calculated that a drainage basin of 6,000 km^2 is required in order to accumulate such a volume of precipitation (not taking into account percolation and evaporation processes). Therefore, it is possible to suggest that the sources of the natural lakes in the Sarykamysh are ground water and springs coupled with the contribution of precipitation.

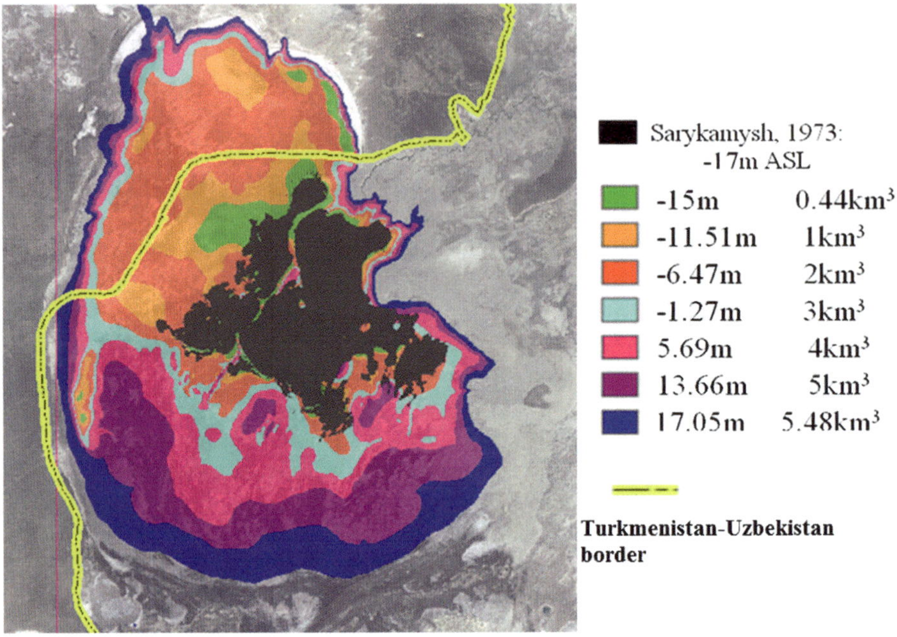

Fig. 15 Sarykamysh depression. Water area expected in the year 2100 according to the different inflow scenarios. In the middle (in *black*), the lake as seen by Landsat MSS (1973)

5.3.2 Lake's Expansion Scenarios

In a scenario of inflow of 5 km^3 per a year with the same inflow's salinity, water level and salinity in the lake will reach 13.66 m ASL and 27.6 g/L, respectively, by 2100. At this height the lake will cover an area of 7,823 km^2.

5.3.3 International Border Exposure

If the inflow to the Sarykamysh will stand on 0.44 km^3/year, the water level will drop down to −15 m ASL. This implies that the entire borderline with Uzbekistan, which crosses the Sarykamysh depression, will dry out. The salinity in this case will reach 159 g/L, inevitably defining the Sarykamysh as an unviable ecosystem. Figure 15 presents possible lake dimensions as a result of a few possible scenarios.

5.3.4 Lake's Viability

In a terminal lake the most acute process in the accumulation is salts accumulation. The total salt stocks in the lake will always increase as long as there is inflow. Since the lake's viability depends primarily on the lake's salinity two possibilities to

Table 7 Lake viability duration according to different inflow scenarios

Maintaining salinity lower than 16 g/L		Maintaining salinity lower than 35 g/L	
Total inflow (km^3 per a year)	Viability duration	Total inflow (km^3 per a year)	Viability duration
0	2008[a]	0	2016
1	2008[a]	1	2019
2	2009[a]	2	2026
3	2010	3	2041
4	2013	4	2085
5	2020	5	Beyond 2099
6	2041	6	Beyond 2099
7	2070	7	Beyond 2099

[a]Value has low significance.

preserve the lake as a viable ecosystem were checked: (1) maintaining the lake's salinity under 16 g/L, which will enable the existence of the brackish water ecosystem; (2) maintaining the lake's salinity under 35 g/L, which will turn the lake into an ocean-like ecosystem. Table 7 presents the duration of lake's viability according to different scenarios of inflow. The basis for all the calculation was inflow salinity of 3.5 g/L which is the average reported inflow's salinity.

According to the results, a salinity increase above the current values will take place in a couple of years. Maintaining the current inflow to the Sarykamysh (around 5 km^3/year) will keep the lake under a salinity level of 16 g/L until 2020. Beyond this stage salinity is expected to pass the salinity threshold. An attempt to maintain the Sarykamysh as an ocean-like ecosystem (in terms of salinity) is feasible for longer duration. Maintaining the current inflow to the Sarykamysh will enable salinity values lower than 35 g/L even beyond 2099.

6 Conclusions

In order to forecast the behavior of the Sarykamysh Lake, a model was produced based on 1940s topographic maps. Using satellite images, Satellite Altimetry data and the Lake's morphological characteristics collected over the years in situ, the morphological changes of the lake in the past 40 years have been studied. By applying water and salt balances the evaporation rates from the lake during this period was calculated. Evaporation rates calculated by the model were compared with evaporation rates based on local measurements and with rates calculated by other scientists. These comparisons proved the reliability of our model. The model was applied to several controlled inflow scenarios to the lake, and lake's parameters were derived. Furthermore, the lake's behavior was examined in scenarios of different parameters such as inflow's salinity, annual precipitation, and annual evaporation rates.

In almost all scenarios water level stability is achieved around 40 years from today in different levels. A total disconnection of the lake from its feeding canals will probably not bring the lake into total desiccation, at least not until the year of 2100, but salinity values will be so high (268 g/L) that life could not exist in it. The environmental impact of the salinity issue is severe since the expected loss is of species that were forced out of the desiccating ecosystem of the Aral Sea area toward the Sarykamysh where they found a suitable ecological niche.

The water level according to the different scenarios is not perfectly stable. It is moderately fluctuating around certain levels. These fluctuations occur because there is constant inflow of salt (though total inflow and total evaporation balance each other).

Multiannual salt residuals average was calculated and stands on 1.5% of the lake's salinity. Due to the complications in calculating each of the residuals elements, and due to the minor size of the residuals, we chose to refer the residuals part as negligible in our salt balance.

The expected drop in the Sarykamysh water level hides severe consequences regarding the exposure of the lake's bed, which carry dust and sands sediments together with chemicals and salts. The additional Aeolian exposed substance might lead to severe regional environmental and health hazards [39–41]. Possible future research should focus on an analysis of soil exposure, based on the presented model, and an analysis of the potential contribution to Aeolian activity for risk assessment process.

Maintaining the current inflow (around 5 km^3/year) will lead to a rise of the water level up to 13.66 m ASL, and an expansion of the lake's area up to 7,823 km^2, more than twice of the lake's area today. By the year 2100 the salinity will rise up to 27.6 g/L.

The meaning of the salinity increase (even with maintaining the lake with ocean like salinity) is the loss of the Sarykamysh as an ecological refuge for many species that have migrated from the Aral Sea during the last four decades.

A scenario in which an inflow of 0.44 km^3/year is kept will lead to a decrease of the water level down to −15 m ASL. The importance of this level is a complete exposure of the borderline between Turkmenistan and Uzbekistan in the Sarykamysh depression.

References

1. CAWATERinfo – Aral Sea Basin (2012) http://www.cawater-info.net/aral/geo_e.htm. An internet site of CAREWIB. Accessed 03 July 2012
2. Ziyadullaev SK (1990) Use of land and water resources in Central Asia and Southern Kazakhstan. Probl Desert Dev 2:1–3 (in Russian)
3. Chembarisov E (1988) Hydrochemistry of irrigated areas (on example of the Aral Sea). FAN, Tashkent, Uzbekistan, 104 p (in Russian)
4. Nikitin A (1987) The lakes of Central Asia. Gidrometeoizdat, Leningrad, 105 p (in Russian)

5. Tolstov SP, Kes' AS (eds) (1960) Low Amudarya River, Sarykamish, Uzboi. History of formation and population. Materials of Khoresm Expedition, vol. 3. Academy of Sciences of the USSR, Moscow, 347 p (in Russian)
6. Kiyatkin A, Sanin MV (1989) The Lake Sarykamish – the largest accumulator of collector and drainage water. Amelioration and Water Industry 1:20–24 (in Russian)
7. Mansimov M, Glazovsky V (1988) Sarykamysh depression: problems and perspectives. Probl Desert Dev 4:71–72 (in Russian)
8. Anisimov V, Staroverova B (1982) The modern state and perspective of fishery in the Sarykamysh Lake. Vestnik Karakalpakstan branch of Academy of Sciences of Uzbekistan 3 (89):47–52 (in Russian)
9. Shaporenko SI (1987) The effect of hydrological regime on fish productivity of salinized water bodies (using the example of the Aral Sea, lakes of the Arnasai system and Sarykamysh) [Vliyaniy gidrologicheskogo rezhima na ryboropoduktivnost' osolonyayushikhsya vodoyemov (na primere Aralskogo morya, ozer Arsanajskoi sistemy i Sarykamysh)]. *Avtoref na soisk uch. step.k.g.n.* (Moscow)
10. Djumaniyazov K (1978) Lake Sarykamish. Proc SANIGMI 59(140):67–73 (in Russian)
11. Vitkovskay T, Mansimov M, Shekhter L (1985) Dynamics of Sarykamysh Lake by remote sensing data. Probl Desert Dev 6:38–43 (in Russian)
12. Gorelkin N, Mansimov M (1990) Hydrometeorological regime and water/salt balance of the lake Sarykamysh. Proc SANIGMI 133(214):70–84 (in Russian)
13. Mansimov M (1987) Lake Sarykamish and its influence on the adjacent environment. Probl Desert Dev 2:65–68 (in Russian)
14. Mansimov M (1994) Dynamics of landscape and ecological conditions in the zone of influence of the Sarykamysh Lake. Probl Desert Dev 3:32–38 (in Russian)
15. Saparov U, Golubchenko V (2001) Turkmen Lake in the Karakum Desert. Probl Desert Dev 1:46–51 (in Russian)
16. Kes' AS (ed) (1991) Lake Sarykamysh and water bodies collectors of drainage water. Nauka, Moscow, 150 pp (in Russian)
17. Aladin N (2000) Water and environment health – historic evolution of the Aral Sea system. Water security – opportunity of development and cooperation in the Aral Sea area. A SIWI/RSAS/UNIFEM Seminar, Stockholm, 2000
18. Grigoryev AA (1955) The Sate of Soviet Geography. Geogr J 121–124:429–439 (in Russian)
19. Countryturkmenistan (2005) Information portal accessed on http://countryturkmenistan.tripod.com/index.blog?start=1118403201
20. Aladin N, Letolle R, Micklin P, Plotnikov I (2005) Uzboy and the Aral regressions: an hydrological approach. International conference on rapid sea level change – a Caspian perspective, Rashat, Iran
21. Sanin M et al (1991) Sarykamysh Lake and water bodies-collectors of drainage waters. Nauka, Moscow, 149 p (in Russian)
22. Pavlovskaya LP (1995) Fishery in the lower Amu-Darya under the impact of irrigated agriculture. FAO Fisheries Circular No. 894, 42–57. FAO, Rome
23. Petr T (ed) (2003) Fisheries in irrigation systems of arid Asia. FAO Fisheries Technical Paper. No. 430. Rome, FAO, 2003, 150 p
24. Alimokhamedov MA, Lesnik IUN, Timoshenko AM (1982) The use of space photographs in the mapping of glaciers and water bodies. Geodeziia i Kartografiia, May 1982, 24–30 (in Russian)
25. Nuriddinov OS (1989) Use of remote sensing in the study of the changing shoreline of Sarykamysh Lake. Sci Remote Sens 26(1):74–77 (in Russian)
26. Kikichev KG, Kostyyukovsky VI, Sanin MV, Shevtsov LA, Shkarin BI (1990) Assessment of evaporation off the surface of the Sarykamysh Lake from Radioisotope data. Int J Radiat Appl Instrum Part E Nucl Geophys 4–1:91–98

27. Nezlin PN, Kostianoy AG, Lebedev SA (2004) Interannual variations of the discharge of the Amu Darya and Syr Darya estimated from global atmospheric precipitation. J Mar Syst 47:67–75
28. Cretaux JF, Birkett C (2006) Lake studies from satellite altimetry. CR GeoSci 338:1098–1112
29. Asmar BN, Ergzinger P (1999) Estimation of evaporation from the Dead Sea. Hydrol Process 13:2743–2750
30. Brutsaert W (1982) Evaporation into the atmosphere: theory, history and applications. D. Reidel Publishing Company, Dordrecht
31. Penman HL (1948) Natural evaporation from open water, bare soil and grass. Proc Royal Soc Lond Series A 193:120–145
32. Amer MH (1992) Fayoum Oasis of Egypt. A Ph.D Thesis work. Water Research Center, El Kanater, Egypt
33. Glazovsky NF (1995) The salt balance of the Aral Sea. GeoJournal 35(1):35–41
34. Benduhn F, Renard P (2004) A dynamic model of the Aral Sea water and salt balance. J Mar Syst 47:35–50
35. Orlovsky NS (1971) Evaporation from the small water bodies of Turkmenistan. Ashgabat, Ylym, 96 pp (in Russian)
36. WHO (2003) Background document for preparation of WHO guidelines for drinking-water quality. Geneva, World Health Organization (WHO/SDE/WSH/03.04/17)
37. US-EPA (2003) Drinking water contaminants list on the US Environment Protection Agency web site: http://www.epa.gov/safewater/contaminants/index.html#mcls. Accessed on 18 Oct 2007
38. Pescod MB (ed) (1992) Wastewater treatment and use in agriculture no. 47 Series title: FAO Irrigation and Drainage Papers. Food and Agriculture Organization of the United Nations, Rome, 1992
39. Gill TE (1996) Aeolian Sediments generated by anthropogenic disturbance of Playas: human impacts on the geomorphic system and geomorphic impacts on human system. Geomorphology 17:207–228
40. O'Hara SL, Wiggs GFS, Mamedov B, Davidson G, Hubbard RB (2000) Exposure to airborne contaminated with pesticide in the Aral Sea region. Lancet 355:627–628
41. Orlovsky L, Orlovsky N, Durdyev A (2005) Dust storms in Turkmenistan. J Arid Environ 60:83–97
42. IFAS (2000) State of environment of the Aral Sea Basin: regional report of the Central Asian States. Portal accessed on http://enrin.grida.no/aral/aralsea/index.htm. 13-4-2005
43. LEGOS, Hydrology from space, Cretaux JF (ed). Database portal accessed on http://www.legos.obsmip.fr/en/soa/hydrologie/hydroweb/12-8-2007
44. Allmetsat – online weather and climate database. http://www.allmetsat.com. Accessed 23 July 2007
45. Earth-Explorer, Earth Resources Observation & Science (EROS) USGS. http://www.earthexplorer.usgs.gov/. Accessed during Mar 2005 to Oct 2007
46. Roll G, Alexeeva N, Aladin N, Sokolov V, Sarsembekov T (2004) The Aral Sea – management, experience and lessons learned. A status report on the portal: International Waters Learning Exchange and Resource Network. http://www.iwlearn.net/publications/ll/aralsea_2005.pdf/view

Groundwaters and Salinization of Soils in Turkmenistan

Paltamet E. Esenov

Abstract The characteristics of groundwaters, their salinity, and chemical composition of the upper water-bearing horizon are the important factors for development of land reclamation of oasis and pasturelands in the plain part of Turkmenistan. This chapter is focused on the main characteristics of groundwaters and salinization of soils in the Dashoguz, Lebap, Mary, Akhal, and Balkan Velayats in Turkmenistan.

Keywords Altyn Asyr, Groundwater, Karakum River, Karakums, Salinization, Soil, Turkmenistan

Contents

P.E. Esenov (✉)
National Institute of Deserts, Flora and Fauna of the Ministry of Nature Protection of Turkmenistan, 15, Bitarap Street, Ashkhabad 744000, Turkmenistan
e-mail: desert@online.tm

I.S. Zonn and A.G. Kostianoy (eds.), *The Turkmen Lake Altyn Asyr and Water Resources in Turkmenistan*, Hdb Env Chem (2014) 28: 141–150, DOI 10.1007/698_2012_215,

1 Introduction

The soil and land reclamation conditions may be assessed after study of groundwaters (GW) and soil salinity. The characteristics of GW, their salinity, and chemical composition of the upper water-bearing horizon are the important factors for development of land reclamation of oasis and pasturelands in the plain part of Turkmenistan. The plains in Turkmenistan cover the territories of two large artesian basins of the platform type – Karakum and West Turkmenian [1].

The Karakum Basin embraces water-bearing complexes confined to the Mesozoic-Cenozoic rocks of the sedimentary cover. Groundwaters occurring in the Jurassic to Paleogene rocks are artesian with high salinity, even to brines. The Neogene-Quaternary rocks form the single top water-bearing complex.

By the recharge of GW and peculiarities of hydrogeological conditions we may distinguish several regions in the flat part of Turkmenistan. The Southeastern Karakums feature the single groundwater table. GW flow is directed to the northwest, while in Lower Karakums – to the west. The GW are recharged in the east from the Amudarya valley, in the south by groundwater flow from the Paroparmiz and also from the Karakum River zone. In these regions the waters are mostly saline and heavily saline occurring at depths from 15 to 20 m. In the south, on the slope of Badkhyz and Karabil GW occur at a depth of 100–200 m, while in the Karakum River zone and in deltas with intensive irrigation development GW occur at a depth from 0.5 to 3 m. In the delta margins the GW level is deeper – 10–20 m.

For land reclamation very essential is the data on the upper water-bearing horizon, on the level of GW occurrence and on GW salinity. Accordingly, at evaluation of the GW condition the oases and desert (beyond oases) zones should be studied separately.

The plain zone generally reveals gradual salinity growth from the Amudarua to Central Karakums. In the areas nearby the Amudarya GW with salinity less than 3 g/L prevail. The low- and medium-saline waters (to 10 g/L) cover over 60% of the territory practically in all regions of the country. The heavily saline GW (10–50 g/L) and brines are most widespread in the Central Karakums and on the Caspian Sea coast proving general flow of GW in this direction.

By their chemical composition GW differ significantly by regions of the country. GW of the hydrocarbonate-calcium to chloride calcium type are widespread (they make about 50% of all waters in the country) which confirms the fact that this territory belongs to the sulfate–chloride province of salt accumulation [2].

By the anion–cation combination in groundwaters over 200 chemical compositions were found beyond the irrigated zone [3]. They were arranged in 11 groups. Waters from one group of chemical composition include 4–5 gradations by the salinity level and vice versa waters with identical salinity may be referred to 4–5 groups by chemical composition.

Analysis of GW of irrigated lands in the main delta areas of Turkmenistan revealed a close relationship between general salinity of GW and their chemical composition, mostly with sodium ion. The formulas and graphic methods for calculation of salts by sodium are suggested in [2–7].

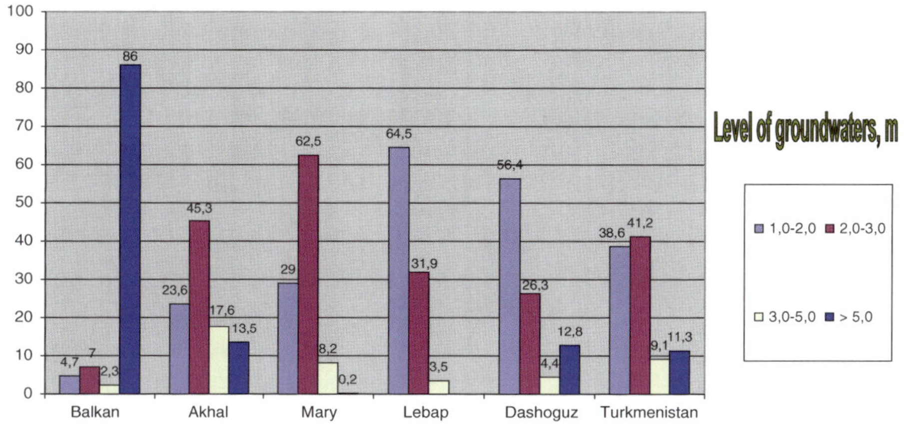

Fig. 1 Distribution of irrigated farmlands in velayats of Turkmenistan on the level of groundwater in %

2 Groundwaters

2.1 Amudarya Area

In the middle and lower reaches of the Amudarya GW occurring mostly to the north are characterized by lower salt concentrations compared to the delta areas in the southern subzone of deserts (deltas of the Murghab and Tedjen rivers and the Khauzkhan area) where they reveal higher salinity. Thus, in the Amudarya middle reaches (Lebap Velayat) GW with salinity less than 3 g/L makes 86%, while in the Amudarya lower reaches – over 70% of the total number of analyzed samples (over 900 samples were analyzed). These figures for the Murghab River are about 40% and for the Tedjen delta – only 2%. This confirms once more the conclusions of V. A. Kovda [2] that the combination of physiographical conditions in each area determines the relationship between the composition and content of salts in GW and characterizes the regional peculiarities of the relationships between the salinity and chemical composition of GW.

2.2 Dashoguz Velayat

The irrigated zone in Dashoguz Velayat (over 400 thousand ha) features high occurrence of GW: more than 82% of the area of irrigated lands is characterized by GW occurrence at a depth less than 3 m (Fig. 1). High occurrence of GW is connected with inadequate drainage, high water horizons in collectors and water head in them formed due to the action of the transit drainage flow from Khorezm Velayat in Uzbekistan.

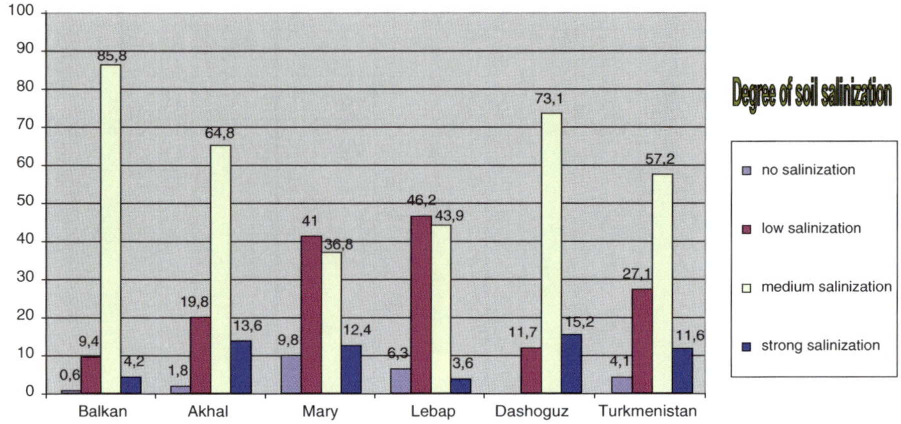

Fig. 2 Distribution of irrigated farmlands in velayats of Turkmenistan on the level of salinization of soils in %

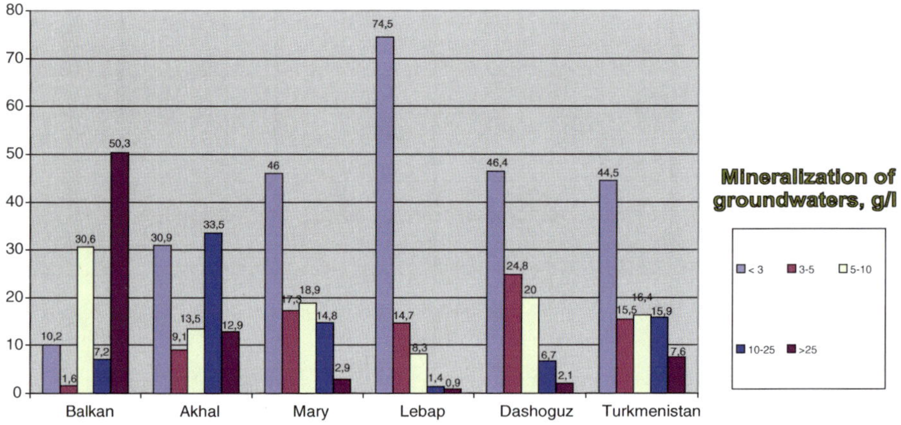

Fig. 3 Distribution of irrigated farmlands in velayats of Turkmenistan on the level of mineralization of groundwaters in %

GW in the irrigated lands of this velayat are less saline in summer in the period of intensive irrigation, while in spring and autumn the water salinity is the maximum. GW with high salinity occur mostly in the regions located in the lower and poorly drained western part of the velayat where the share of soils with very high and high salinity is greater than in the regions locating at higher altitudes. In the velayat GW in irrigated lands on over 46% of the area have salinity to 3 g/L, on about 25% of the area – 3–5 g/L and on the rest area – over 5 g/L (Fig. 2).

GW of the velayat are diverted beyond the oasis via the large interstate collector systems (Daryalyk, Ozerny, etc.) to the Sarykamysh Lake and from 2009 a part of the collector-drainage waters (CDW) is diverted into the Turkmen Lake "Altyn-Asyr." The many-year annual CDW flow in Dashoguz Velayat is 0.98–1.60 billion m^3, while the CDW transit flow from Khorezm Velayat is 2.61–4.54 billion m^3 which affected seriously the condition of irrigated lands in the velayat (Fig. 3).

It is expected that diversion of a part of the CDW flow of the velayat and the transit CDW flow from Uzbekistan to the Turkmen Lake will improve significantly the ecological and reclamation condition of lands in Dashoguz Velayat.

2.3 *Lebap Velayat*

In the Lebap Velayat GW in irrigated lands occur close to the surface. The Amudarya River crossing this velayat drains the irrigated lands here and determines the depth of occurrence and salinity of GW. In more than 64% of the irrigated land area GW occur at a depth of 1–2 m, and in 40% of the irrigated lands – at a depth of 2–3 m.

The salinity of GW in more than 74% of the irrigated lands is less than 3 g/L. For a long time GW of various salinity of the Lebap Velayat and also CDW of the Bukhara and Karshi oases in Uzbekistan were disposed as return drainage flow into the Amudarya River, thus, creating the unfavorable ecological and reclamation conditions in the downstream areas. Depending on the water abundance in a year only in the Lebap Velayat about 2.5–4.8 billion m^3 of drainage waters were discharged annually into the Amudarya from Turkmenistan and Uzbekistan for many years.

After commissioning in 2009 of the first stage of the Turkmen Lake "Altyn-Asyr" project the drainage waters from the left-bank areas of the Lebap Velayat with salinity about 2.5 g/L were diverted from the Amudarya River and directed via the Main Collector to the Karashor Depression. At the second stage of this project the drainage waters from the right-bank areas of the Lebap Velayat and drainage waters from Uzbekistan will be also diverted from the Amudarya and via two aqueducts will be transferred to the left bank and directed via the Main Collector to the Turkmen Lake.

2.4 *Mary Velayat*

GW of irrigated lands of the Mary Velayat consist of two complexes – groundwaters of the Murghab oasis and groundwaters of the Khauzkhan irrigated area.

The Murghab oasis covers about 5,000 km^2 and it is located entirely within the borders of modern delta. The absolute altitudes of the oasis decrease from 260 to 200 m to the north and northwest with very mild surface gradients (0.001–0.005). GW of the oasis are reached by the water seeped from the Murghab and Karakum canals and, to a lesser degree, by atmospheric precipitations. GW of the Murghab oasis form filtration mounds along the river, Karakum Canal and new irrigation canals. The depth of GW occurrence of 1–2 m is found in about 30% of the irrigated area, while the depth of 2–3 m – in more than 60% of the area. GW salinity in the oasis is very diverse – the salinity of filtration mounds is mostly to 5 g/L (in over 60%

of the irrigated lands). In non-irrigated areas and in peripheral irrigated lands GW salinity is usually more than 10 g/L and controlled by infiltration and evaporation.

In the Khauzkhan irrigated area GW are mostly recharged by water filtrating from the Karakum and Khauzkhan canals and partially from the Khauzkhan Reservoir and atmospheric precipitations. In the greater part of the Khauzkhan irrigated area (over 90%) GW occur at a depth of 3–5 m. GW salinity here varies within wide ranges – from 2 to 90 g/L. Low GW salinity is witnessed in the zone of large canals and also in intensively irrigated areas where GW are formed due to infiltration from canals and irrigated fields. GW with salinity to 5 g/L occur in about 15% of the area, 5–10 g/L – 34% and 1- g/L – over 50% of the area. Beginning from 1969 GW were diverted from the Khauzkhan irrigated area via drainage network. The average salinity of diverted CDW after 1969 gradually decreased from 40 to 10–12 g/L.

In general, from the Mary Velayat, including Khauzkhan irrigated area, over 1.2 billion m^3 of drainage waters are diverted annually into the Turkmen Lake. Salinity of these waters varies from 5 to 10 g/L.

2.5 Akhal Velayat

The irrigated lands in the Akhal Velayat extend over the plain at the foot of the Kopetdag Mountains and modern delta of the Tedjen River. The total area of irrigated lands in the velayat is over 500,000 ha. The main sources of GW recharge in the near-Kopetdag zone are the Karakum Canal, mountain rivers, karizes, and wells. In this zone GW occur at a depth less than 3 m in more than 60% of the irrigated area and 3–5 m and more in the rest area. GW salinity here is mostly 3 g/L (about 60% of lands), 3–5 g/L and 5–10 g/L (about 30% of lands), and more. GW are diverted from irrigated lands via large interregional collectors into the Main Collector of the Turkmen Lake.

The main sources of GW recharge in the Tedjen zone is the Tedjen River and the Karakum Canal. In this zone in about 80% of irrigated lands GW occur at a depth of 103 m. The salinity of CDW diverted from this zone varies within 10–18 g/L being the largest in the country.

3 Soil Salinity

3.1 Salinization of Soils

Out of the total land area of Turkmenistan being 49.40 mln ha the land reclamation stock (the lands suitable for irrigated farming) is evaluated at 7.01 mln ha. The land reclamation stock is available in all velayats, but it is the greatest in the Balkan and Akhal velayats.

By the degree of their salinity the lands of the reclamation stock are grouped as follows: non-saline 1659.5 thousand ha or 23.7%, slightly saline 1098.8 thousand ha of 15.6%, medium saline 1183.7 thousand ha of 16.9%, heavily saline 2251.3 thousand ha or 32.1%, highly saline and solonchaks 820.0 thousand ha or 11.7%.

By their texture they are divided into: light soils (sand, sandy loam) – 2560.0 thousand ha or 36.5%; medium (loam, light loam) – 1933.0 thousand ha or 27.6%; heavy (heavy loam, clay) – 2520.3 thousand ha or 35.9%.

The efficiency of land use in Turkmenistan is preconditioned by the natural-reclamation and irrigation-economic conditions of individual regions. At present two-thirds of the irrigated lands in the country need reclamation improvements.

The existing salinity level is the main indicator of soil degradation. Its types and forms of its development are dependent on many natural and man-made factors among which drainage conditions are the key ones.

Salinity as a result of meadow processes of soil formation is most often formed in poorly drained areas where groundwaters occur at depths of 2–3 m and more with the seasonal amplitude of variations from 0.8 to 1.0 m. And the salt differentiation is observed herewith not only in the profile of irrigated fields but also in the space across agricultural lands (fallow, abandoned lands, dry drainage) where with the increase of soil wetting and groundwater salinity the soils add up in salinity.

Salinization as a companion of irrigation is manifested at all stages of land cultivation due to rise of the level of saline groundwaters. Accordingly, the main sources of salinity in oases are groundwaters. The closer to the ground surface the saline groundwaters, the more intensive is the salt accumulation in soils. Therefore, the depth of groundwater occurrence is the key factor of soil degradation; thus, there is a need to regulate artificially the groundwater level by construction of the collector-drainage network.

The irrigated lands in Turkmenistan are confined to non-drained or poorly drained deltaic plains, river terraces, proluvial piedmont plains and lower part of debris cones that, in general, complicate the creation of optimal land reclamation conditions.

In the conditions of the hot and dry climate of Turkmenistan the yields of agricultural crops are limited by many factors, but the factor of soil salinization is the key one. Therefore, the idea of land reclamation is principally in removal of water-soluble salts toxic for plants from the root soil horizon, thus, improving its physical and chemical properties and fertility.

The salinization control measures include application of agrotechnical and agro-reclamation techniques for desalinization of irrigated lands with the help of leaching and land leveling operations in autumn and winter; introduction of plants growing on saline soils, application of new advanced methods of irrigation and drainage; improvement of the existing methods of land cultivation, application of more effective crop rotations and optimal norms of application of organic and mineral fertilizers, etc. These measures may be corrected for particular soil and reclamation conditions.

The target of land reclamation at radical improvement of saline soils is, first of all, the medium and heavily saline irrigated and newly developed lands. Their total area in the country is 1185.41 thousand ha or 68.8%.

3.2 *Dashoguz Velayat*

The area of lands in the Dashoguz Velayat requiring flushing is 359.5 thousand ha. Here the water application rates are determined differently depending on the groundwater level (GWL), degree of salinization, and soil texture. For leaching the medium saline soils which area if 297.5 thousand ha the water application rate should be 2,500–6,000 m^3/ha. For high and very high saline soils covering an area of 62,000 ha the greater water application norms for leaching are needed 5,000–10,000 m^3/ha. Such lands are usually improved only by growing of rice on them followed by fodder crops.

In the Dashoguz Velayat the water resources in autumn and winter are limited; the irrigation waters start inflow in early spring and in this case the flushing irrigation is often combined with preliminary watering. As a result, in some places the soils remain non-flushed by the beginning of the vegetation period.

In the Dashoguz Velayat the rice cultivation is practiced for reclamation of saline soils. Here the long-time flooding of rice checks with a water application rate of 15,000–20,000 m^3/ha is alternated with periodical discharge of water into the drainage network. As a result, already on the second or third year of rice cultivation the favorable conditions for growing of fodder crops and even cotton are created.

3.3 *Akhal Velayat*

In the Akhal Velayat the planning of land reclamation measures should take into consideration the significant sloping of irrigated lands located in the piedmont plain of Kopetdag Mountains. Unlike other regions the flushing here is conducted by feeding water along furrows 60–90 cm wide cut across the slope. The furrow length depends on the slope steepness.

In the modern Tedjen delta the greater part of lands in the old irrigated zone got saline and was abandoned, thus, the farms started developing virgin lands beyond the cultural zone. The existing drainage network (20 linear m/ha) was unable to divert heavily saline soils (22–24 g/L).

The irrigated lands in the zone of the third phase of the Karakum Canal construction (in particular, in its northern part) revealed grave land reclamation situation due to GWL rise, salinization, overgrowing which, in its turn, required performance of complicated land improvement operations, including leveling (400–2,000 m^3/ha), flushing (3,000–12,000 m^3/ha), drainage (from 20–30 to 75 linear m/ha) and application of mineral fertilizers.

Irrigated lands in the Pre-Kopetdag zone by their physical, chemical, and water properties have more advantages compared to the irrigated lands in deltaic plains. The specific feature of the soil-formation process is the growing fine-texture of soils from the Kopetdag towards Karakums. Locally it is visible in the growing share of fine-textured soils from the top of debris cones to inter-cone hollows.

In general, 402,770 ha of irrigated lands in the Akhal Velayat require priority reclamation improvements.

3.4 Mary Velayat

In the Mary Velayat the areas of saline lands requiring urgent reclamation us 215,070 ha. In the zone of old irrigation about 50% of lands are affected by medium, heavy, and high salinization. Their condition may be improved by annual leaching with large water application rates and lowering of GWL. It is very difficult to increase soil fertility in the regions of old irrigation demonstrating rather vividly the dissected irrigation relief.

In the Khauzkhan region about 80% of irrigated lands reveal medium, heavy, and high salinization. Low water permeability, the layered nature of soils and occurrence of heavily saline groundwater close to the surface make reclamation of saline soils here a rather complicated task.

3.5 Lebap Velayat

In the Lebap Velayat 134,000 ha of irrigated lands require urgent land improvement measures. The unsatisfactory reclamation condition of lands here is connected with high GWL and soil salinity. The newly developed lands where reclamation is rather difficult locate in the Yulangyz area in the southeast of velayat. Their improvement requires leaching with large water application rates and adequate discharge of drainage waters.

3.6 Balkan Velayat

In the Balkan Velayat out of 82,300 ha of irrigated lands 74,000 ha require reclamation improvements.

4 Conclusions

Two-thirds of the irrigated lands in Turkmenistan need reclamation improvements. Apart from flushing, these lands require construction of new and repair of the existing collector-drainage network, comprehensive rehabilitation of the irrigation network, capital leveling, and improvement of water availability (Fig. 3).

Apart from the above-mentioned specific features there are some factors contributing to salinization and waterlogging of irrigated lands. The most important of them is the "reclamation geochemical load," i.e., inflow of salts with irrigation waters. This is applicable, primarily, to the Amudarya zone as great volumes of drainage waters are disposed into this river in its upper and middle reaches.

The salinity of drainage waters discharged into the upper reaches of the Amudarya varies from 0.34 to 5.4 g/L, in its middle reaches – from 2.8 to 9.1 g/L. Accordingly, the salinity of rivers waters in the middle reaches of the Amudarya, including within the Dashoguz Velayat, may reach in some years 2.0 g/L.

After irrigation with such waters each irrigated hectare of lands in the Dashoguz Velayat receive annually 10–32 t/ha of salts, out of which 6–23 t/ha – in the seasons of intensive irrigation. For comparison we may stress that at a greater water application rate and less salinity of irrigation water (0.48–0.81 g/L) in the zone of Phase II of the Karakum Canal construction the salt input per each hectare is much less. This factor should be taken into consideration while planning actions on the rational use of irrigation water.

An important factor of land salinization in the Turkmen Pre-Aral area is the salt drift from the dry bed of the Aral Sea. At present each hectare of lands in the Dashoguz Velayat receive annually 200–800 kg of solid aerosols, over 70% in the zone of present-day irrigation. The effect of this factor on the natural environment of the Pre-Aral region is constantly growing.

References

1. Hydrogeology of the USSR (1972) Vol 38: Turkmen SSR. Nedra, Moscow (in Russian)
2. Kovda VA (1946) Origin and regime of saline soils, vol 1. Leningrad, Moscow, in Russian
3. Muratova VS (1974) Salinity and chemical composition of groundwaters in Turkmenia. Scientific Center for Biological Research of Academy of Sciences of the USSR, Pushchino
4. Minashina NG (1970) Toxic salts in soil solution, their calculation and soil classification by salinity. Soil Science 8 (in Russian)
5. Esenov P (1978) Content and calculation of toxic salts in virgin and irrigated lands of Tedjen. In: The biological productivity of delta ecosystems in the Caspian Lowland of Caucasus. Makhachkala, pp 106–107
6. Esenov P (1996) Assessment of the components of irrigation landscapes of Turkmenistan. Probl Desert Dev (3):43–46
7. Esenov P (2006) Application of nomograms for determination of soil and ground water salinity. Probl Desert Dev (3):59–61

Forecast of the Water–Salt Balance in the Turkmen Lake Altyn Asyr

Eduard T. Pyagay

Abstract This chapter provides the forecast of formation of the artificial lake Altyn Asyr after transfer here of the collector and drainage waters from the irrigated lands of Turkmenistan and nearby territories. The balance estimates have shown that about 50 years will be required to fill this lake with the design disposal of the collector and drainage waters. At the same time approximately one billion tons of salts are expected to get into the lake with these waters.

Keywords Balance, Collector and drainage waters, Evaporation, Forecast, Infiltration

Contents

1 Introduction

Summing up the results of the regime and balance investigations in Turkmenistan has shown that long-time irrigation of oasis lands leads to considerable increase in the volume of collector and drainage waters. Until recently, saline collector and drainage waters from irrigated fields were disposed directly into rivers and natural depressions in the Karakum Desert. As a result, the near-oasis pasture lands

E.T. Pyagay (✉)
V.V. Dokuchaev Soil Study Institute, Russian Academy of Agricultural Sciences, Moscow, Russia
e-mail: epyagay@gmail.com

I.S. Zonn and A.G. Kostianoy (eds.), *The Turkmen Lake Altyn Asyr and Water Resources in Turkmenistan*, Hdb Env Chem (2014) 28: 151–158, DOI 10.1007/698_2012_199,

surrounding the irrigation systems have already degraded or are threatened with waterlogging and salinization.

In this context the problem of regulation of the water–salt balance in irrigated lands of oases should be addressed in combination with the forecasted alterations in the natural land reclamation and environmental situation in desert pastures near oases. Such approach seems helpful to optimize the water–salt balance in the oasis lands if the technical feasibility and environmental admissibility of improvements, including land reclamation with the assigned level of economic return, satisfy the present-day requirements.

In the face of the growing volumes of saline collector and drainage waters in Turkmenistan, it seems advisable, apart from their utilization, to recycle them for irrigation. As a result of withdrawal of a part of drainage waters disposed into the desert, the crop cultivation on the near-oasis lands will permit, on one hand, to improve significantly the forage base of the region and, on the other hand, to alleviate the environmental stress in the near-oasis zone of the Karakum Desert.

2 The Altyn Asyr Lake Project

For augmentation of water resources and improvement of the desert environment, Turkmenistan initiated a magnificent project being in close control of the Turkmen President on construction of the artificial lake Altyn Asyr or Turkmen Lake. This is the grandest project of the century envisaging transfer of great volumes of collector and drainage waters into the natural depression Karashor, thus resolving the whole complex of scientific and economic problems. As concerns land reclamation and wise nature management it seems very important to evaluate, primarily, the likely hydrological changes in Northwestern Turkmenistan, including water–salt balance of pastures in the watershed zone and feeding networks.

It should be noted that Karashor Depression locates between two lakes – the Kara-Bogaz Gol Bay of the Caspian Sea and Sarykamysh [1]. The depression area is no more than 2,000 km^2 (length 100 km, width 20 km) with the maximum depth being 75 m. The water volume in the lake, the area of its water surface, and the average depth of the water body under design filling will be about 130 km^3, 2,000 km^2, and 70 m, respectively. At present the annual flow of all collector and drainage waters in Turkmenistan is approximately 7 km^3, out of which more than the half accounts for Lebapsky and Dashoguzsky velajats (provinces). Adding here the flow of collector and drainage waters from the nearby territory of Uzbekistan, the total flow may reach 11 km^3 [2, 3].

The flow transfer project envisages that the collector and drainage waters will flow into the terminal point – the Turkmen Lake via the Northern Route – Dashoguz canal 350 km long and via the Southern Route – Main Header 720 km long. Herewith the Dashoguz canal will divert a part of the flow from the Ozerny and Daryalyk headers which will result in drawdown of groundwater level by approximately one meter and a half, and this, in its turn, will normalize the water outflow from irrigated fields and soil salt regime.

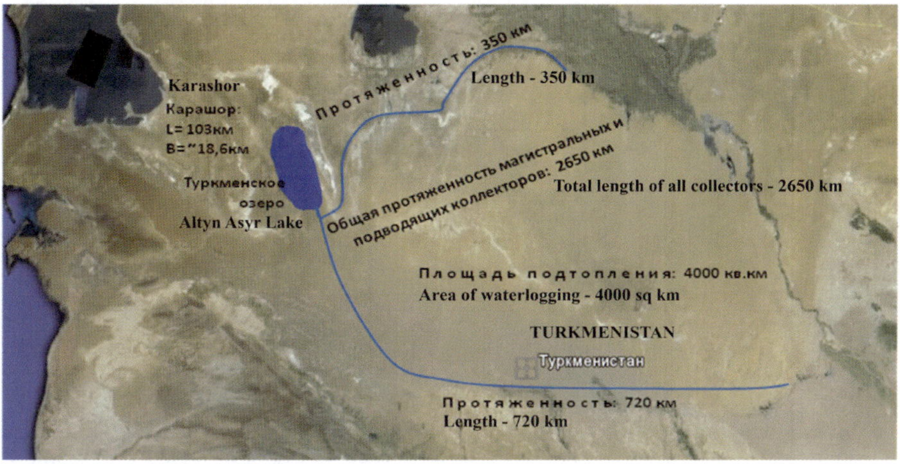

Fig. 1 Scheme of transfer of the collector and drainage waters into the Altyn Asyr Lake

At present diversion of collector and drainage waters from Dashoguz and Khorezm velajats into the Sarykamysh Lake leads to overwetting and waterlogging of agricultural lands, increases the risk of destruction of engineering structures and communications, including gas pipelines, communication lines, and power transmission lines [4]. According to the project, the Main Header will be used for diversion of drainage waters from irrigated lands in the Mary and Akhalsky velajats and also flood waters of the Tedjen and Murghab rivers that are presently disposed directly into the sand desert which results in waterlogging and salinization of valuable pasture lands.

It should be said that very low air humidity and high temperatures in Central Karakums may cause great water losses to evaporation. Even the water balance of the Caspian Sea possessing such potent source of recharge as the Volga River remains at present negative and loses annually over 20 km^3 of water. More than 400 km^3 are lost here only to evaporation. It means that still greater water losses may be expected also in the Turkmen Lake.

The results of long-time monitoring of the water flow into the Sarykamysh Lake and the condition of nearby lands permit to forecast development of similar natural and soil-hydrogeological processes in the zone of Karashor Depression filling and water feeding routes.

In estimates of formation of artificial Altyn Asyr Lake, the inflow routes and the volume of saline collector and drainage waters diverted from oases are assumed the same as in the project, i.e., via the northern and southern route over 1,000 km long and in the amount of 10 km^3 a year. In this case, according to our estimates, the water resources of the lake should grow about 2 km^3 a year or 20% of the annual flow of collector and drainage waters. The remaining part of water will be invariably lost to evaporation and infiltration from the lake and feeding network of headers and canals. These losses are distributed rather evenly over the transit zone forming in the future a vast space of overwetted and saline lands with an area of approximately 4,000 km^2 (Fig. 1).

Table 1 Epignostic estimates of the water balance of the Sarykamysh Lake for the period 1970–2010

	Design time, years				Total for 40 years
Water balance elements	1–10	11–20	21–30	31–40	1–40
CDW flow into the lake, km^3	46	52	58	65	221
Total evaporation, km^3	23	30	40	48	141
Infiltration, km^3	17	13	7	3	40
Inflow into the lake, km^3	6	9	11	14	40
Lake surface area, km^2	1,600	2,100	2,800	3,400	3,400
Lake average depth, m	4	7	9	12	12

Such redistribution of the collector and drainage waters may bring about certain changes in the natural, soil, hydrogeological, ecological, and economic situation practically across the whole territory of the Central Karakums. Let us demonstrate the likely changes of the hydrological regime in the Turkmen Lake zone taking as an example the estimations of the water balance for the Sarykamysh Lake.

3 The Sarykamysh Lake Water Balance

The forecast of formation and development of the Sarykamysh Lake was prepared taking into account the present condition of the lake applying the following scheme. The year of 1970 was adopted as the initial reference time. From this time on during the first decade no less than 4.6 km^3 of collector and drainage waters should flow into the Sarykamysh Depression via two headers Daryalyk and Ozerny. In the subsequent years the volume of the collector and drainage waters diverted into the lake increased by 0.6 km^3 a year in each 10 years, so that by the end of the design period of the Sarykamysh Lake formation the real volume of water inflow into it is attained. In the estimates the water losses to evaporation from the lake surface and from headers were assumed equaling the mean daily evaporation from the water surface in the hot season of a year (April–September) making some 7–8 mm, while losses to infiltration equaling 1.3–0.3 km^3 a year in the conditions of backup infiltration. For the period with unsteady regime of free filtration, the infiltration losses from the lake were assumed 35% of the total water flow into the lake which fits the real water losses from irrigated lands in the Dashoguz velajat (Table 1).

Given these initial values of the water balance components, the estimates have shown that the total inflow into the lake for 40 years was about 40 km^3. The average water body depth in the period from 1980 to late 2010 increased from 4 to 12 m, while the lake area from 1,400 to 3,400 km^2 (Table 1).

Table 2 Forecast estimates of the water balance of the Turkmen Lake for the period 2009–2059 with the permanent annual CDW flow

	Design time, years					Total for 50 years
Water balance elements	1–10	11–20	21–30	31–40	41–50	1–50
CDW flow into the lake, km^3	100	100	100	100	100	500
Total evaporation, km^3	34	35	36	37	39	181
Infiltration, km^3	45	42	39	36	25	187
Inflow into the lake, km^3	21	23	25	27	36	132
Lake surface area, km^2	1,900	1,910	1,920	1,930	1,950	1,950
Lake average depth, m	11	23	36	50	68	68

4 The Altyn Asyr Lake Water Balance

Similar forecast estimations were conducted for the Turkmen Lake. In these estimates the year of 2009 was taken as a reference time point. From that time on about 10 km^3 of the collector and drainage waters were supplied annually into the Turkmen Lake via the northern Daryalyk and Ozerny headers and the southern Main Header, out of which every year over 3.5 km^3 were disposed into the lake via the northern route and about 7.4 km^3 via the southern route. The total water losses from the canal network with the backup filtration regime will make about 0.23 km^3 a year; the water losses to evaporation from the surface of the lake and open headers were taken equaling the mean daily evaporation of 8–9 mm in the hot season of a year (April–September) making the year average of 5–6 mm a day for climatic conditions of Central Karakums. At the initial filling of the lake, the infiltration losses accounted for 10–15% of the total water inflow at a free filtration regime. Later on the infiltration losses decreased gradually reaching 20–25% of the total volume of water in the lake.

The balance estimates have shown that with the unchanged flow of the collector and drainage waters in the forecasted period (full filling of the lake) about 500 km^3 of water will be transferred from oases (Table 2). Out of this amount the annual flow for the "net feeding" of the lake will average 2.64 km^3. The greater part of the flow (about 73%) gets lost to infiltration and evaporation from the water surface of the lake and feeding canals. The Table 2 also shows that the water losses to evaporation are 3.6 km^3 a year with the annual evaporation equaling approximately 1,800 mm, while the water losses to infiltration and underground outflow – about 3.8 km^3 a year or approximately 37% of the total flow of the collector and drainage waters.

If we take that the maximum capacity of the whole system of depressions in the Karashor area is equal to 132 km^3, it will be required about 50 years to fill it completely, in other words, the Karashor Depression will be filled completely in 2060.

For stabilizing the water level in the lake with the area of its water surface being 1,950 km^2, we will have to withdraw part of the water (about 2 km^3/year) to irrigate lands in the canal zone. This amount will be sufficient to irrigate over 200,000 ha of

Table 3 Forecast estimates of the salt balance in the Turkmen Lake for the period 2009–2059 with the permanent annual CDW flow equaling 10 km^3 and salinity of 3 g/L, in million tons

	Design time, years					Total for 50 years
Salt balance elements	1–10	11–20	21–30	31–40	41–50	1–50
Total salt inflow into the lake with CDE	300	300	300	300	300	1500
Salt outflow into ground waters	135	123	117	108	75	558
Salt stock in the lake, km^3	165	177	183	192	225	942

soils with the light granulometric composition along the route of the northern and southern headers.

It should be noted that apart from water this transfer project involves salt, the level of which varies broadly in CDW. Depending on the salt content in water its salinity varies from weak to heavy. Even assuming that the salinity of collector and drainage waters does not exceed the critical threshold of 3 g/L, then every year about 30 million tons of salts will be imported into the lake with water and by year 2060, by the time of the Turkmen Lake filling about 1.5 billion tons of salts will be transported from oases into the desert (Table 3). Out of the total amount of toxic salts brought into the lake, over of 500 million tonnes of salts are evacuated with water infiltrating into ground waters, and the remaining amount of salts (approximately one billion tons) dissolves in the lake water.

5 Conclusions

It is also worth noting that the Turkmen Lake Project envisages development of the lands along the northern and southern water feeding canals. In this context it becomes necessary to conduct here various investigations, including monitoring of the land condition.

These investigations and observations should be based on simulation modeling in combination with land and remote sensing monitoring. This will permit to get answers to many questions connected with transfer of collector and drainage waters from oases into the desert. In particular, it is possible to substantiate:

- Areas of waterlogged and saline lands in the zone of transit of collector and drainage waters into the Altyn Asyr Lake
- Assessment and allocation of lands fit for growing fodder crops under irrigation with saline waters
- Estimates of the water application rates with regard to the soil cover structure and structure of the zone of aeration
- Identification of the case-and-effect relationships in the system soil–environment–man
- Forecast of changes in land fertility affected by natural and anthropogenic factors
- Operative managerial solutions and others

Therefore, even the roughest estimates indicate that we deal with the grand project on transfer of the enormous amounts of water and salts from oases into the desert during 50 years and more which will invariably induce global natural, ecological, and socioeconomic changes in the whole region of Northwestern Turkmenistan.

This circumstance requires the scientific community and the engineering-technical services to focus their attention on the Altyn Asyr Project in Turkmenistan. Special attention should be paid to the methodology and technique of monitoring and forecast of the water–salt balance applying modern mathematical models and computation technique. It is also advisable to build the permanently operating mathematical model of the Turkmen Lake and water supply canals to "play" different situations that may arise at various stages of collector and drainage water transfer from oases into the desert. Such model will be helpful for prediction of the object behavior and, accordingly, for taking appropriate managerial solution, thus optimizing the regime of the Altyn Asyr Lake formation.

References

1. Saparov UB, Golubchenko VG (2001) The Turkmen Lake in the Karakum Desert. J Desert Dev. Ashgabad, No. 1 (in Russian)
2. Abdirakhmanov HA, Malabaev NI (1978) Irrigation of fine-fiber cotton with drainage waters on takyr light loams in the Sherabad Steppe. Problems of application of saline drainage waters. Central Asian Sci. Conf. Tashkent: SoyuzNIHI (in Russian)
3. Atalyev KH, Saparov UB (2007) Management of transborder drainage waters in the Aral Sea basin. J Desert Dev. Ashgabad, No. 4 (in Russian)
4. Esenov P (2006) Problems of land degradation and some ways of their resolution. J Desert Dev. Ashgabad, No. 4 (in Russian)

The Turkmen Lake Altyn Asyr

Igor S. Zonn and Andrey G. Kostianoy

Abstract Currently, the development of water economy has become one of the priorities of the state policy in Turkmenistan. For Turkmenistan, which does not have a large amount of water (about 25 km^3 per year), the main task is the efficient use and conservation of water resources, despite the fact that most of them (90%) is used for irrigation. One of the main problems of the country has been the depletion of water resources due to the complex natural and anthropogenic factors, including a rise in air temperature and decrease in rainfall as a result of regional climate change under global warming. Deficit irrigation water for agriculture may be partially compensated for the expense of drainage water. Until now, almost all the collector-drainage water (CDW) without treatment is discharged into the Karakum Desert and partially to the rivers, thereby worsening water quality. The total amount of drainage water discharged annually to Sarykamysh Lake and sands of the Karakum Desert ranges from 6 to 8 km^3 of water with an average mineralization of 3–5 g/l. Discharge of drainage water in the desert caused underflooding, waterlogging and salinization of lands, and pollution of groundwater in an area of 700,000 ha. In this regard, it was decided to streamline the collection and discharge of drainage water into the Karashor Depression, and creating a large Turkmen Lake “Altyn Asyr” (Golden Age). In the future, it should be a source of water for livestock, fisheries development, and become an area of recreation and nature tourism.

I.S. Zonn
Engineering Research Production Center for Water Management, Land Reclamation and Ecology “Soyuzvodproject”, 43/1 Baumanskaya ul, 105005 Moscow, Russia

A.G. Kostianoy (✉)
P.P. Shirshov Institute of Oceanology, Russian Academy of Sciences, 36 Nakhimovsky Pr., Moscow 117997, Russia
e-mail: kostianoy@gmail.com

I.S. Zonn and A.G. Kostianoy (eds.), *The Turkmen Lake Altyn Asyr and Water Resources in Turkmenistan*, Hdb Env Chem (2014) 28: 159–176, DOI 10.1007/698_2012_216,

Keywords Altyn Asyr, Collector-drainage water, Karashor depression, Salinization, Turkmen Lake, Turkmenistan, Water resources

Contents

1 Introduction

Water in Turkmenistan is the basis of livelihood of the people of this country and its economic development. In a country where more than 80% of the territory is the Karakum Desert agriculture is based on irrigation farming. The main source of surface water resources in the country is the Amu Darya River, which, through the Karakum Canal (also called Karakum River), brings together in a single complex irrigation drainage systems of four of the five velayats (provinces) – Lebap, Mary, Akhal, and Balkan. It also provides water supply for Dashoguz velayat (Fig. 1). The use of water resources in irrigated areas of the country leads to a formation of 6 km^3 of drainage water, and with the collector-drainage water (CDW), formed in the neighboring Uzbekistan, situated on the right bank of the Amu Darya River, and discharged on the territory of Turkmenistan, the total CDW volume may exceed 10 km^3.

Until now, a small part of the CDW with low mineralization (about 2.5 g/l) is discharged into the Amu Darya River and the other part to natural depressions in the Karakum Desert, polluting the river water and salting the land. To prevent these processes a big water project is implemented in Turkmenistan. It is aimed at a construction of a single drainage network for collection of CDW, formed as a result of irrigated agriculture in Turkmenistan, and its discharge into a natural depression Karashor (Fig. 1), where the Turkmen Lake Altyn Asyr (Golden Age) will be created [1]. The project is valued at more than $ 6 billion and is one of the largest projects in the world in the field of development of land reclamation.

2 Collector-Drainage Waters

The history of the main changes in the hydrography of the country started in 1950s with the beginning of a construction of the Karakum Canal (1950–1988), which allowed to increase several times the area of irrigated lands in Turkmenistan.

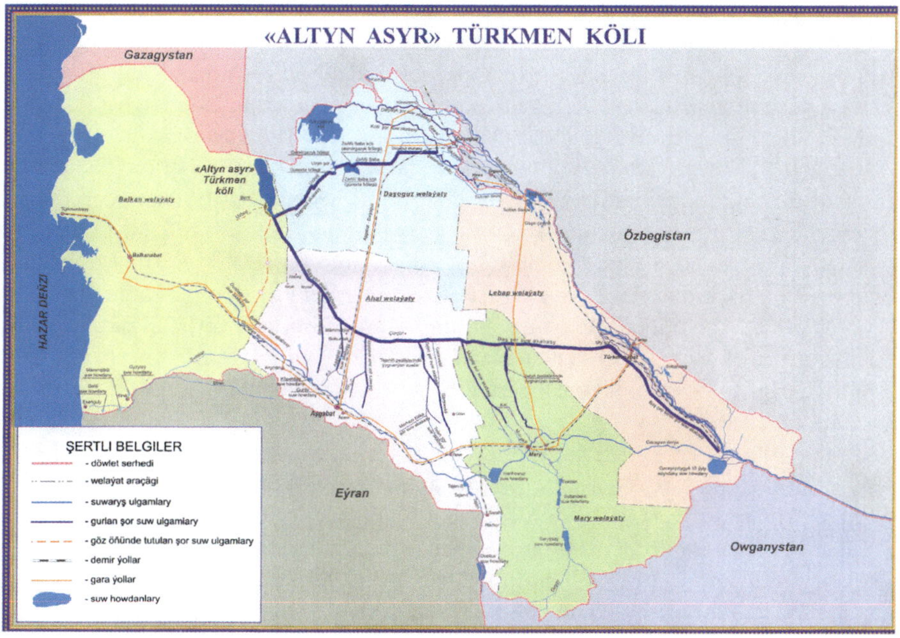

Fig. 1 Administrative division of Turkmenistan, collector-drainage network, and the Turkmen Lake "Altyn Asyr" (Turkmen Lake. . ., 2010)

In 1960s the northern collectors network started to operate, which included Lake Sarykamysh as a terminal water reservoir for CDW. In 2000 a new project Turkmen Lake "Altyn Asyr" begun, which involved the construction of the Transturkmen Main Collector, northern branch of CDW collectors, collection of drainage water from the irrigated lands in the country, and their discharge into the Karashor Depression, which will lead to the formation of the Altyn Asyr Lake.

In the early 1960s, there was an increase of area under cotton and rice in the lower reaches of the Amu Darya, which has increased the volume of irrigation water, and therefore the volume of drainage water, which since 1961 has been dumped in the Sarykamysh Depression, continues today. As a result, since the middle of 1960s Lake Sarykamysh has increased its area significantly: in 1975 – 1,500 km^2, in 2000 – 2,500 km^2, and today about 3,900 km^2. It became one of the largest artificial water bodies in Central Asia.

Back in Soviet times, all the irrigated lands of Turkmenistan had collector-drainage network, which all this time continued to develop as on the old and new lands for irrigation. By the beginning of 1982 the collector-drainage system was built on an area of 800,000 ha. This network allowed to divert CDW from the irrigated areas by a system of canals to remote natural depressions. The first built was the Main Murghab Collector (Fig. 1), which discharged about 1 km^3/year of CDW, collected from the irrigated lands of Mary Velayat, to the Unguz lowlands.

The annual volume of CDW drained from the irrigated lands was 11.45 km^3, including: (1) Chardzhou (now Turkmenabat), Bukhara, Karshi (in the Amudarya River) – 4.05 km^3; (2) Tashauz (now Dashoguz), Khorezm – 5.8 km^3; and (3) Mary, Tedjen, Kopetdag Oasis (in the Karakum Desert) – 1.6 km^3. In addition to CDW, the flow of mudslides from Kopetdag Mountains, floods from Murghab and Tedjen rivers, and discharges from the Karakum Canal are moved to the Central Karakums. As a result, highly mineralized CDW deteriorate water quality characteristics in receiving water reservoirs, and when they accumulate in the Karakum Desert, the pastures are flooded. In the beginning of 1982 about 65,000–75,000 ha of pasture lands in the Karakums was flooded. In the flooded and under flooded zones, an active process of salt accumulation, waterlogging, salinization of soils, and deterioration (salinization) of water in the distant pasture wells is going on. Furthermore, transport communications are damaged; operation of gas pipelines and some of gas fields become difficult.

In the 1980s some predictions showed that the growth of irrigated lands in the region (Turkmenistan, Khorezm, Karshi, and Bukhara regions of Uzbekistan) will lead to an increase of CDW, which by 2000 will reach 20 km^3/year. In 1992, the volume of wastewater and drainage water in Turkmenistan amounted to 8.7 km^3, including 6.6 km^3 of CDW. Almost all of these waters without treatment were discharged into the desert, rivers and sources of irrigation, thereby worsening water quality and polluting the environment. Only in the Lebap Velayat in 1990–2000, about 2.57–4.86 km^3 (depending on the dryness of the year) of CDW was discharged to the Amu Darya River from Turkmenistan and Uzbekistan territories. These discharges reduce the quality of the Amu Darya water, increasing its mineralization from 0.9 to 1.44 g/L [2].

To protect the Amu Darya River in the Lebap Velayat, in the late 1990s, it was planned to move the flow of the Main Collector at the left bank (which discharges into the river) in the amount of 1.2 km^3/year to the Yaradzha depression and further. CDW formed within Mary, Akhal, and Balkan velayats are discharged into the lowerlands of the Karakums, where they fill and flood pasture lands, reducing their size and productivity. Dashoguz Velayat is in a more difficult situation. Over 65% of the annual CDW flow is formed in the Khorezm Region of Uzbekistan and transits via the Ozerny and Daryalyk collectors to Sarykamysh Lake, which is a natural drainless water reservoir (Fig. 1). Currently, its volume reached 59 km^3 at the water surface area of 3,900 km^2 [2].

The problem of eliminating the negative impact of discharge of drainage water and its reuse for the needs of the economy could be solved by a construction of the unifying Transturkmen Main Collector, which was planned to be held on the ancient beds of the Amu Darya. It was planned that the final destination of CDW will be either the Caspian Sea or the Karashor Depression. Ancient Amu Darya channels go from east to west across the country, and the planned position of the Transturkmen Collector allowed to connect to it all the drainage, flood and mud discharges in Turkmenistan.

Implementation of the construction of the unifying Transturkmen Collector already in the Soviet times would:

(a) Reduce the negative impact of saline drainage water on the environment.
(b) Replace unproductive desert natural associations by more productive complexes of flora and fauna.
(c) Irrigate pastures, create reserve stocks of feed for the needs of transhumance.
(d) Use the Transturkmen Collector for fishery.
(e) Create about 100,000 ha of spawning of carp and roach (example of Adzhiyab spawning in Gasan-Kuli) at the confluence of the collector with the Caspian Sea.
(f) Be useful to use the freshwater of rivers of the Caspian Sea basin to the needs of agriculture in the amount of CDW discharge into the Caspian Sea.

Later, the following technical scheme of the construction of a unifying Transturkmen Collector was proposed: to bring limit discharges of the Tedjen and Murghab oases to ancient channels of the Amu Darya and to allow water pass to the Caspian Sea by creating a canal at a rate of 40 m^3/s, and thus creating a flowage of the Central Karakums. As for receiving drainage water from Chardzhou, Tashauz, and Khorezm oases, later they could also be connected to this unification path. Thus, the Transturkmen gravity (self-moving) collector could be created, collecting and discharging waste, drainage, flood and mud water from the irrigated lands of Turkmenistan and the Khorezm Region of Uzbekistan. Later, it was planned to connect the right bank lands in the middle and lower reaches of the Amu Darya (part of Chardzhou, Bukhara, and Karshi) to the Transturkmen Collector also using the ancient riverbed.

The main advantage of the construction of the Transturkmen unifying collector was a possibility of reuse of CDW for agricultural needs, which was very important and urgent at the growing shortage of irrigation water. Finally, the Karashor Depression was chosen as a receiver for drainage water [3]. The project involved the creation of small water reservoirs along and around the collector for growing aquatic plants (cattails, rushes, reeds, etc.), whose yield could then be used in the construction industry.

Currently, the total amount of the CDW in the country is estimated at 6 km^3/year, of which 3–3.5 km^3/year have salinity less than 5 g/l and are suitable for growing salt-tolerant crops and pastures [2]. The operation of all the drainage network will allow to transport to Altyn Asyr Lake about 23–28 million tons of salts from all irrigated lands in Turkmenistan.

3 Karashor Depression

Depression Karashor (Garashor) ("Black Saline") or Shor (a saline) Geklenkui is located between the Kara-Bogaz-Gol Bay of the Caspian Sea on the west and Lake Sarykamysh on the northeast (Fig. 2). It is located west of the well Charyshly in the north of the Karakum Desert in the Balkan and Dashoguz velayats in the northwest of Turkmenistan. From the west the depression is surrounded by Uchtagan Sands

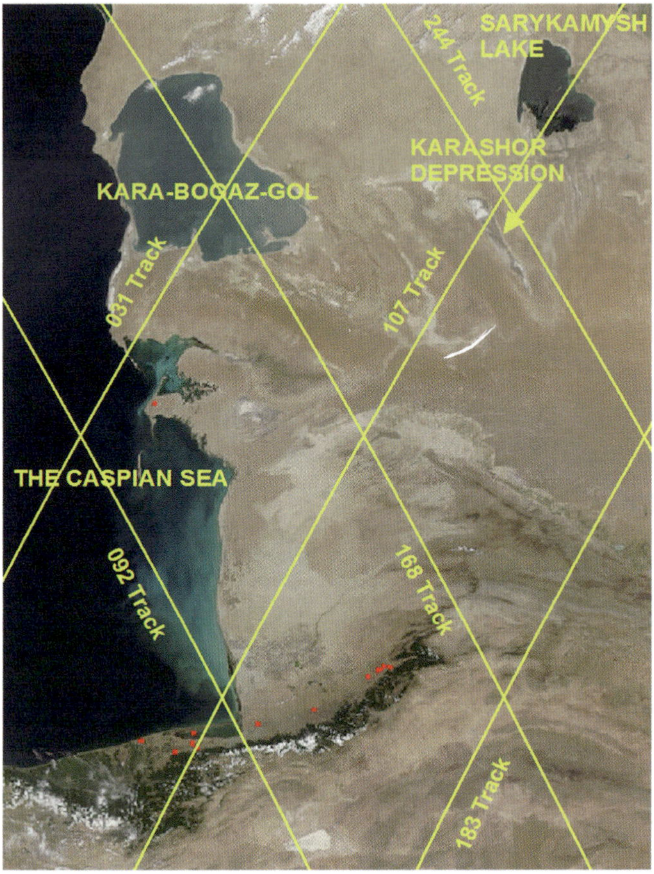

Fig. 2 Location of the Karashor Depression, the Southeastern Caspian Sea, Kara-Bogaz-Gol Bay, Sarykamysh Lake, and Western Turkmenistan on MODIS image acquired on 28 May 2002 with superposition of the TOPEX/Poseidon and Jason-1/2 satellite ground tracks

which close the Chulyungkyr Plateau. From the south, the ancient riverbed of Uzboy River limits a trough and plateau Kaplankyr. It belongs to the Turkmen Central zone of uplift, which covers a broad area of Kara-Bogaz-Gol and Krasnovodsk Peninsula, Tuarkyr, Kumsebshen and Zaunguz Karakums along with the northern Karakums lowlands. To the northwest of the Karashor Depression there is a smaller dry saline depression Kazakhlyshor (Kazakh Shor), a little is known about this trough. The Karashor Depression is compared with the California Death Valley in the USA [4].

Karashor is a deep dry depression. Its length along the long axis is about 120 km, it is 20 km wide in northwest, and in a south-easterly direction, gradually narrowing, it is closed (Fig. 3). In the northeast it is bordered by high (up to 319 m) chink of the Kaplankyr Plateau. From the west it adjoins a large array of

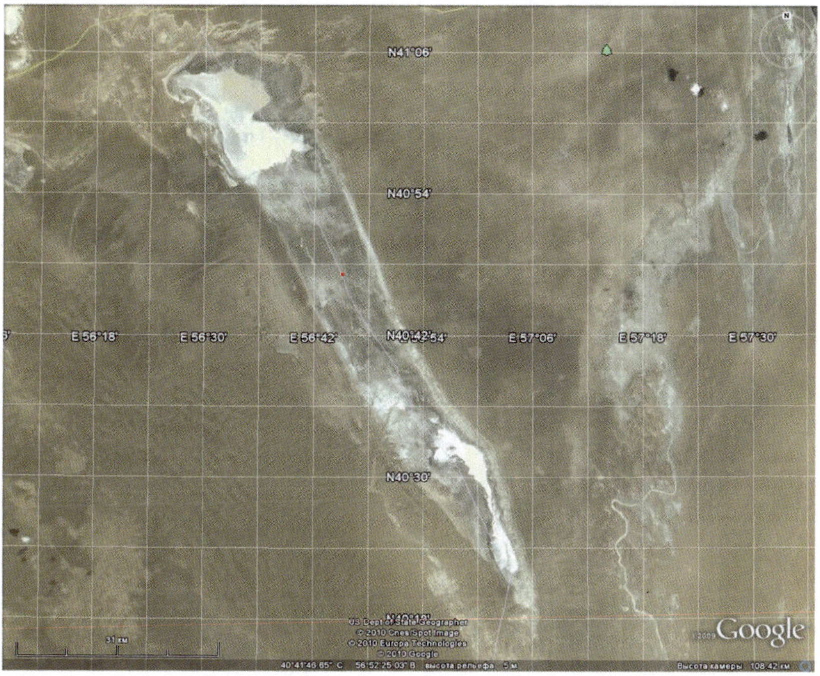

Fig. 3 The Karashor Depression in the high-resolution optical satellite image presented by Google Earth

hilly sands. The bottom of the depression is located at a level of −28 m (28 m below the ocean level), which corresponds approximately to the present position of the Caspian Sea level (−27.6 m by the end of 2011).

At the central axis of the depression there is a channel, diverting surface and partly ground water. In the northwest, in the deepest part of the depression, there is “dry” salt lake, which is more than 5 km wide and 15 km long (Fig. 4) [5]. Layer of halite in the lake is very tight. In 1956, when the survey was performed it was drilled to a depth of 0.75 m only. According to some data the thickness of salt deposits here may reach 5 m. Here halite is contaminated by silt. Halite and salt brine from the sulfate salts contain only calcium sulfate. Brine is referred to the chloride type. Halite contains 88–96% of sodium chloride.

Surrounding chinks 150–320 m high, are pronounced, stacked by stone-lime boards, cut by numerous gullies and erosion, with ledges, piles, closets, and cracks. Local chinks feature are almost horizontal, slightly inclined into plateau dense surface layers overlying soft rock. In the northeastern slopes, very high and steep, almost vertical in places, carbonate rocks are exposed.

On the plateau there is a gray-brown soil. Numerous logs, where the flow is formed only in the rainy season, there is a mosaic of land cover. Salt marshes in depressions with very rare halophytes, between them on the gentle slopes there is a thick sandy soil with small shrubs, which occupy a small area compared to the salt marshes. Area

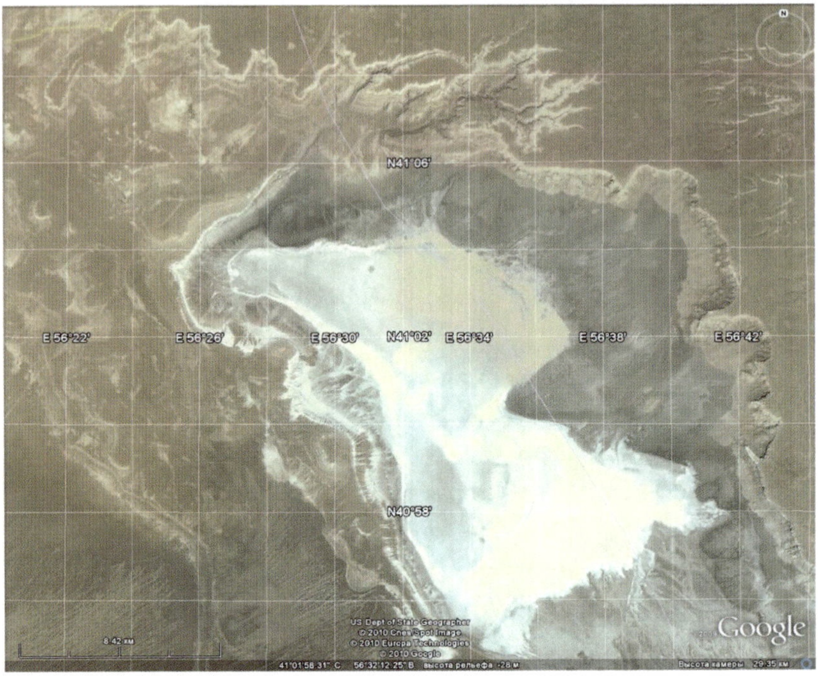

Fig. 4 Dry salt lake (*white area* is salt deposits) in the northwestern part of the Karashor Depression in the high-resolution optical satellite image presented by Google Earth

around the Karashor Depression is not populated, there is a private transhumance (sheep, camels). The region is an important bird area, situated within Kaplynkar State Reserve (see the chapter by Rustamov and Belousova in this book) [6, 7].

The average annual temperature in the area is about 12°C, a maximum of 43°C and a minimum of −30°C. Amount of precipitation is about 100 mm/year, which is a characteristic feature for the deserts [6].

Historically, the formation of the relief in the Karashor area was as follows. In the early Cenozoic, the whole territory of Turkmenistan was covered by the sea. Modern macro relief began to form at the end of the Paleogene. With the onset of the Neogene tectonic movements have intensified. Great Balkhan has uplifted. In the platform area two folded structures – Tuarkyr and Karashor anticlines, and between them Uchtagan deflection are formed. In the early Miocene, only several bays of the former Paleogene sea remained in northern Turkmenistan and Near-Kopetdag Strait. In the Late Miocene the Sarmatian Sea flooded much of the territory of Turkmenistan. In the early Pliocene isolation of the Caspian basin, lowering of its bottom led to the retreat of the sea. In the middle Pliocene a very strong deflation has opened structures and created giant depressions – Aral, Karashor, Sarykamysh, and Kara-Bogaz-Gol. From the former bottom of the Sarmatian Sea a number of areas survived, in particular, Ustyurt, Chelinkyr, and Kaplankyr – areas of flat elevated plains [8].

Later Akchagyl Sea covered Krasnovodsk Peninsula, the area between the Great Balkhan and Tuarkyr, Uchtagan deflection, Karashor and Kumsebshen depressions, and the Karakums lowlands up to Repetek and Iolotani. On the northeast Zaunguz area rivers flew into the Aral-Sarykamysh Depression, from which flew further through the Upper-Uzboy corridor to the Caspian Sea. At the turn of the Pliocene and Early Quaternary time surface areas east of Lake Yaskhan were raised, and Ancient Amu Darya during the Quaternary surrounded them from the south. In the early-Quaternary Amu Darya came into Karashor and formed a vast lake [8].

4 Project "The Turkmen Lake Altyn Asyr"

The purpose of the project – streamlining collection of CDW, discharged so far in the river basins and in the Karakums, and dumping them via the collector network into the Karashor Depression – the future Turkmen Lake Altyn Asyr (Fig. 1). Implementation of the project requires a complex of water economy, melioration, environmental, social and economic activities, and construction work to be performed. Creation of a lake has an important role in solving problems of salinization of soils, water pollution, groundwater level rise and underflooding of irrigated lands and desert grasslands. Total capacity of natural water reservoir is estimated in 132 km^3. The project will be performed during 20 years.

According to the Project, the Turkmen lake will drain water via two systems of drainage channels – the Dashoguz branch (northern route) and Transturkmen Main Collector (southern route). Dashoguz branch, 470 km long, crosses the territory of Turkmenistan from northeast to the western part of the Karakums (Fig. 1). The northern drainage network will take much of the CDW from the territory of Dashoguz velayat, from the Ozerny and Daryalyk interstate main collectors, as well as CDW from irrigated lands in Uzbekistan. Maximum CDW flow of 210 m^3/s in the Dashoguz branch consists of two CDW discharges from the Daryalyk collector – 60 m^3/s and the Ozerny collector – 150 m^3/s. Dashoguz branch begins at the 57th km of the Ozerny collector and in a little over 140 km will bring water to Zengibaba Depression. From this depression water will go into the ancient bed of Uzboy River via a 45 km long canal. Then for about 160 km water will go up to Kurtysh Baba, where the river bed turns to the west. At this point there is its connection with the Transturkmen (Main) collector. Then, CDW passes through the Uzboy still about 26 km before the canal turns to the south, where a connecting canal 54 km long continues to the northwest till Karashor Depression (Figs. 5 and 6).

The Transturkmen (Main) collector begins from the Main left bank collector in Lebap velayat, then it crosses the territory of Turkmenistan in submeridional direction from Deinau village in the east to Karashor Depression in the northwest. The total length of this collector will be about 720 km, flow in its head is estimated at 123 m^3/s, from which 58 m^3/s is the CDW flow from the right bank of Amu Darya River in the range of Lebap velayat. It will fully drain water from irrigated lands of Lebap, Mary, Akhal, and Balkan velayats at a rate of 240 m^3/s. The collector system

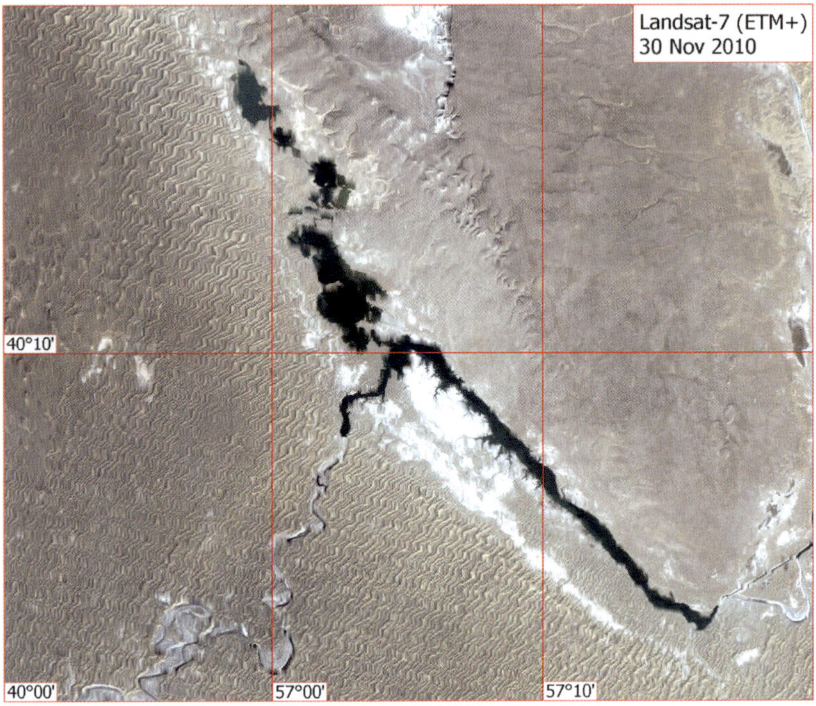

Fig. 5 Satellite image of the CDW collector southward of the Karashor Depression acquired by Landsat-7 (ETM+) on 30 November 2010

will drain water from the left and right banks of the middle reaches of the Amu Darya, thus will prevent their discharge into the river (1.8–2.8 km^3/year). From the right bank of the river the CDW will be transported via a pipeline laid under the bottom of the Amu Darya. The connecting collector network of velayats has a total length of 824.4 km [1].

Construction of the lake will be done in three stages. On October 20, 2000, 50 km north of Ashkhabad, the Turkmen capital, a start of the construction of one of the collectors that will fill the Turkmen Lake was done at a special ceremony. The work was done by 175 excavators and 125 bulldozers, bought for $54 million from the "Komatsu" Company (Japan) and "Caterpillar" (the USA). The main objective of the first phase of construction was laying pioneer canal along the pass of the Main Turkmen Collector, construction of a connecting canal from Uzboy to Karashor Depression 54 km long for ensuring the passage of drainage water of 20 m^3/s for 200 km distance (Ozerny collector – Zengibaba – Uzboy). Construction of bridges and special structures was completed in 2004. By 2008, the Dashoguz branch was built with a total length of 381 km. A water dam 600 m long and 22 km long sandy dam were constructed. The first phase was put into operation in 2009. Now work is underway on the supplying collectors at the Murghab, Tedjen, and Djarsk networks.

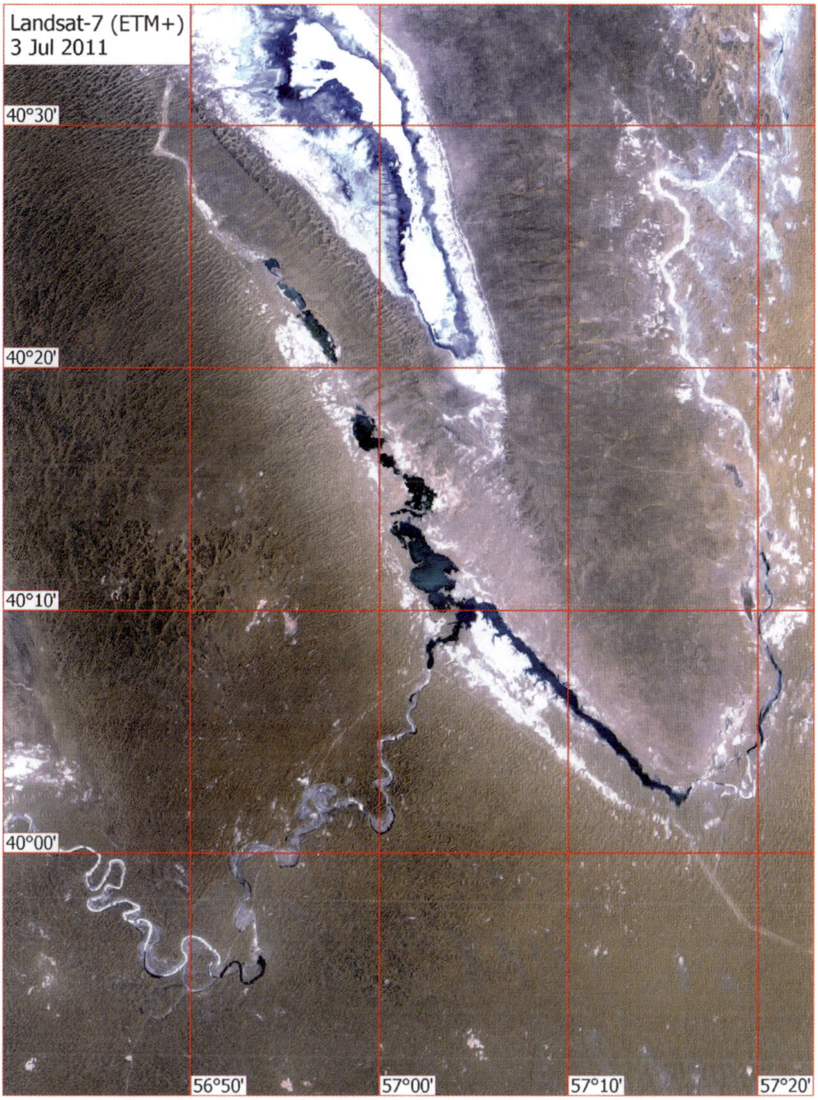

Fig. 6 Satellite images of the southern part of the Karashor Depression and the surrounding collectors acquired by Landsat-7 (ETM+) on 3 July 2011

The total length of the main and supplying collectors will reach 2,654 km. According to the Project, Lake Altyn Asyr will be 103 km long, 18.6 km wide, and 69 m deep. It will have a surface of about 1915.8 km^2 and a volume of 132 km^3 [1]. It is planned that it will receive annually 10 km^3 of CDW, thus it will be filled during about 15 years. When accumulation of water will be completed, the lake will look similar to the modern western part of the Large Aral Sea [9] (Fig. 7). Recently,

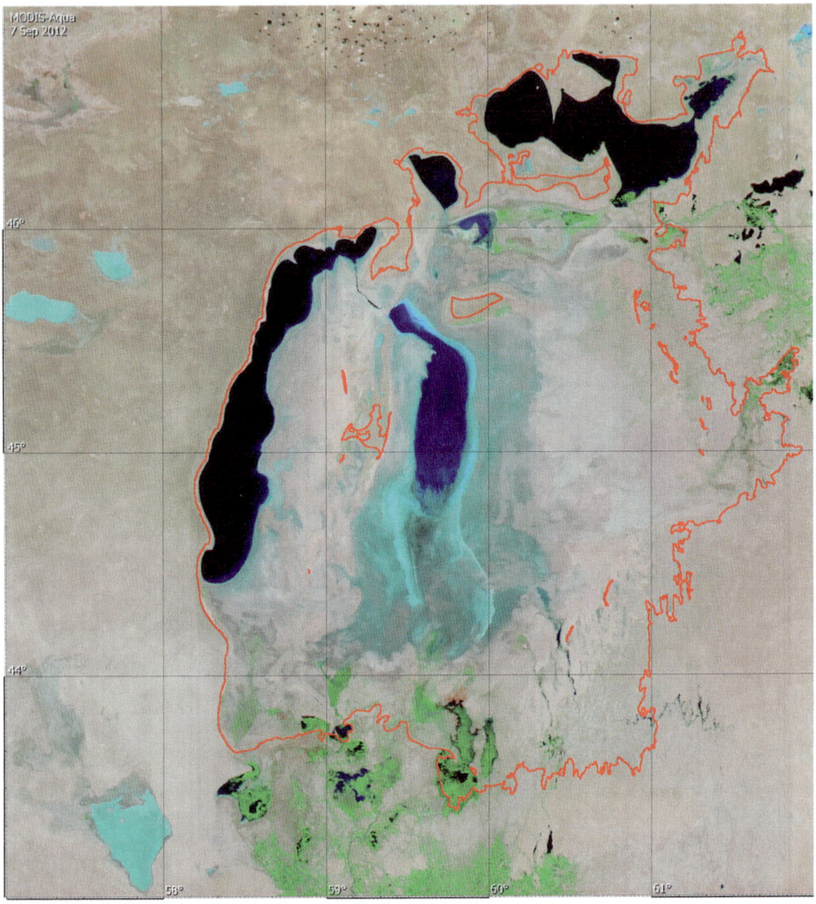

Fig. 7 Satellite images of the Aral Sea acquired by MODIS-Aqua on 7 September 2012. *Red line* shows the coastline of the former Aral Sea in 1960

digital elevation model of the Karashor Depression of 90 m spatial resolution and 1 m vertical resolution was prepared and investigated [10]. The precise topographic model of the reservoir showed significant discrepancy in morphometric characteristics of the lake in comparison with the project.

5 Value and Prospects of the Project

Turkmen scientists have repeatedly emphasized the fact that the creation of the lake is designed solely to collect CDW and does not mention diversion of the Amu Darya water (or water from other sources) to fill the lake. Construction of the Turkmen Lake Altyn Asyr is a unique project that requires special attention due to the unpredictable consequences of the impact of saline water on the natural environment along the route of the drainage canals and lake itself. This is required by the environmental safety in the operation of the constructed water canals.

It is planned that with the realization of this ambitious project many of the social and economic problems in Turkmenistan will be solved. Turkmen lake will have a positive impact on the quality of water in the collectors and lake, flora and fauna of the surrounding area, biological productivity of streams and lake. And, most importantly, the wastewater will not be spent on useless evaporation and infiltration but will be sent to the lake for accumulation and further reuse. Construction of the Turkmen Lake will allow to use new water reservoirs and collectors for development of fishery in the country. Construction of the Main Turkmen Collector will largely contribute to the improvement of the pond ranches for marketable fish. Realization of the project envisages the development of recreation facilities, eco-tourism, transport, communications, and infrastructure in the Karakum Desert.

Completion of the Turkmen Lake construction will solve the following environmental and economic problems of the country:

1. Collection of all CDW from the irrigated lands into a single stream will significantly increase the water reserves for reuse them in water economy in the country.
2. Return to the use of 4,000 km^2 of desert pastures now underflooded.
3. Decrease of the water level in the Ozerny and Daryalyk collectors by 1.0–1.5 m will ensure normal operation of drainage systems, improve salt balance in the irrigated zone of Dashoguz Velayat, and reduce the threat of destruction of the transport network.
4. Improvement of the quality of irrigated lands, the area of which is projected to reach 2.24 million ha [11].
5. Dumping of CDW from the irrigated lands in Lebap Velayat into the Amu Darya will be stopped. As a result, Turkmenistan will be the first country in Central Asia, which will stop pollution of this river.
6. Implementation of these measures will significantly improve environmental safety and create conditions for the further sustainable development and improvement of water supply and, thus, the health of the population of the region.

When implementing such a large project a set of questions, comments, and doubts arise among the international scientific community, especially concerning the impact of the collector network and the lake on the environment. The answers for these questions can be given basing on the comprehensive monitoring of the construction and exploitation of the system in the framework of international collaboration. Satellite monitoring has to be incorporated into the in situ monitoring system because it can provide effective (a set of land and water parameters with high resolution), operational (daily), continuous (years), objective (the same satellite sensors are used in other parts of the world), and independent (from in situ measurements) control of the state of the environment and to answer questions that are starting to excite the public.

6 Satellite Monitoring of the Altyn Asyr Lake Construction

On 15 July 2009 a filling of collectors by water begun in the vicinity of the Karashor Depression. Since that time, P.P. Shirshov Institute of Oceanology, Russian Academy of Sciences (Moscow, Russia) has combined efforts with Geophysical Center, Russian Academy of Sciences (Moscow, Russia) and (Marine Hydrophysical Institute (Sevastopol, Ukraine) in order to monitor the Karashor Depression and surrounding collectors and canals by means of satellite remote sensing. For this task we used optical scanners of medium spatial resolution (250 m) MODIS-Terra and -Aqua, as well as of high resolution (20 m) Landsat-5 (TM) and Landsat-7 (ETM+) [12–19]. Satellite images of medium resolution are available daily and that of high resolution – about once a month, which is sufficient for the observed rate of construction. In the above-mentioned papers, as well as in this book a chronology of the construction of the collectors in the vicinity of the Altyn Asyr Lake is followed by means of satellite imagery.

Figure 8 shows the satellite image of the southern part of the Karashor Depression and the surrounding collectors received by Landsat-5 (TM) on 21 July 2009, 6 days after start-up of water supply. A southern part of the Karashor Depression is clearly visible at the top of the frame, the bottom of which is partially covered with salt (white color). The ancient riverbed of the Uzboy River zigzags from northeast to southwest. In the center of the frame we see 20 km long section of the collector filled by water (black color). It was constructed on the straightened Uzboy riverbed. In its northwestern part the collector bends to the southwest and ends at 6 km by a dam, which prevents the spread of CDW down the ancient riverbed.

Figure 5 shows a zoom on the same area on 30 November 2010, i.e. 16 months after a start of water supply in this area. Comparison of Figs. 5 and 8 shows the progress in the construction at this site of the project. Patchiness of the water table (dark color) in this area suggests that water flows in the direction of the Karashor Depression by gravity and consistently fills the natural cavities in the relief. Figure 6 shows the same area on 3 July 2011, i.e. 24 months after the beginning of water supply. Comparison of Figs. 6 and 5 suggests that during this time period there were observed minor changes, water propagated ahead another 19 km in the direction of the Karashor Depression. The total length of canals filled by water reached 50 km. Due to intensive deposition of salt (white color) in a dry season we clearly detect the ancient riverbed of the Uzboy. In the southeastern corner of the satellite image we can see the Transturkmen (Main) collector which is dry. In the northwestern corner of the frame we can observe a new 18 km section of the collector which joins the northernmost flooded area with the Karashor Depression. Due to a strong evaporation much more salt is visible in the satellite image in July than in November, especially in the southern part of the Karashor Depression as well as along and around the Uzboy.

High-resolution Landsat optical imagery (see a zoom in Fig. 5) showed its effectiveness in the monitoring of the Altyn Asyr Lake construction. Many details, including the wave-like structure of sandy dunes in the desert, are clearly visible in

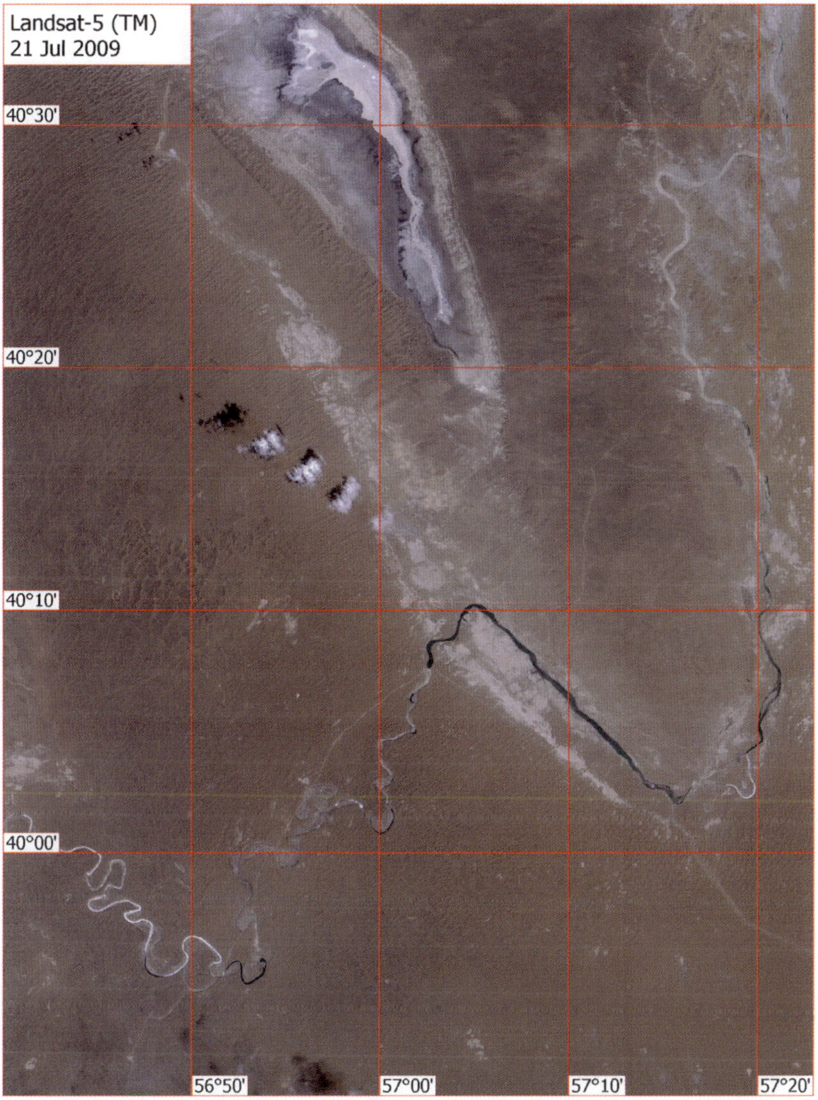

Fig. 8 Satellite images of the southern part of the Karashor Depression and the surrounding collectors acquired by Landsat-5 (TM) on 21 July 2009

the satellite images. Further progress in the construction in the framework of the project "The Turkmen Lake Altyn Asyr" is given in a special chapter by Kostianoy et al. [20].

7 Conclusions

In a country where more than 80% of the territory is the Karakum Desert, water in Turkmenistan is a basis of livelihood of the people, agriculture, and economic development. The main source of surface water resources in the country is the Amu Darya River, which, through the Karakum Canal and a canals network, supplies water for irrigation and industry. The observed depletion of water resources is due to a complex of natural and anthropogenic factors, including a rise in air temperature and decrease in rainfall as a result of regional climate change under the global warming. Deficit of water for irrigation of lands may be partially compensated for the expense of drainage water. Until now, almost all the CDW without treatment is discharged into the Karakum Desert and partially to the rivers, thereby worsening water quality. The Turkmen Lake Altyn Asyr Project is aimed to streamline collection of CDW in the country and to dump them via the collector network into the Karashor Depression – the future Altyn Asyr Lake. Implementation of the project requires a complex of water economy, melioration, environmental, social and economic activities, and construction work to be performed. Creation of a lake (in 10–15 years) has an important role in solving problems of salinization of soils, water pollution, groundwater level rise and underflooding of irrigated lands and desert grasslands. It is supposed that realization of the Turkmen Lake Project will provide a development of recreation facilities and eco-tourism, transport, communications, and will change the entire infrastructure of now vacant areas of the Karakum Desert.

Recently, a number of questions and comments, in particular about environmental issues of the project, arised, which requires answers, based on the interdisciplinary comprehensive research and monitoring of the construction and operation of the Turkmen Lake "Altyn Asyr." As a friendly initiatives, since July 2009 we organized a permanent satellite monitoring of the collector's construction in the vicinity of the Karashor Depression, which combined efforts of P.P. Shirshov Institute of Oceanology and Geophysical Center of the Russian Academy of Sciences (Moscow) and Marine Hydrophysical Institute, National Academy of Sciences of Ukraine (Sevastopol). This important part of any modern monitoring system was lacking in this great irrigation project. Since the beginning of the CDW filling into the Karashor Depression satellite altimetry measurements of water surface level in the lake will be added to the monitoring scheme. Since this time, every 10 days we will have independent measurement of the Altyn Asyr Lake level with accuracy of 2 cm. Some of the preparatory works were already done. For example, a digital elevation model of the Karashor Depression of 90 m spatial resolution and 1 m vertical resolution was prepared [10], which will be crucial for the analysis of satellite altimetry data.

We hope on a close collaboration with Turkmen scientists and international community on the organization of a permanent monitoring system, analysis of its results, and assessment of the impact of the Turkmen Lake Altyn Asyr Project on the environment of Turkmenistan.

References

1. Turkmen Lake "Altyn Asyr", Ashgabat (2010), p 104 (in Turkmen, Russian and English)
2. National Environmental Action Plan by the President of Turkmenistan Saparmurat Turkmenbashi on Environment Protection in Turkmenistan, Ashgabat (2002), pp 235
3. The State of Environment of Turkmenistan, Ashgabat (1999). Kurbanov P (ed) Ministry of Nature Protection of Turkmenistan and UNDP, pp 154
4. Stone R (2008) A new great lake – or dead sea? Science 320(5879):1002–1005
5. Grokhovsky LM (1963) Salt lakes in the arid zones of the USSR and their development. In: Bogomolov GV (ed) Problems of integrated research of arid zones in the USSR. USSR Academy of Sciences, Moscow, pp 69–97 (in Russian)
6. Rustamov EA, Welch GR, Brombacher M (eds) (2009) Important bird areas in Turkmenistan. Priority sites for conservation. MNPT, Ashgabat
7. Rustamov EA, Belousova AV (2013) Waterbirds of hydrographic network of Altyn Asyr. In: Zonn IS, Kostianoy AG (eds) The Turkmen Lake Altyn Asyr and water resources in Turkmenistan. Springer-Verlag, Berlin, Heidelberg
8. Dobrin LG (1984) Relief of nature in Turkmen SSR. In: Atamamedov NV (ed) Turkmen SSR. Encyclopedia. Ashkhabad. Chuvash Regional Committee of CPSU, Cheboksary, pp 24–30 (in Russian)
9. Kostianoy AG, Kosarev AN (eds) (2010) The Aral Sea environment, vol 7. Springer, Berlin, p 332
10. Kostianoy AG, Lebedev SA (2013) Three-dimensional topographic model of the Altyn Asyr Lake. In: Zonn IS, Kostianoy AG (eds) The Turkmen Lake Altyn Asyr and water resources in Turkmenistan. Springer-Verlag, Berlin, Heidelberg
11. Sustainable Development of Turkmenistan, RIO +10. National overview (2002) Ministry of nature protection of Turkmenistan, Ashgabad, pp 94
12. Kostianoy AG, Lebedev SA, Solovyov DM (2010) Satellite monitoring of the Altyn Asyr Lake and water resources of Turkmenistan. In: Abstracts, international conference on science, technique, and innovation technologies in the epoch of great revival, vol 1, Ashgabad, 12–14 June 2010, pp 388–391 (in Russian, Turkmen, English)
13. Kostianoy AG, Lebedev SA, Solovyov DM (2010) Satellite monitoring of water resources in Turkmenistan. In: Abstracts, 4th Coastal altimetry workshop, Porto, 14–15 October 2010
14. Kostianoy AG, Lebedev SA, Solovyov DM (2011) Satellite monitoring of water resources in Turkmenistan. In: Proceedings on CD, XV International water technology conference, Alexandria, 28–30 May 2011
15. Lebedev SA, Kostianoy AG, Solovyov DM (2011) Integrated satellite monitoring of the Turkmenistan water resources. In: Abstracts, 34th International symposium on remote sensing of environment "The GEOSS era: towards operational environmental monitoring", Sydney, 10–15 April 2011
16. Kostianoy AG, Lebedev SA, Solovyov DM (2011) Satellite monitoring of water resources in Turkmenistan. Int Water Technol J 1(1):4–15
17. Kostianoy AG, Lebedev SA, Zonn IS, Lavrova OYu, Solovyov DM (2011) Satellite monitoring of Turkmenistan, Moscow, Signal, p 16 (in Russian and English)
18. Kostianoy AG, Zonn IS, Solovyov DM (2012) Construction of the Turkmen lake "Altyn Asyr" and the results of its satellite monitoring in 2009–2011. Int Water Technol J 2(1):59–66
19. Kostianoy AG, Zonn IS, Solovyov DM (2012) The Turkmen Lake "Altyn Asyr": satellite monitoring of its construction. Mod Probl Remote Sens Earth Space 9(1):219–227 (in Russian)
20. Kostianoy AG, Lebedev SA, Solovyov DM (2013) Satellite monitoring of the Caspian Sea, Kara-Bogaz-Gol Bay, Sarykamysh and Altyn Asyr Lakes, and Amu Darya River. In: Zonn IS, Kostianoy AG (eds) The Turkmen Lake Altyn Asyr and water resources in Turkmenistan. Springer-Verlag, Berlin, Heidelberg

Three-Dimensional Digital Elevation Model of the Karashor Depression and Altyn Asyr Lake

Andrey G. Kostianoy and Sergey A. Lebedev

Abstract Digital elevation model of the Karashor Depression was constructed on the base of the Shuttle Radar Topography Mission with the horizontal resolution of 90 m and the vertical resolution of 1 m. The model allowed to get a highly precise three-dimensional topography of the Karashor Depression, to analyze it at several sections crossing the depression, to investigate morphometric characteristics of the future Altyn Asyr Lake at all stages of its filling with a step of 1 m depth, and to establish real values of the lake maximum allowed depth, surface, and volume, as well as estimate the time of the lake filling.

Keywords Altyn Asyr Lake, Digital elevation model, Karashor Depression, Shuttle Radar Topography Mission, Turkmenistan

Contents

A.G. Kostianoy (✉)
P.P. Shirshov Institute of Oceanology, Russian Academy of Sciences, 36, Nakhimovsky Pr., Moscow 117997, Russia
e-mail: kostianoy@gmail.com

S.A. Lebedev
Geophysical Center, Russian Academy of Sciences, 3, Molodezhnaya Str., Moscow 119296, Russia

Russian Space Research Institute, Russian Academy of Sciences, 84/32, Profsoyuznaya Str., Moscow 117997, Russia

I.S. Zonn and A.G. Kostianoy (eds.), *The Turkmen Lake Altyn Asyr and Water Resources in Turkmenistan*, Hdb Env Chem (2014) 28: 177–196, DOI 10.1007/698_2013_238,

1 Introduction

Today there is an increasing demand for satellite and in situ environmental data for scientific and practical purposes. In many cases the analysis of these data requires high-resolution models of the surface topography. Digital terrain modelling is a very useful tool for professionals working on the applications of terrain models in support for policies and decision making [1]. Digital elevation models (DEM) are widely used in many hydrogeological applications, such as delineation and analysis of watersheds and drainage networks, flood simulation, flood risk mapping and the design of flood emergency reservoirs, as well as agriculture, melioration of lands, construction of canals, roads, pipelines, and tunnels, civil engineering, military applications, etc. [1]. Turkmenistan is a country where the "Altyn Asyr" Project is developing, which includes a constriction of a long chain of drainage canals, which will forward drainage water from arable lands to the Karashor Depression, where during 15 years Altyn Asyr Lake will be formed. DEM technologies should be used implementing this greatest national project.

According to the Project, Altyn Asyr Lake will be 103 km long, 18.6 km wide, and 69 m deep. It will have a surface of about 1,915.8 km^2 and a volume of 132 km^3 [2, 3]. It is planned that the depression will receive annually 10 km^3 of collector-drainage water (CDW); thus it will be filled during about 15 years. It seems that these characteristics of the lake represent only some rough estimates, which are far from real values, knowledge of which is very important for calculation of the water budget of the lake, the amount of year-to-year water supply required for filling the depression, and the forecast of the time of the lake filling. This is explained by the fact that the high-resolution DEM of the Karashor Depression was not constructed and used for the analysis of morphometric characteristics of the future lake. Instead, conventional topographic cartography of uneven quality was used, and the surface and volume of the lake were calculated by a simple multiplication of the length, width, and depth of the lake which is not correct, because the Karashor Depression is not a rectangular body.

In this chapter we describe the DEM of the Karashor Depression we constructed on the base of digital surface topography of the area with the spatial resolution of 90 m and the vertical resolution of 1 m, provided by the Shuttle Radar Topography Mission (SRTM) [1, 4–6]. We hope that the obtained results will help decision-makers in Turkmenistan to take correct decisions at the final stage of the Altyn Asyr Lake construction.

2 Shuttle Radar Topography Mission

The SRTM acquired surface topography elevation data on a near-global scale (80 % of Earth's land surface from 60°N to 54°S, see Fig. 1) to generate the most complete high-resolution digital topographic database of Earth. SRTM consisted of a specially modified radar system (interferometric Synthetic Aperture Radar) that flew onboard

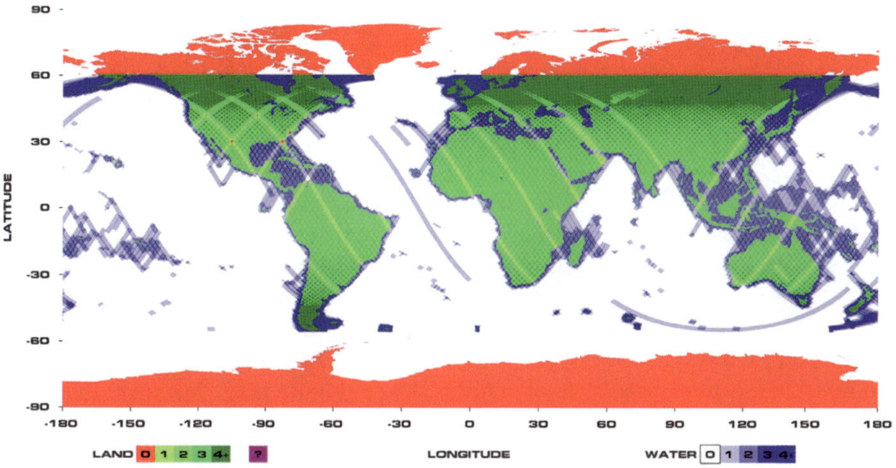

Fig. 1 The SRTM global coverage map [5]

the Space Shuttle "Endeavour" during an 11-day mission in February of 2000. In order to gather topographic (elevation) data of Earth's surface, SRTM used the technique of interferometry, when two images are taken from different vantage points of the same area. The slight difference in the two images allows scientists to determine the height of the surface with high accuracy. SRTM is an international project spearheaded by the National Geospatial-Intelligence Agency (NGA) and the National Aeronautics and Space Administration (NASA) of the USA [4–6].

The SRTM radar contained two types of antenna panels, C-band and X-band. The near-global topographic maps of Earth called DEMs are made from the C-band radar data. These data were processed at the Jet Propulsion Laboratory (USA) and are being distributed through the US Geological Survey's EROS Data Center. The data are available from the USGS server at http://dds.cr.usgs.gov/srtm/. The data are also available through the USGS seamless server at http://seamless.usgs.gov/. Data from the X-band radar are used to create slightly higher resolution DEMs but without global coverage of the C-band radar. SRTM X-band radar data are being processed and distributed by the German Aerospace Center (DLR) [4, 5].

Users should be aware that these data are intended for use with the Geographic Information System (GIS) or other special application software, and are not directly viewable in a browser. Also, users should be aware that the digital topographic data and images are unedited and are intended for scientific use and evaluation. The SRTM data may contain numerous voids (areas without data), water bodies that may not appear flat, and coastlines that may be ill defined [4, 5]. NASA has released version 2 of the SRTM digital topographic data (also known as the "finished" version). Version 2 is the result of a substantial editing effort by the NGA and exhibits well-defined water bodies and coastlines and the absence of spikes and wells (single-pixel errors), although some areas of missing data ("voids") are still present. The version 2 directory also contains the vector coastline mask derived by the NGA during the editing, called the SRTM Water Body Data (SWBD), in ESRI Shapefile format [4, 5].

Figure 1 shows global coverage of the SRTM mission. The colors of the swaths indicate the number of times the area was imaged by SRTM. For land, one-time coverage is green, twice is yellow-green, and so on, as shown in the color palette at lower left. Over water, the color code is in shades of blue as shown in the color palette at lower right. Areas in red could not be mapped. SRTM is a topography mission, so data were mostly acquired over Earth's land. Small amounts of data were collected over water for calibration purposes [4]. The cartographic products derived from the SRTM data were to be sampled over a grid of 1 arc-s by 1 arc-s (approximately 30 m by 30 m), with a linear vertical absolute height error of less than 16 m, and a linear vertical relative height error of less than 10 m. A comparison with available ground truth data showed that the absolute vertical error is better than 9 m globally (better than 5 m for the Turkmenistan territory) and the vertical resolution is about 1 m [5].

Raw data, full-resolution terrain height data, and strip DEMs with the 1 arc-s (30 m) spatial resolution for areas outside the territory of the USA are under the control of the Department of Defense. NASA and the NGA are trying to work out a policy allowing access to 1 arc-s non-US SRTM data for scientific use. The same types of data for areas within the USA and its possessions are not subject to restrictions. Distribution of terrain height data with the spatial resolution larger than or equal to 3 arc-s (90 m) is not subject to restrictions [5]. In the present analysis we used this restriction-free type of data.

3 Digital Elevation Model of the Karashor Depression

The Karashor Depression (Altyn Asyr Lake) is located in the northwestern part of the country between the Kara-Bogaz-Gol Bay of the Caspian Sea on the west and Lake Sarykamysh on the northeast (see Fig. 2 for location). From the west the depression is surrounded by Uchtagan Sands which close the Chulyungkyr Plateau. From the south, the ancient riverbed of Uzboy River limits a trough and plateau Kaplankyr. To the northwest of the Karashor Depression there is a smaller dry saline depression Kazakhlyshor (Kazakh Shor). Karashor is a deep dry depression with some salt deposits at the bottom. Its length along the long axis is about 110 km, it is 20 km wide in the northwest, and in the southeasterly direction, it gradually narrows (Fig. 3). In the northeast it is bordered by a high (up to 319 m) chink of the Kaplankyr Plateau. From the west it adjoins a large array of hilly sands. The bottom of the depression is located at a level of −28 m (28 m below the ocean level) [3].

The Karashor Depression DEM was constructed from the SRTM data with the spatial resolution of 90 m and the vertical resolution of about 1 m. We limited the model by the following geographical coordinates – 40.25–41.15°N and 56.35–57.15°E (Fig. 3). The depression is stretched from the southeast to the northwest with the deepest part there. The bottom of the depression is quite flat and inclined into the northwestern direction, where it reaches −32 m level, i.e., 32 m below

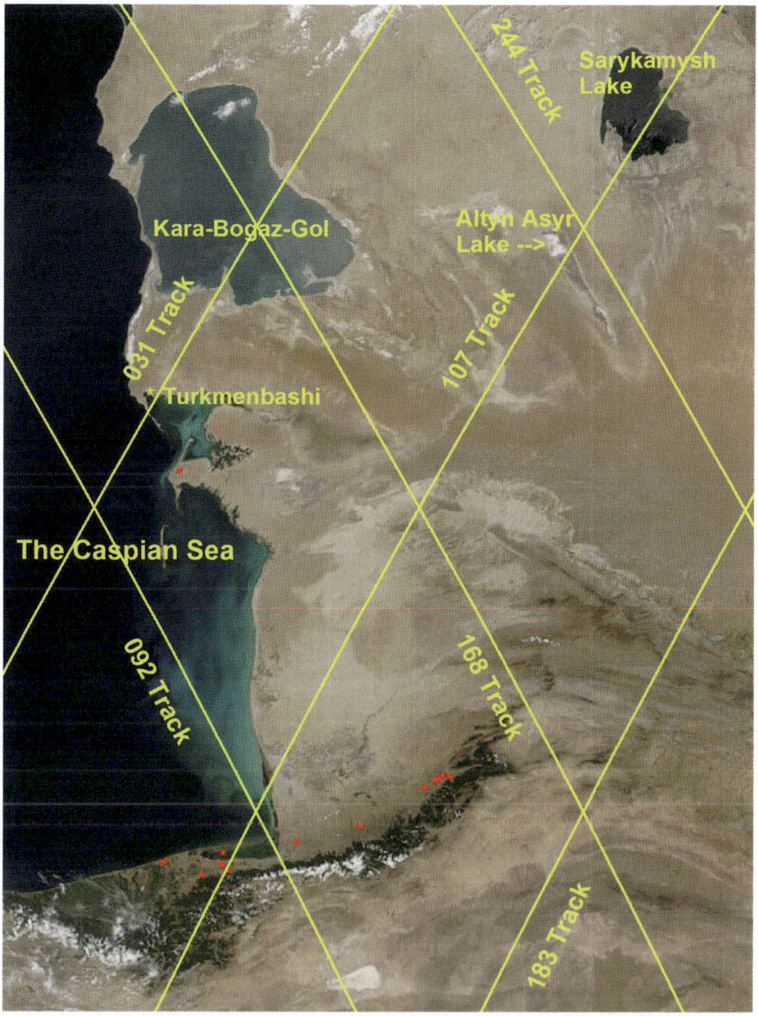

Fig. 2 The Caspian Sea, Kara-Bogaz-Gol Bay, Sarykamysh Lake, Altyn Asyr Lake (the Karashor Depression) on MODIS image acquired on 28 May 2002 with the superposition of the TOPEX/Poseidon and Jason-1/2 satellite ground tracks

the ocean level, which corresponds well with in situ topographic measurements (−28 m) [3]. The difference in 4 m abs. level may be explained by the absolute vertical error of about 5 m for Turkmenistan. The walls of the depression look to be very steep, as the gradient of the level changes very sharply at the borders of the depression. The maximum length of the depression is 110 km and the maximum width is 21.5 km in its northwestern part (Fig. 3). The 107 ground track of the TOPEX/Poseidon, Jason-1, and Jason-2 satellites crosses the northwestern part of the depression (Figs. 2 and 3). The length of the track inside the depression is 21 km. In the future it will be possible to

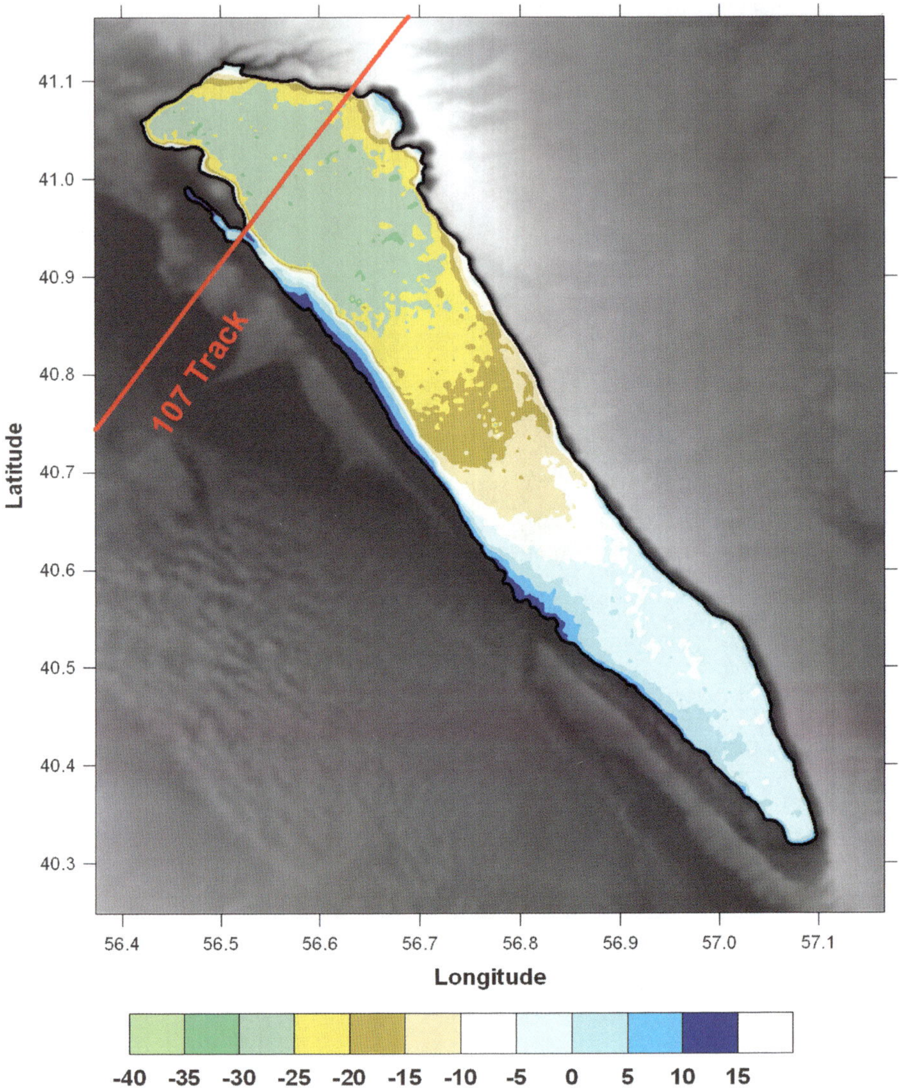

Fig. 3 The topography map of the Karashor Depression. *Color palette* shows the absolute level in meters in the depression. Areas outside the depression were shaded by *grey* colors. *Red line* shows the 107 ground track of the TOPEX/Poseidon and Jason-1/2 altimetry satellites

follow water filling of the lake basing on satellite altimetry data along this track as it has been done since 1993 for the Kara-Bogaz-Gol Bay case study [7].

Three-dimensional topography of the Karashor Depression is presented in Fig. 4, where there are two airviews – from the southwest (transversely to the depression) and from the northwest (along the axis of the depression). The first view shows that the depression has high steep walls at its back northeastern side with the highest

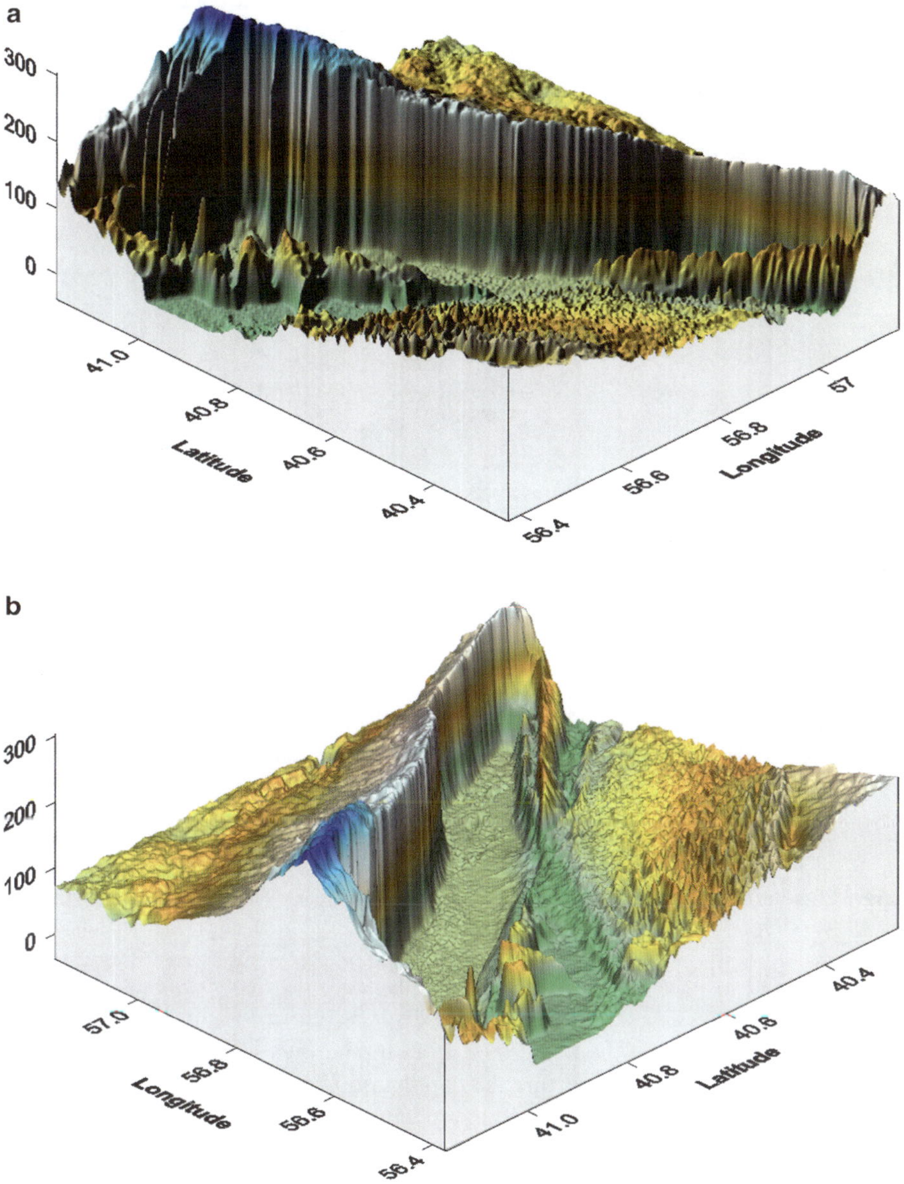

Fig. 4 Three-dimensional topography of the Karashor Depression: An airview from the southwest (**a**) and from the northwest (**b**)

point more than 300 m at the Kaplankyr Plateau closer to the northwestern end of the depression (Fig. 4a). The height of the chink (which looks like a chain of mountains) gradually decreases to the left and right, encircling the depression from both sides. The front southwestern side of the depression is partially open; therefore

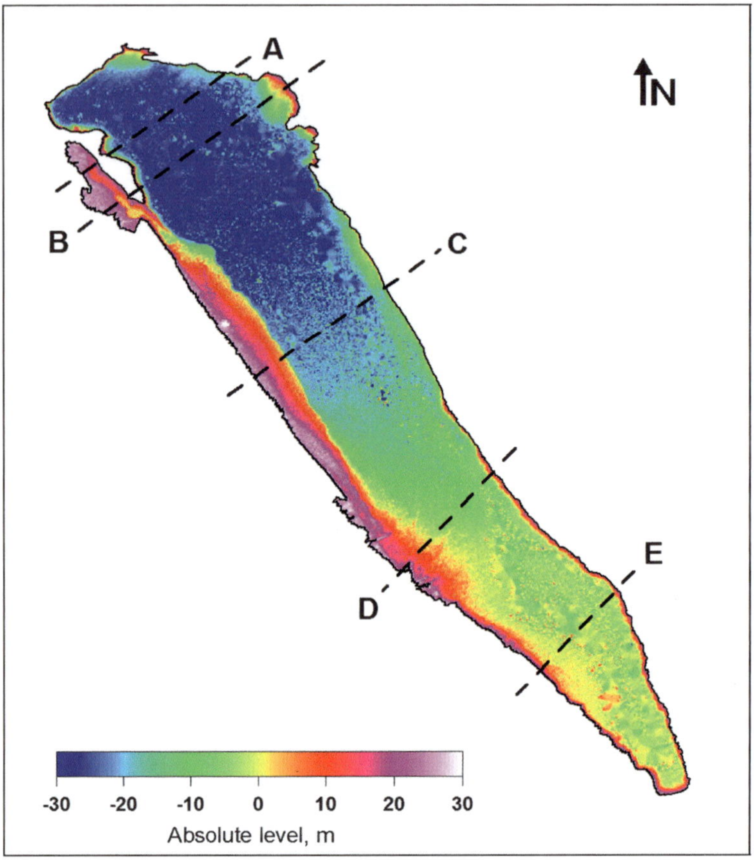

Fig. 5 Bottom topography of the Karashor Depression (absolute level in meters) and the location of cross sections A, B, C, D, and E

the canal will bring CDW to Altyn Asyr Lake via local gates somewhere in this place.

A bird's eye view on the depression from the northwestern corner is also very spectacular (Fig. 4b). It shows that the depression is deep, the bottom is wide and flat, and the walls of the depression are high and almost vertical on the left (northeastern) side and low on the right (southwestern) side of the depression. In the middle of the southwestern side there is a natural decrease in topography which will be used as a point for CDW delivery to the depression. It seems that the future Altyn Asyr Lake will have a fantastic landscape, and a magnificent view from any point will reward travelers and tourists. Also, the easiest accessible place at the coast of the lake will be located in the middle of its southwestern side, where the lake management infrastructure, probably, will be constructed.

Let's look at the vertical cross sections in different parts of the depression. The location of five sections A, B, C, D, and E is shown in Fig. 5. Sections A and B are located in the widest northwestern part, C and D – in the middle part, and E in the southeastern narrowest part of the depression. Figure 6a–e shows vertical profiles of topography in the Karashor Depression and surrounding borders at different sections. The dashed line denotes the maximum allowed level of the future Altyn Asyr Lake, which was defined by the analysis of the closed contours at different levels of land topography with a step of 1 m (see Table 1). The last closed contour was at the level of +28 m, which gives the maximum allowed depth of the lake of 60 m. The first open contour was at the level of +29 m, which means that excess water will flow outside of the lake and flood the nearest areas located farther northwestward.

Figure 6a (Section A) shows a flat bottom, a depth of about 55 m, a width of about 20 km, an almost vertical wall 220 m high over the lake level on the right side of the depression axis, a future narrow peninsula, and a small and shallow bay close to the mainland on the left side (see also a plan view in Fig. 5). Figure 6b (Section B) shows more or less the same characteristics, but the wall is higher – 260 m. Figure 6c (Section C) shows that the wall height over the lake level will be 190 m, the depth in the middle of the lake – 50 m, and the width – 18 km. The left coast will be almost flat because the land will rise to 40 m gradually along a distance of 1 km from the shoreline. Section D (Fig. 6d) shows the lake width of about 15.3 km, the maximum depth less than 40 m, the eastern wall as high as 190 m, and the western coast to be flat – 30 m for 3 km distance from the shoreline. The southeasternmost Section E (Fig. 6e) crosses the lake with the depth of about 30 m, its width here is 13 km, the wall height is of 130 m, and left coast topography on the distance of 4 km is irregular with maximum heights of 80 m (see also Fig. 4a for a plan view).

The morphometric characteristics (depth, surface, volume) of Altyn Asyr Lake were calculated on the base of the Karashor Depression DEM with a step of 1 m abs. level (Table 1). This information is of vital importance for the Altyn Asyr Project, as well as for other scientific and practical purposes. The morphometric characteristics are given for the range of absolute levels between –30 m and +28 m, which is a maximum possible level of the lake. At this level mark the maximum depth of the lake will be about 60 m (mean depth – 37 m), surface – about 1,325 km^2, and volume – about 50 km^3. These values significantly differ from those suggested in the Altyn Asyr Project – 69 m, 1,915.8 km^2, and 132 km^3 [2]. The difference is explained by the fact that the Project values were obtained by a simple multiplication of the planned length (103 km), width (18.6 km), and depth (69 m) of the lake, which is not correct.

DEM of the Karashor Depression in the northwestern part of its bottom gives topography level values deeper than −30 m as well (down to −38 m), but the analysis of spatial distribution of these depths shows that these are small-scale separated ponds. A much greater number of the smallest "ponds" at levels below −32 m is probably related with horizontal and vertical errors of the SRTM database. From the other hand, if we divide the lake volume of 0.002 km^3 on its surface of 1.45 km^2 (characteristic values for −30 m abs. level) (see Table 1), we

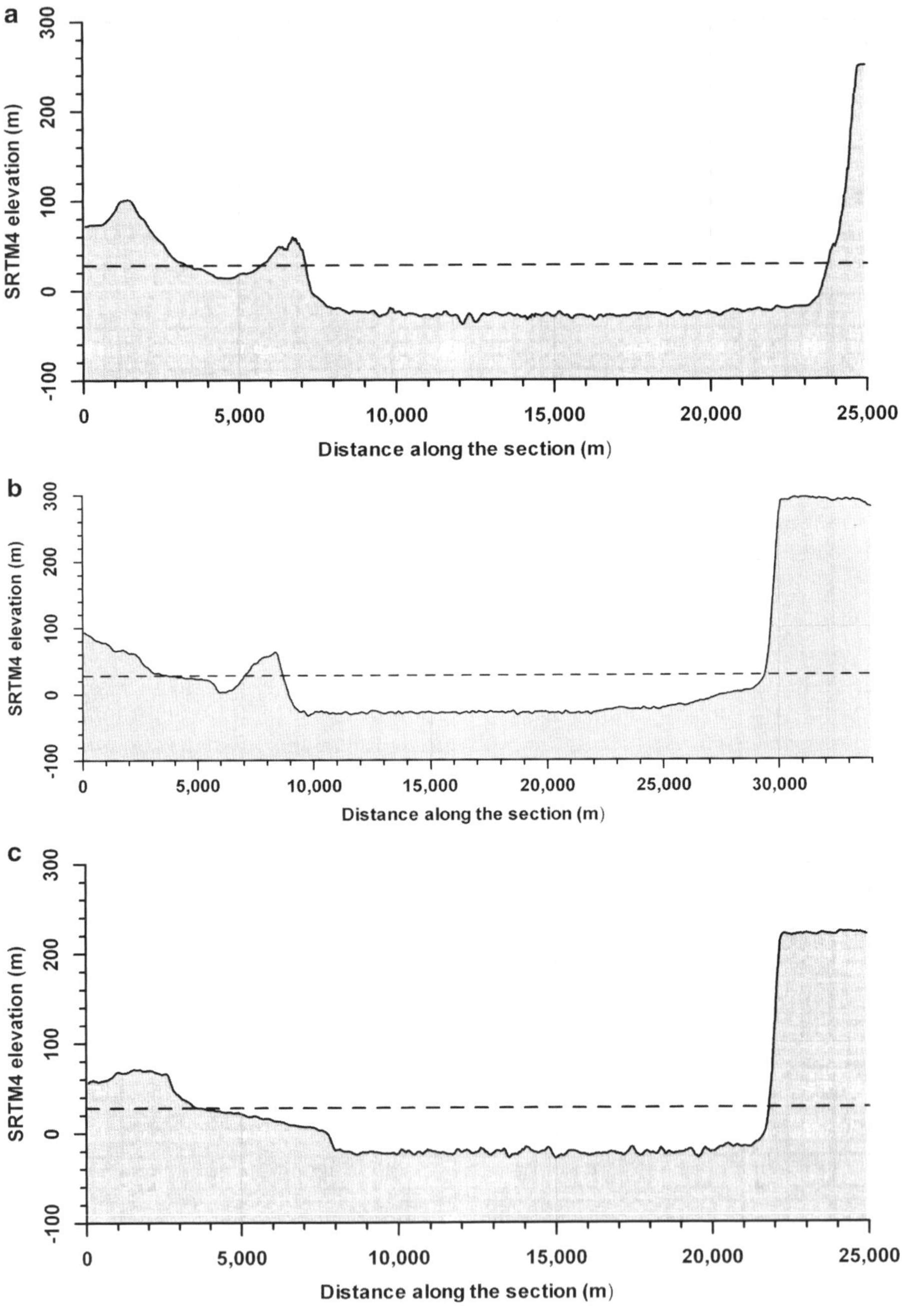

Fig. 6 (continued)

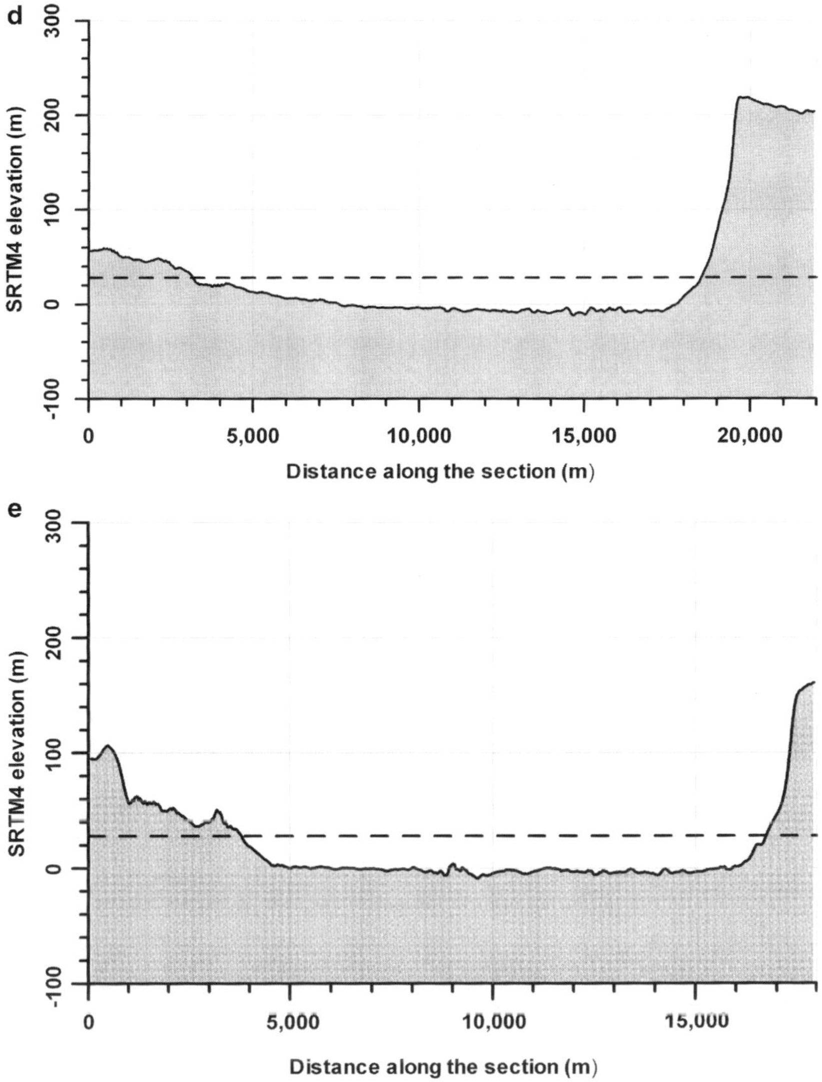

Fig. 6 Vertical profile of the topography across the Karashor Depression at sections A (**a**), B (**b**), C (**c**), D (**d**), and E (**e**). See Fig. 5 for location of sections. Vertical height is in meters of absolute level. The *dashed line* shows the maximum water level in Altyn Asyr Lake. *Grey* color shows land

will get 1.4 m as an averaged depth (assuming that the lake already represents a single water body). Thus, for simplicity we can suggest that the bottom of the lake is located at −32 m abs. level. This is the reason why we started the table with −30 m abs. level (2 m depth).

Table 1 Morphometric characteristics of Altyn Asyr Lake at different stages of its filling

Lake abs. level (m)	Lake surface (km^2)	Lake volume (km^3)	Lake depth (m)	Lake abs. level (m)	Lake surface (km^2)	Lake volume (km^3)	Lake depth (m)
−30	1.450	0.002	2	0	1,011.681	16.729	32
−29	3.062	0.007	3	1	1,042.332	17.754	33
−28	20.932	0.044	4	2	1,058.429	18.804	34
−27	220.655	0.361	5	3	1,070.937	19.870	35
−26	281.584	0.608	6	4	1,082.894	20.951	36
−25	318.386	0.886	7	5	1,094.182	22.045	37
−24	345.765	1.200	8	6	1,104.882	23.151	38
−23	376.518	1.559	9	7	1,115.116	24.269	39
−22	411.269	1.958	10	8	1,124.630	25.398	40
−21	446.782	2.389	11	9	1,136.451	26.550	41
−20	469.450	2.833	12	10	1,145.142	27.700	42
−19	494.174	3.312	13	11	1,153.819	28.861	43
−18	513.610	3.810	14	12	1,162.189	30.029	44
−17	532.538	4.333	15	13	1,169.939	31.203	45
−16	553.940	4.878	16	14	1,179.778	32.388	46
−15	575.246	5.445	17	15	1,189.221	33.583	47
−14	600.727	6.036	18	16	1,198.849	34.784	48
−13	616.974	6.645	19	17	1,208.674	35.996	49
−12	636.356	7.273	20	18	1,217.602	37.218	50
−11	655.206	7.925	21	19	1,226.217	38.449	51
−10	671.744	8.591	22	20	1,235.801	39.689	52
−9	691.232	9.276	23	21	1,246.285	40.940	53
−8	707.943	9.976	24	22	1,254.653	42.197	54
−7	721.580	10.689	25	23	1,263.599	43.463	55
−6	741.686	11.427	26	24	1,273.271	44.738	56
−5	756.948	12.178	27	25	1,286.912	46.027	57
−4	797.461	12.991	28	26	1,299.132	47.329	58
−3	860.029	13.861	29	27	1,308.265	48.641	59
−2	930.179	14.794	30	28	1,324.898	50.039	60
−1	979.590	15.743	31				

To illustrate these values we presented Table 1 in different graphical forms. In Fig. 7 we draw contours of the lake surface (shaded by black color) at different absolute (sea) levels, which correspond to different stages of the lake filling. In total there are 12 frames corresponding to the absolute levels between −30 m (2 m max depth) and +25 m (57 m max depth) with a step of 5 m. A contour of the maximal lake surface is shown in every frame for spatial comparison of the shapes. Figure 7 shows that the filling of the lake will start from the deepest part of the depression in the northwestern part, and the configuration of the lake will change significantly with an increase of the water level from −30 m to 0 m abs. level (max depth of 32 m). Then the surface of the lake will increase slowly and the lake configuration will not change significantly till it reaches the maximum level of +28 m abs. level.

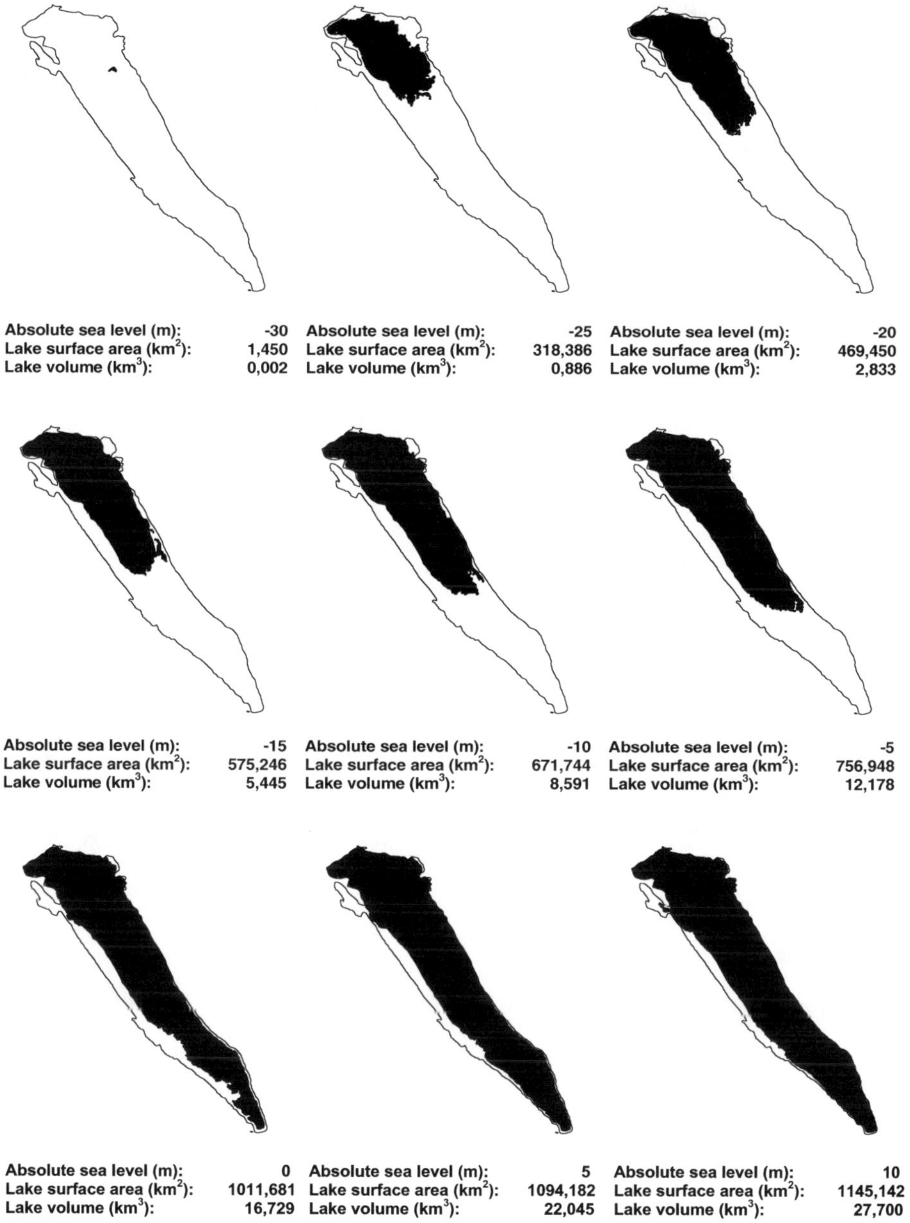

Fig. 7 (continued)

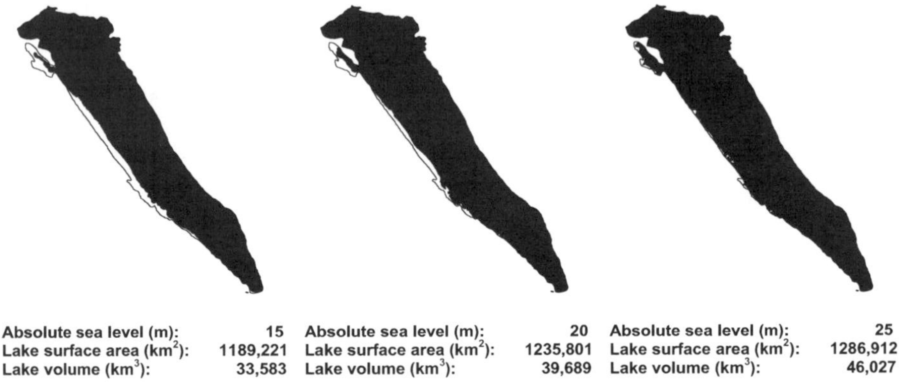

Fig. 7 Filling of Altyn Asyr Lake at different absolute (sea) levels. Surface of the lake is shaded by *black* color. Contour of the maximal lake surface is shown by a *solid line*. Every frame has indications of the absolute (sea) level, and the corresponding values of the lake surface and volume

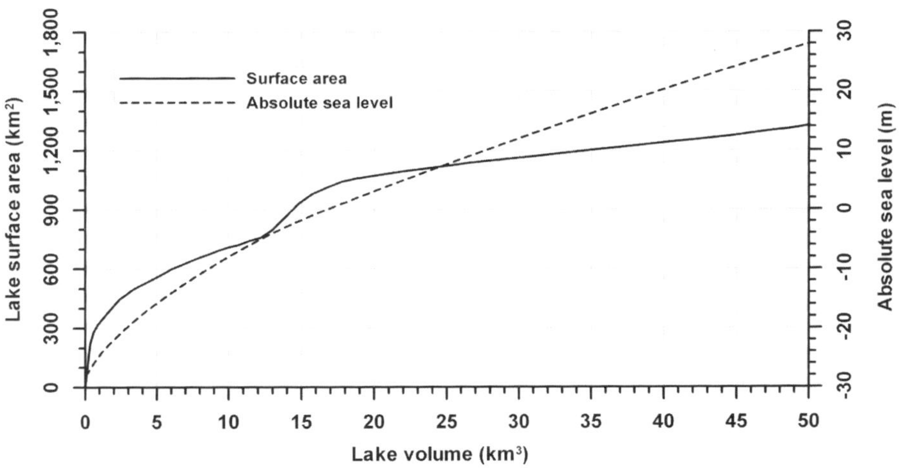

Fig. 8 The Altyn Asyr Lake surface and absolute (sea) level vs. the lake volume

Figures 8, 9, and 10 present the relationship between the absolute (sea) level, surface, and volume of Altyn Asyr Lake. The lake surface and level as a function of the lake volume are given in Fig. 8, and the lake surface and volume as a function of the lake level – in Fig. 9. Figure 10 shows the increment of the lake surface and volume vs. the absolute lake level, i.e., a value of an increase of the lake surface and volume with a step-by-step increase of the lake level by 1 m. Figures 8, 9, and 10 show that the lake surface will raise quickly till the absolute level of about +2 m, and then it will increase slowly. The lake volume will increase almost monotonically with a raise of the level.

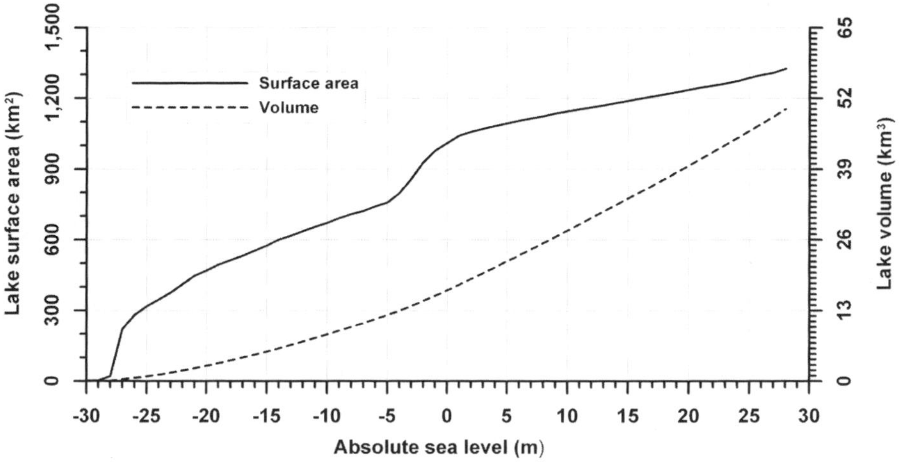

Fig. 9 The Altyn Asyr Lake surface and volume vs. the lake absolute (sea) level

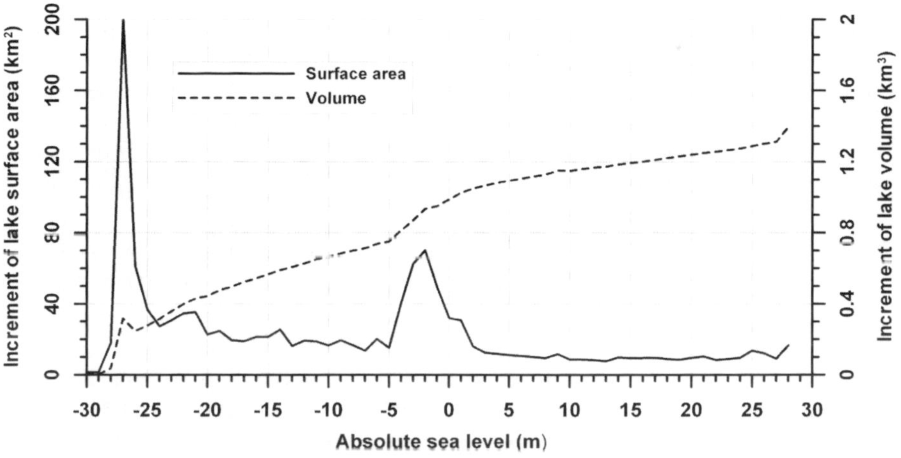

Fig. 10 The increment of the Altyn Asyr Lake surface and volume vs. the absolute lake level

Now, let's discuss two scenarios of the lake filling with the rates of CDW supply of 10 km^3/year, envisaged in the Project, and 5 km^3/year, which seems to be more realistic. Let's assume that the depression is a rigid box (no infiltration), the lake has no evaporation/precipitation, and we fill it with a rate of 10 km^3/year. In this case the lake volume will rise linearly and in 5 years the lake will be filled (Fig. 11). During the first two years (total volume of CDW of 20 km^3) the lake level will rise till 3 m abs. level (max depth of 35 m), and then it will rise slowly – 25 m during 3 years. If we add a strong evaporation of about 1–1.5 m/year, which is a characteristic feature for the other water bodies around – Kara-Bogaz-Gol Bay, Sarykamysh Lake, and the Aral Sea – we

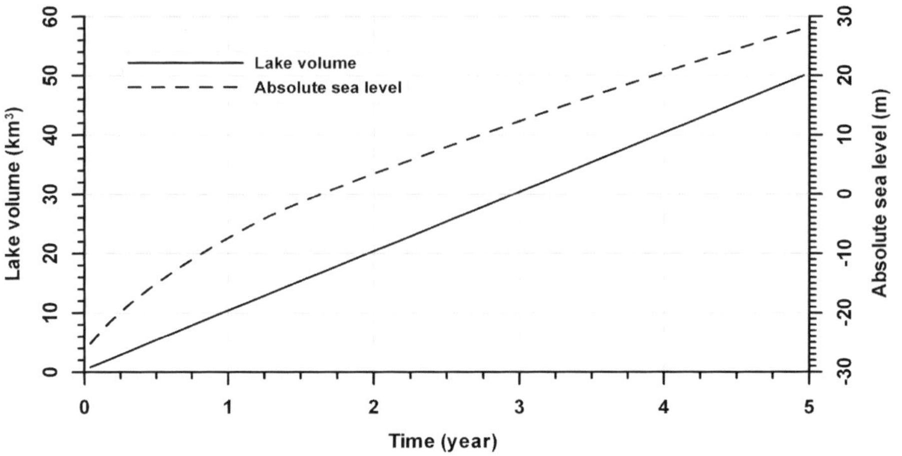

Fig. 11 The Altyn Asyr Lake volume and absolute (sea) level vs. time (years) of filling with a rate of 10 km^3/year

will have a loss of water of about 1 km^3 during the first year, 3 km^3 after 3 years, and 10 km^3 in total after 5 years (see Table 1 and Fig. 11). It means that one more year of water filling will be required, and the total time of the lake filling will increase till 6 years.

If the lake is filled with a rate of 5 km^3/year, in the ideal case 10 years will be required, but evaporation will "consume" more than 15 m of the water layer, which corresponds roughly to 20 km^3 of water. It means four additional years will be required, which in their turn will "consume" another 6 m of water or about 8 km^3, which will require additional 2 years, etc. Thus, the total time of the lake filling may be estimated as 17 years. These rough estimates showed that a decrease of water supply to the lake by a factor of 2 from the rate mentioned in the project will lead to a huge water loss for evaporation and increase of the total time of the lake filling by three times. Our calculations show that the lake being filled, only 2 km^3/year will be required to compensate yearly evaporation and keep Altyn Asyr Lake at a constant level (Table 1).

4 Conclusions

According to the Project, Altyn Asyr Lake will be 103 km long, 18.6 km wide, and 69 m deep. It will have a surface of about 1,915.8 km^2 and a volume of 132 km^3 [2, 3]. It is planned that the depression will receive annually 10 km^3 of CDW; thus it will be filled during about 15 years. These characteristics of the lake represent only some rough estimates, which are far from real values, knowledge of which is very important for calculation of the water budget of the lake, the amount of year-to-year water supply required for filling the depression, and the forecast of the time of the lake filling. This is explained by the fact that the high-resolution DEM of the Karashor Depression was not constructed and used for the analysis of morphometric characteristics of the future lake.

In this chapter we have described the DEM of the Karashor Depression we constructed on the base of digital surface topography of the area with the spatial resolution of 90 m and the vertical resolution of 1 m, provided by the SRTM. The absolute (sea) level, surface, volume, and maximum depth of Altyn Asyr Lake were calculated with a step of 1 m. This information was given in a table and different graphical forms, which would be useful for scientific and practical purposes related to the Altyn Asyr Project. The results showed that the maximum depth of the lake will be about 60 m (mean depth – 37 m), surface – about 1,325 km^2, and volume – about 50 km^3. Two scenarios of the lake filling showed that a decrease of the rate from 10 to 5 km^3/year will increase the total time of the lake filling from about 6 to 17 years due to the enormous loss of water for evaporation. When the lake is filled out, only 2 km^3/year will be required to compensate yearly evaporation and keep Altyn Asyr Lake at a constant level. We hope that the obtained morphometric characteristics of the lake will help decision-makers in Turkmenistan to take correct decisions at the final stage of the Altyn Asyr Lake construction.

Our future research will include modelling of the water budget of Altyn Asyr Lake basing on the analysis of the potential water supply, precipitation, and evaporation from the changing lake surface, taking into account typical seasonal meteo conditions (air and water temperature, wind speed, and humidity). It will be done basing on the reanalysis meteo data, but in the future an establishment of weather and limnological station at the coast of the lake will require. In situ data will have a great importance for the project. All this will allow to provide a more accurate relationship between the amount of water supply to the lake and its level (surface and volume) at different stages of its filling. This will also allow a correct forecast of the time required to fill the lake by CDW.

Ten years after the SRTM (February 2000), in 2010 a new German space experiment was performed aimed to produce better DEM of Earth's land. TanDEM-X (TerraSAR-X add-on for Digital Elevation Measurements) is a second, very similar spacecraft successfully launched on 21 June 2010 from Baikonur Cosmodrome in Kazakhstan [8]. Since October 2010, TerraSAR-X and TanDEM-X fly in a close formation at distances of 250–500 m and synchronously record data (Fig. 12). This unique twin-satellite structure allows the generation of global DEM of Earth's land surface of an unprecedented accuracy, coverage, and quality. It was planned that the data will be acquired and the DEM generated within 3 years after the launch. This DEM will have a vertical accuracy of 2 m (relative) and 10 m (absolute), and the horizontal resolution of 12 m [8]. The quality of this new product is shown in Fig. 13 in comparison with the SRTM DEM 90 and 30 m models.

Fig. 12 TerraSAR-X and TanDEM-X satellite mission. © Astrium GmbH

Fig. 13 Comparison of the quality of the TanDEM-X, SRTM DEM 90 m and 30 m data sets for generation of land topography in the Death Valley. © DLR

We hope that these data will be soon available to the scientific community, which, in particular, will improve the Karashor Depression DEM. Meanwhile, we already constructed the SRTM DEM (90 m) for the whole territory of Turkmenistan, which should be widely used for a broad spectrum of tasks in water management, agriculture, and other sectors of economy in the country.

References

1. Peckham RJ, Jordan G (eds) (2007) Digital terrain modelling. Springer, Berlin, 313pp
2. (2010) Turkmen Lake "Altyn Asyr". Ashgabat, 104 pp (in Turkmen, Russian and English)
3. Zonn IS, Kostianoy AG (2013) The Turkmen Lake Altyn Asyr. In: Zonn IS, Kostianoy AG (eds) The Turkmen Lake Altyn Asyr and water resources in Turkmenistan. Springer, Berlin
4. SRTM Home Page. (2013) http://www2.jpl.nasa.gov/srtm/. Accessed 30 March 2013
5. Farr TG, Rosen PA, Caro E, Crippen R, Duren R, Hensley S, Kobrick M, Paller M, Rodriguez E, Roth L, Seal D, Shaffer S, Shimada J, Umland J, Werner M, Oskin M, Burbank D, Alsdorf D (2007) The Shuttle Radar Topography Mission. Rev Geophys 45:33, RG2004. doi: 10.1029/2005RG000183

6. Vrscaj B, Daroussin J, Montanarella L (2007) SRTM as a possible source of elevation information for soil-landscape modelling. In: Jordan G, Peckham RJ (eds) Digital terrain modelling. Springer, Berlin, pp 99–120
7. Kostianoy AG, Lebedev SA, Solovyov DM (2013) Satellite monitoring of the Caspian Sea, Kara-Bogaz-Gol Bay, Sarykamysh and Altyn Asyr Lakes, and Amu Darya River. In: Zonn IS, Kostianoy AG (eds) The Turkmen Lake Altyn Asyr and water resources in Turkmenistan. Springer, Berlin
8. DLR (2013) http://www.dlr.de/eo/en/desktopdefault.aspx/tabid-5727/10086_read-21046/. Accessed 4 April 2013

Satellite Monitoring of the Caspian Sea, Kara-Bogaz-Gol Bay, Sarykamysh and Altyn Asyr Lakes, and Amu Darya River

Andrey G. Kostianoy, Sergey A. Lebedev, and Dmytro M. Solovyov

Abstract Satellite monitoring of water resources and land is of great importance for Turkmenistan located in the arid zone especially now when significant changes in the regional climate are observed in Central Asia. Modern capabilities of satellite remote sensing technologies in environmental monitoring and examples of use of satellite data and imagery for the analysis of morphometric characteristics, sea/lake level, sea/lake surface temperature, sea/lake wind and waves, oil pollution of the main water bodies in Turkmenistan are shown. Special attention is paid to the construction and water filling of Altyn Asyr Lake water network. Examples of the processed satellite imagery for Sarykamysh Lake and Amu Darya River are given.

Keywords Altyn Asyr Lake, Amu Darya River, Kara-Bogaz-Gol Bay, Sarykamysh Lake, Satellite monitoring, The Caspian Sea, Turkmenistan

---

A.G. Kostianoy (✉)
P.P. Shirshov Institute of Oceanology, Russian Academy of Sciences, 36, Nakhimovsky Pr., Moscow 117997, Russia
e-mail: kostianoy@gmail.com

S.A. Lebedev
Geophysical Center, Russian Academy of Sciences, 3, Molodezhnaya Str., Moscow 119296, Russia

Russian Space Research Institute, Russian Academy of Sciences, 84/32, Profsoyuznaya Str., Moscow 117997, Russia

D.M. Solovyov
Marine Hydrophysical Institute, National Academy of Sciences of Ukraine, 2, Kapitanskaya Str., Sevastopol 99011, Ukraine

I.S. Zonn and A.G. Kostianoy (eds.), *The Turkmen Lake Altyn Asyr and Water Resources in Turkmenistan*, Hdb Env Chem (2014) 28: 197–232, DOI 10.1007/698_2013_237,

Contents

1 Introduction

Turkmenistan is a Central Asian country of about 490,000 km^2, the fourth largest by area in the Former Soviet Union (FSU) after Russia, Kazakhstan, and Ukraine (Fig. 1). It is slightly smaller than Spain and a bit larger than California State in the USA. Over 80% of the country is covered by the Karakum Desert, one of the largest and driest sand deserts in the world. Some regions in the country have an average annual precipitation of only 12 mm. The highest temperature recorded in Turkmenistan was 51.7°C in July 1983. The habitable area is strictly limited, and this huge country has a small population of about six million people, which rapidly grows.

Turkmenistan possesses the world's third-largest proven reserves (8.7%) of natural gas after Russia and Iran, but land and water are the two scarcest and most precious resources in this country. Turkmenistan, like all other Central Asian countries, is critically dependent on water because of its arid desert climate, which is becoming warmer and drier. Amu Darya River, flowing from the Pamir and Tien-Shan Mountains along the entire length of the northeastern border with Uzbekistan to the tragically dying Aral Sea [1], is the main source (84%) of water for all agricultural and non-agricultural uses in Turkmenistan. Water has become the principal strategic resource that determines the region's economic development.

Water allocation from Amu Darya River is governed by regional agreements between all Central Asian states. Turkmenistan's share is 22 km^3/year or 36% of the river's total runoff. Agriculture is the main water user in Turkmenistan, consuming 95% of the available resources. Water intake from Amu Darya River is supplemented with surface runoff from three other rivers – Murghab, Tedjen, and Atrek, as well as minor quantities from small rivers and springs. Groundwater plays a marginal role in Turkmenistan's water resources [2].

Under conditions of continuous massive irrigation in Turkmenistan, considerable importance is attached to collectors and other drainage facilities intended for the removal of excess water from soil. Without proper drainage, soil may become waterlogged due to the rising water table and its salinity may increase to levels detrimental to crop growing. Expansion of irrigated areas naturally requires expansion of the collector-drainage network. The inadequacy of the collector-drainage network is reflected in severe deterioration of soil quality. In 14% of irrigated

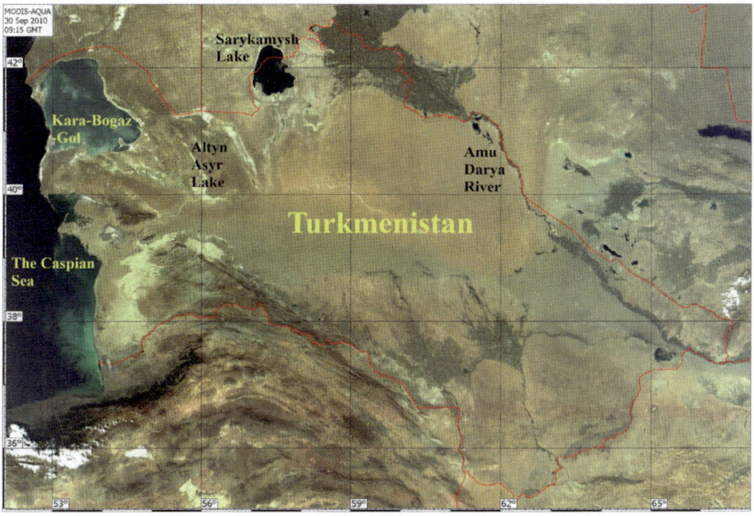

Fig. 1 Satellite view over Turkmenistan (MODIS-Aqua, 30 September 2010)

lands the water table has risen above the critical level, and 1.65 million hectares, or fully 73% of irrigated lands, are salinized [2].

The Turkmen Lake Altyn Asyr (Golden Age Lake) is a new approach to disposal of drainage water from irrigation. Following a decision adopted in August 2000 by the President of Turkmenistan, the country is constructing a huge artificial lake in the middle of the Karakum Desert, on the site of the natural Karashor Depression. The lake is on the border between Balkan and Dashoguz velayats, some 300 km north of the capital Ashkhabad. The lake will be filled with drainage water through a new collector, the Great Turkmen Collector from the south with a combined length of over 1,000 km. It is planned that the collectors will annually divert to the lake up to 10 km^3 of saline drainage water, which is currently discharged into Amu Darya River and Sarykamysh Lake.

Freshwater resources in Turkmenistan should be discussed with other water resources like the Caspian Sea, Kara-Bogaz-Gol Bay, and Sarykamysh Lake which play a very important role in different sectors of the economy of the country (Fig. 2). Turkmenistan is connected with the other Caspian Sea countries by shipping routes. Offshore, there are oil and gas fields, which are developed by national and foreign companies. The sea is rich in bioresources, and fishery is a part of the economy of the country. Kara-Bogaz-Gol Bay during the twentieth century played a key role in the chemical industry of the Turkmen Soviet Republic and today in Turkmenistan. The national tourist and recreation zone "Avaza" near Turkmenbashi town (see Fig. 2) with several modern hotels, restaurants, cafes, a sandy beach, artificial canal, and park zone is progressively developed at the coast of the warm Southern Caspian Sea. The Caspian Sea water (after desalination) is an immense potential source of potable and technical water for the country, living in the desert conditions.

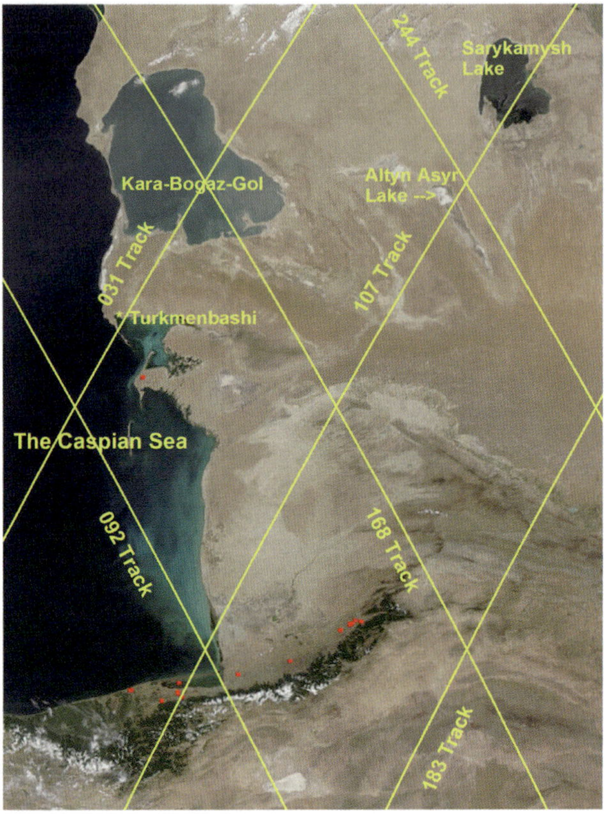

Fig. 2 The Caspian Sea, Kara-Bogaz-Gol Bay, Sarykamysh Lake, Altyn Asyr Lake, and Western Turkmenistan on MODIS image acquired on 28 May 2002 with superposition of the TOPEX/Poseidon and Jason-1/-2 satellites ground tracks

Satellite monitoring of water resources and land is of great importance for Turkmenistan and other Central Asian countries, located in the arid zone, especially now when significant changes in the regional climate are observed. Today, an integral part of any modern environmental monitoring of land, sea, lakes, and rivers is satellite-based monitoring, which has great additional features and advantages over ground-based. First of all, they are: (1) global coverage, (2) instantaneous snapshoting of a vast area, (3) the highest operationality in data acquisition, (4) ability to repeat daily observations, (5) high spatial resolution (from 1 km to 50 cm), (6) receiving of interdisciplinary and multisensor data, (7) ability to ensure comprehensive monitoring at any point of the globe, (8) using the same satellite data for a wide range of supplementary tasks (fires, floods, desertification, vegetation, water resources, etc.), and (9) significantly low cost of satellite monitoring in comparison with in situ observations.

Over the past 20 years in the course of a number of Russian and international projects, P.P. Shirshov Institute of Oceanology of Russian Academy of Sciences

(SIO RAS), Space Research Institute (IKI RAS), Geophysical Center (GC RAS) in cooperation with Marine Hydrophysical Institute of National Academy of Sciences of the Ukraine (MHI NASU) have gained a unique joint experience working with a variety of satellite data on the state of the oceans, seas, lakes, and rivers; have developed and worked out new methods of research that apply for comprehensive environmental monitoring of Russian seas and inland water bodies [3]. We have developed an effective complex (multisensor and interdisciplinary) approach to real-time satellite monitoring of oil pollution of the seas of Russia [3]. This approach was implemented in practice in 2004–2005 in the Southeastern Baltic Sea under the contract with "Lukoil-Kaliningradmorneft," when a full-operational satellite monitoring service was established for oil pollution control. Later, a similar integrated approach has been applied to the Black, Azov, and Caspian seas.

Satellite monitoring of coastal ocean, inland seas, lakes, and rivers is an important method to control their ecological condition. It is based on the reception, processing, and analysis of digital data from different radiometers, scanners, spectrometers, radar altimeters, scatterometers installed mainly on European and USA satellites (NOAA, Terra, Aqua, TOPEX/Poseidon, Jason-1, Jason-2, GFO, ENVISAT, Radarsat-1, Radarsat-2, TerraSAR-X, ERS-2, QuikSCAT, Landsat, IRS, KOMPSAT-2, EROS A, IKONOS, SPOT, QuickBird, FORMOSAT-2, and many others), which give information about fields of the sea surface temperature (SST), suspended matter, chlorophyll concentration, and other optical properties of the water surface and land, oil pollution, as well as anomalies in the sea level, ice cover, variability of currents, wind speed, and wave height with high spatial and temporal resolution. An opportunity to survey huge water and land areas in a short period, as well as a possibility of repeated observations of the same region with a short interval of time (1 day), make the use of remote sensing the cheapest, fastest, and objective method for environmental monitoring. Turkmenistan has supplementary advantages for satellite monitoring, because the country is located in the area with the number of cloudy free days which varies between 240 and 300 and sunshine of 3,100 h/year. These are the highest values in the FSU.

In recent years, with the openning of data banks with global regular satellite information and reanalysis data on the SST, sea level, chlorophyll concentration, ice cover, atmosphere pressure, air temperature, wind, rainfall, snowfall, humidity, heat flux, and other meteorological characteristics (PODAAC JPL, AVISO, UT/CSR, NCEP, GSFC NASA, DAAC GSFC, and many others), there is an opportunity to study not only seasonal but also interannual and even decadal variability of atmosphere, land, and sea parameters. This is particularly important for the study of the regional climate change in Turkmenistan and other Central Asian countries.

We have a long-standing experience in satellite monitoring of water bodies in Central Asia, and first of all this is the Caspian and Aral seas, which are under our permanent attention since 2000 [1, 3–14]. We began satellite monitoring of Turkmenistan with the beginning of water filling the Altyn Asyr water network in July 2009 [15–22]. In this chapter we would like to show only examples of different types of satellite information which can be received, processed, and analyzed, and can be very important for different sectors of economy, science, and education in Turkmenistan. Thus, here we do not intend to analyze satellite images or interannual/seasonal variability of different parameters, but sometimes useful references to the appropriate research will be given. Also, we will focus on the applications generally related to water resources and water quality.

2 The Caspian Sea

The Caspian Sea is the world's largest isolated water reservoir. Its isolation from the ocean and its inland position make the outer thermohydrodynamic factors, specifically, heat and water fluxes through the sea surface, and river discharge the most important for sea level variability, formation of its 3D thermohaline structure, and water circulation [10, 23]. In the twentieth century, there was the Caspian Sea level regression by 2.5 m until 1977 when the sea level lowered to −29 m (29 m below the ocean level). This was a result of a combination of natural factors (decrease of precipitation over Volga River catchment area) and man-made impact (construction of cascade reservoirs in Volga and Kama Rivers). In 1978 the Caspian Sea level started to rise rapidly, reached its maximum in 1995 (−26.4 m), and now it is going down again with some oscillations. By the end of 2012 the sea level already reached a level of −27.7 m. Sea level variability, river runoff, weather conditions, regional climate change, and anthropogenic pressure have a significant impact on the marine environment and ecological state of the Caspian Sea, including waters of Turkmenistan.

The best instrument to monitor the Caspian Sea level is satellite altimetry [7, 8, 24–27]. Satellite altimetry measures the sea surface height (SSH) relative to a reference ellipsoid (or the gravity center) that allows elimination of vertical Earth's crust shifts from interannual level variation. Thus, satellite altimetry has advantages over the old Caspian coastal gauge stations which were not calibrated against the ocean level for several decades and have no precise 3D GPS stations [26]. Figure 2 shows the ground tracks of the TOPEX/Poseidon and Jason-1/-2 altimetry satellites over the Caspian Sea waters of Turkmenistan and Western Turkmenistan, where they also cross Kara-Bogaz-Gol Bay, Sarykamysh Lake, and the Karashor Depression (future Altyn Asyr Lake). Every 10 days the satellite passes along every line, thus in crossover points we have measurements every 5 days. This is enough to investigate seasonal and interannual variability of the Caspian Sea and other water bodies in Turkmenistan. Spatial resolution of altimetry measurements along the tracks is of 7 km, and accuracy is of 4 cm. Figure 3 shows interannual and seasonal variability in the Caspian Sea level basing on satellite altimetry data acquired in 1993–2012.

In Fig. 3 we see that the Caspian Sea level was decreasing from summer 1995 till winter 2001/2002, then it was rising till summer 2005 with a rate of about 10 cm/year, then again it was decreasing till winter 2009/2010 with a rate of 8.5 cm/year, and it accelerated to about 15 cm/year in the past 3 years till winter 2012/2013. Thus, the sea level has reached −27.7 m, which is already 1.3 m less

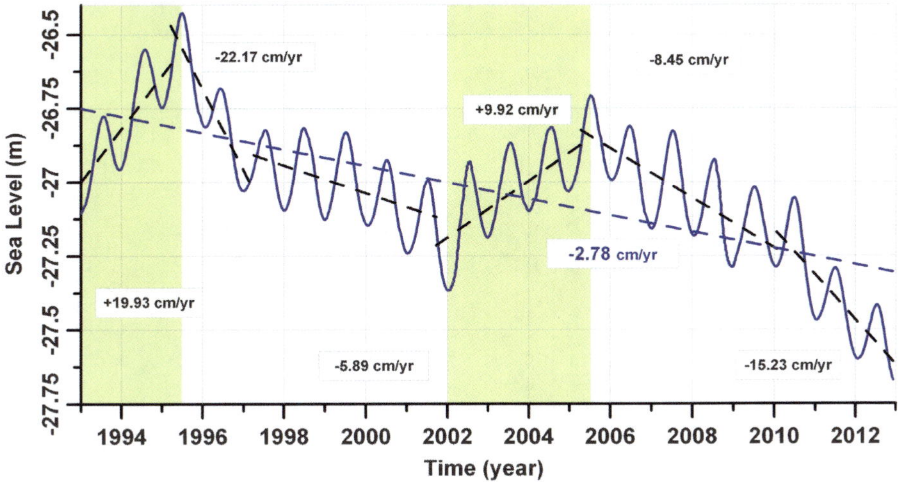

Fig. 3 Seasonal and interannual variability in the Caspian Sea level basing on satellite altimetry data of TOPEX/Poseidon and Jason-1/-2 acquired in 1993–2012. *Yellow* fields show periods when the sea level was rising. *Dashed black lines* show local trends (and trend values in *black*) and *dashed blue line* shows a general trend (and value) for the entire period

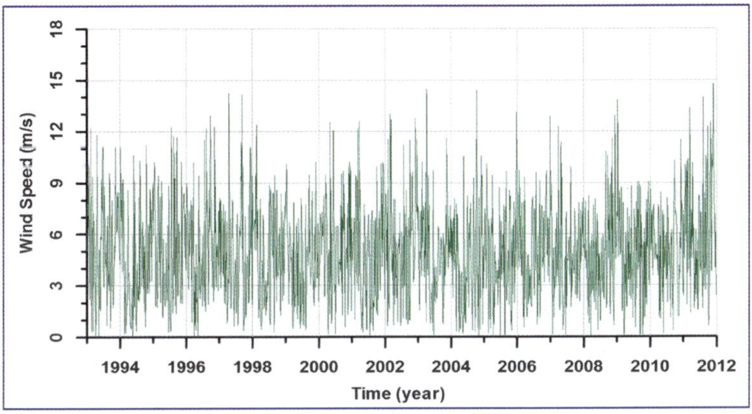

Fig. 4 Seasonal and interannual variability of wind speed at the crossover point of 31/92 tracks (see Fig. 2 for location) in 1993–2011 based on satellite altimetry data

than in summer 1995. Amplitude of seasonal variations of the sea level is of the order of 30–40 cm. All this information is of great importance for Turkmenistan, which has Turkmenbashi port and oil terminals, Turkmenbashi town, and the tourist zone "Avaza" at the shore of the Caspian Sea.

Satellite altimetry also allows to reconstruct values of wind speed (Fig. 4) and wave height (Fig. 5) along the tracks over the sea surface with the same spatial and temporal resolution. This is a very valuable source of information for shipping activities and offshore oil/gas platforms operations at sea. Figures 4 and 5 show

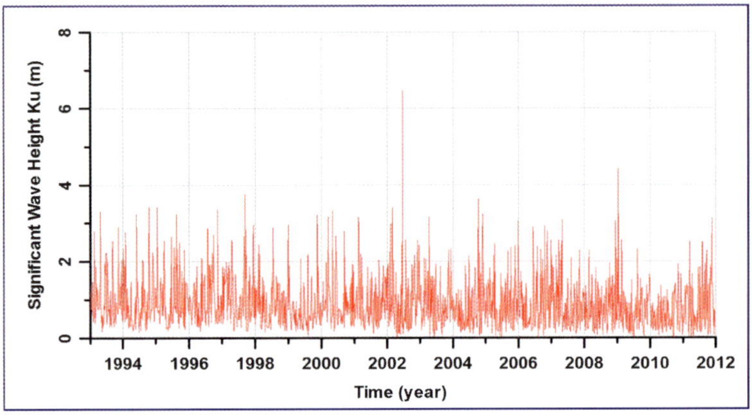

Fig. 5 Seasonal and interannual variability of wave height at the crossover point of 31/92 tracks (see Fig.2 for location) in 1993–2011 based on satellite altimetry data

variability of these parameters at the crossover point of 31/92 tracks (in front of Cheleken Peninsula, see location in Fig. 2) between January 1993 and December 2011 with time step of 5 days.

Satellite synthetic aperture radar (SAR) technologies allow to detect oil spills on the sea surface [3–6, 17]. This satellite technology is widely used for oil pollution control of shipping routes, offshore oil/gas platforms, ports, and oil terminals in the seas. Figure 6 shows a radar image of the coastal waters of Turkmenistan between Turkmenbashi and Cheleken Peninsula on 10 July 2010. In front of Cheleken Peninsula there are dozens of oil/gas platforms which look like a set of numerous white bright dots. Two of them are sources of oil spillages, which are marked by yellow circles. The largest one is stretching from south (location of one of the platforms) to north for 36 km. The spill passed at a distance of 18 km from the national tourist zone "Avaza." The other three oil spills in two circles (closer to the coast) could be released from ships or have natural origin (oil seepages from the bottom). Additional examples of oil pollution in coastal waters of Turkmenistan can be found in [17], where we proposed to organize permanent satellite monitoring of oil pollution in the sea, basing on our experience of operational satellite monitoring of the Lukoil D-6 oil platform in the Southeastern Baltic Sea [3, 28]. This is a very important task to control ecological conditions of the marine environment of Turkmenistan. The same SAR technology can be used for detection of new oil fields in the Caspian Sea basing on statistical data on the location of natural oil seepages from the bottom discovered by specific oil pollution features at the sea surface.

The quality of the marine environment can be controlled by optical imagery acquired by different spectroradiometers like MODIS-Terra and -Aqua and MERIS ENVISAT [3]. These instruments allow to reconstruct fields of suspended matter, chlorophyll concentration, and algal bloom events. High concentrations of these parameters represent another type of chemical and biological contamination of

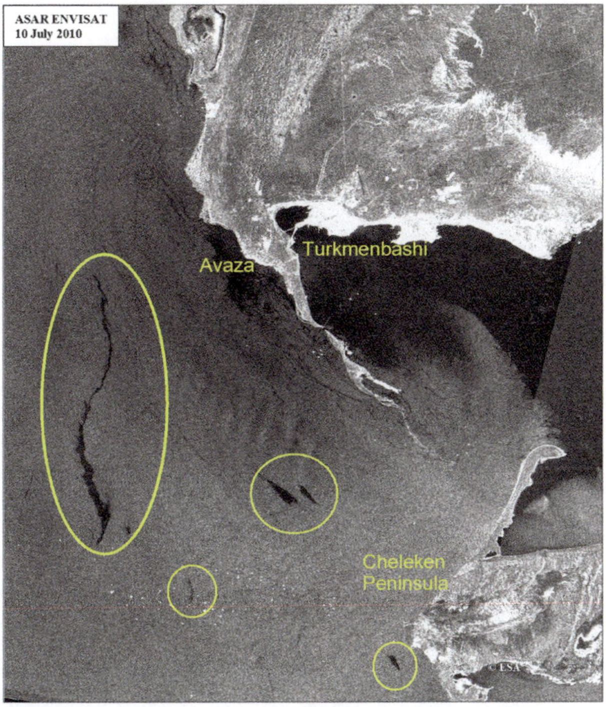

Fig. 6 ASAR Envisat image of coastal waters of Turkmenistan acquired on 10 July 2010. Oil spills (*black patches*) are shown in *yellow circles*. ©ESA, 2010

Fig. 7 Anomalous algal bloom in the Southern Caspian Sea (*green* vortical area) on 1 September 2005 revealed by MODIS-Terra

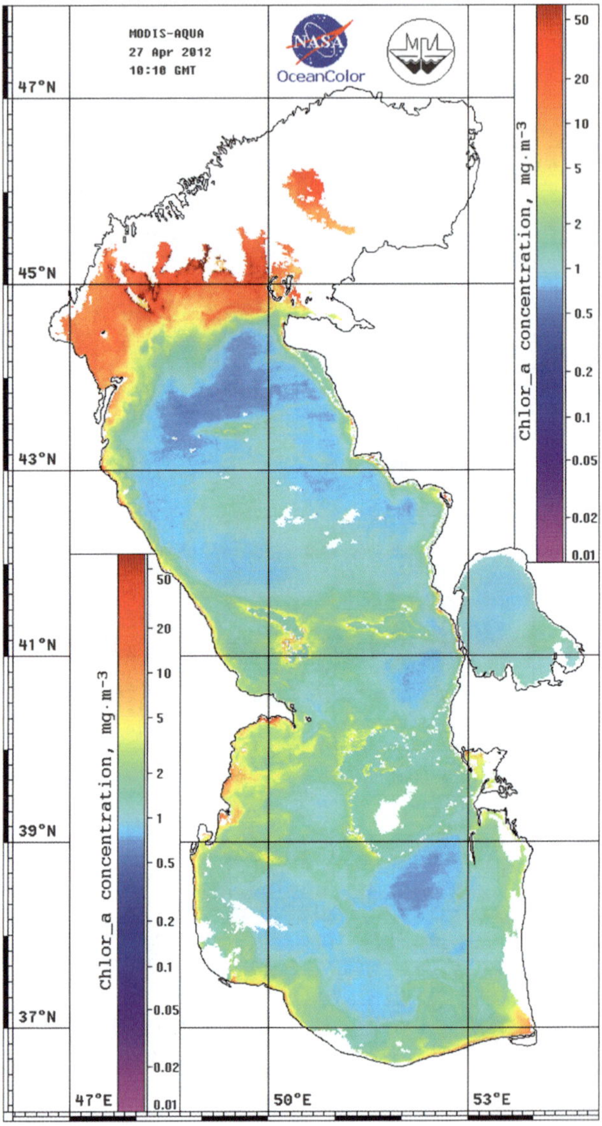

Fig. 8 Chlorophyll concentration in the Caspian Sea basing on the MODIS-Aqua data acquired on 27 April 2012

sea water. Intensive bloom of blue-green algae significantly reduces the quality of sea water and sometimes it may be even dangerous for humans and animals, because a part of these algae may be toxic. This is a well-known phenomenon, for example, in the Baltic Sea, where it occurs yearly in the increasing area and during longer periods of time. Recently, the same events began to appear in the northwestern part of the Black Sea, and a very large anomalous bloom event occured in 2005 in the Southern Caspian Sea (Fig. 7). The bloom area reached 20,000 km^2, and the bloom event lasted almost 2 months during August and

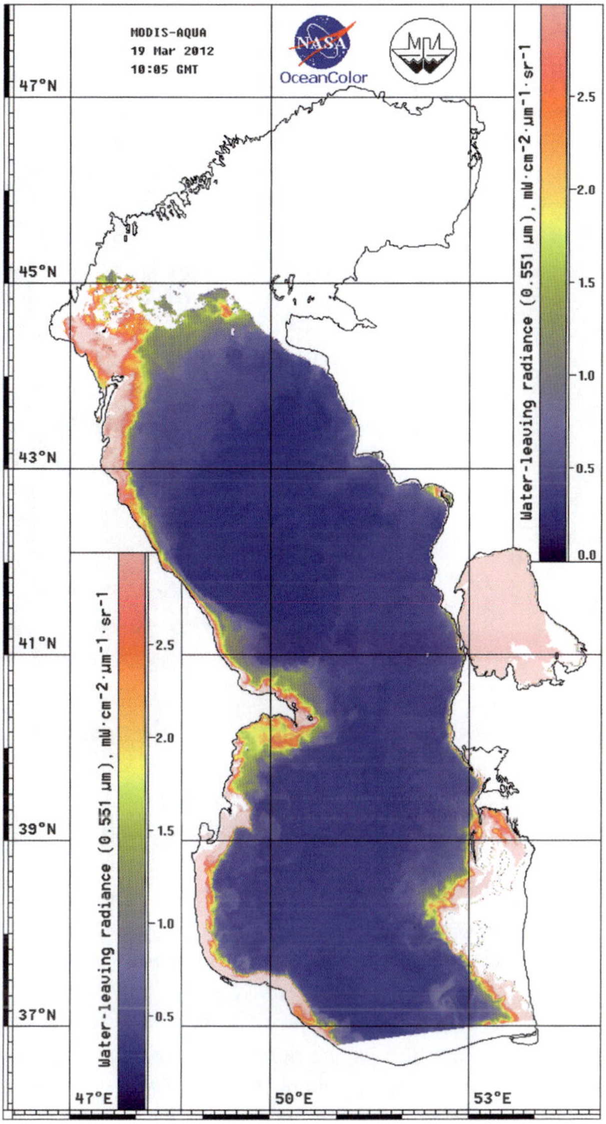

Fig. 9 Spatial distribution of turbid waters in the Caspian Sea displayed by water-leaving radiance acquired by MODIS-Aqua on 19 March 2012. *Blue* color shows clean waters

September. This event was detected and followed by MODIS-Terra daily imagery, which showed that fortunately the bloom area did not reach the coasts of Turkmenistan, Iran, or Azerbaijan (Fig. 7).

There are different algorithms to calculate chlorophyll concentration in the upper layer of the sea from satellite optical data. This technology allows to reconstruct fields of chlorophyll concentration for the whole Caspian Sea and to digitally monitor the quality of sea water. One of the examples is shown in Fig. 8 for 27 April 2012. A typical spatial distribution of chlorophyll suggests larger

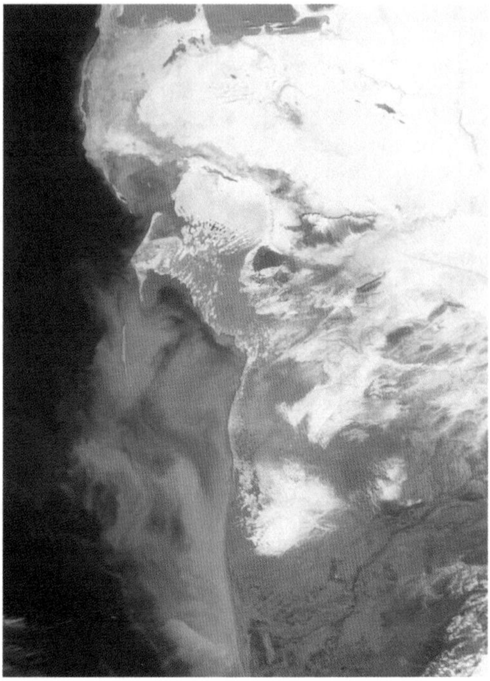

Fig. 10 Spatial distribution of turbid waters along the coasts of Turkmenistan displayed by MODIS-Terra on 15 February 2012. *Light* colors show turbid waters

values in the Northern Caspian, in front of the Volga River Delta, and along the coasts of the sea. Coastal zone of Turkmenistan is normally characterized by low values of chlorophyll concentration (Fig. 8).

The coastal zone from the Turkmenbashi Bay to the southern coast of the Caspian Sea in Iran is characterized by a permanent very large zone of turbid waters (Fig. 9). The turbid area is about 200–250 km long and 100–120 km wide and is displayed on most of the optical satellite images. The shape of the turbid area corresponds exactly to the bottom topography with an isobath of 25–30 m in the case of maximum development. It was found that the reason for generation of these turbid waters is resuspension of bottom sediments due to wind forcing at a very shallow area, moreover the radiances (turbidity) were about twice as high for winds having an offshore component in comparison with the onshore wind conditions [29]. Other wind directions produce also a turbid zone of a different size, this is why it is observed almost anytime [29]. High concentration of suspended matter is typical also for the shallow Turkmenbashi and Kara-Bogaz-Gol bays (Fig. 9).

Total concentration of suspended matter may be characterized by water-leaving radiance measured by SeaWiFS, MODIS-Terra, or -Aqua, or directly calculated in absolute values from the MERIS ENVISAT data, which unfortunately are no more available since April 2012 due to a failure of the ENVISAT satellite. The turbid area is also very well visible in “true colors” displayed by the MODIS instrument with 250 m spatial resolution (Fig. 10). This information is available daily from Terra and Aqua satellites. This turbid area may have a negative impact on fishery,

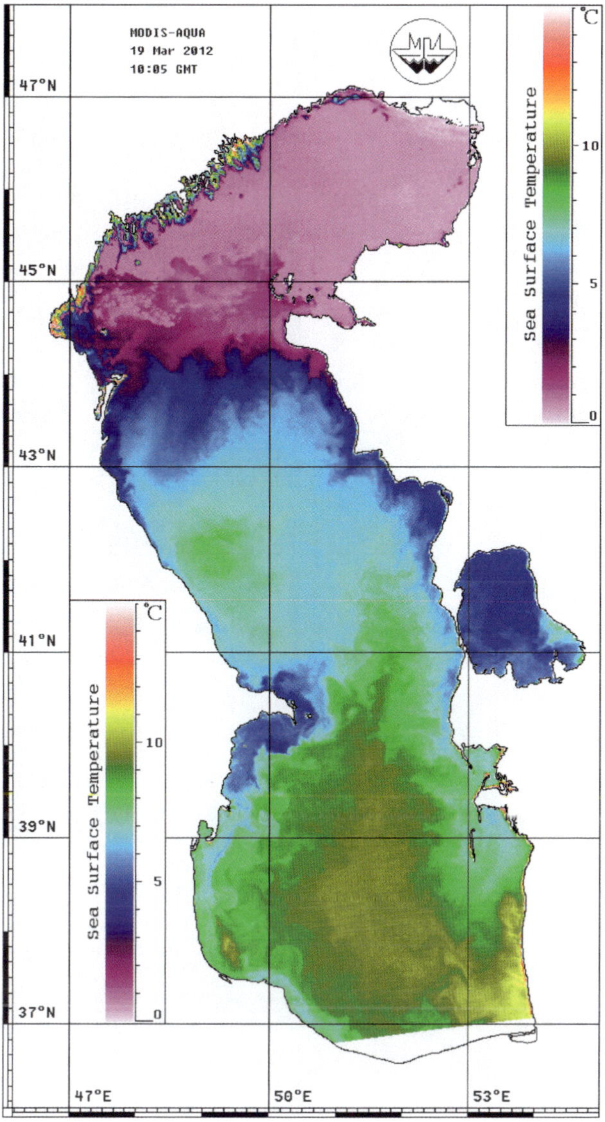

Fig. 11 SST field in the Caspian sea (MODIS-Aqua, 19 March 2012)

and its advection by the coastal current northward may have a negative impact on water quality in the recreation area of the "Avaza" tourist zone.

Radiometers AVHRR NOAA, spectroradiometers MODIS-Aqua and -Terra provide information on the SST, which is very important characteristics for any water body (Fig. 11). SST fields are used to detect meso- and small-scale water dynamics in the sea, which is required, for example, for the analysis and forecast of pollutants transport; to locate and investigate zones of coastal upwellings [30, 31], which play a key role in fishery and have a significant impact on recreation zones

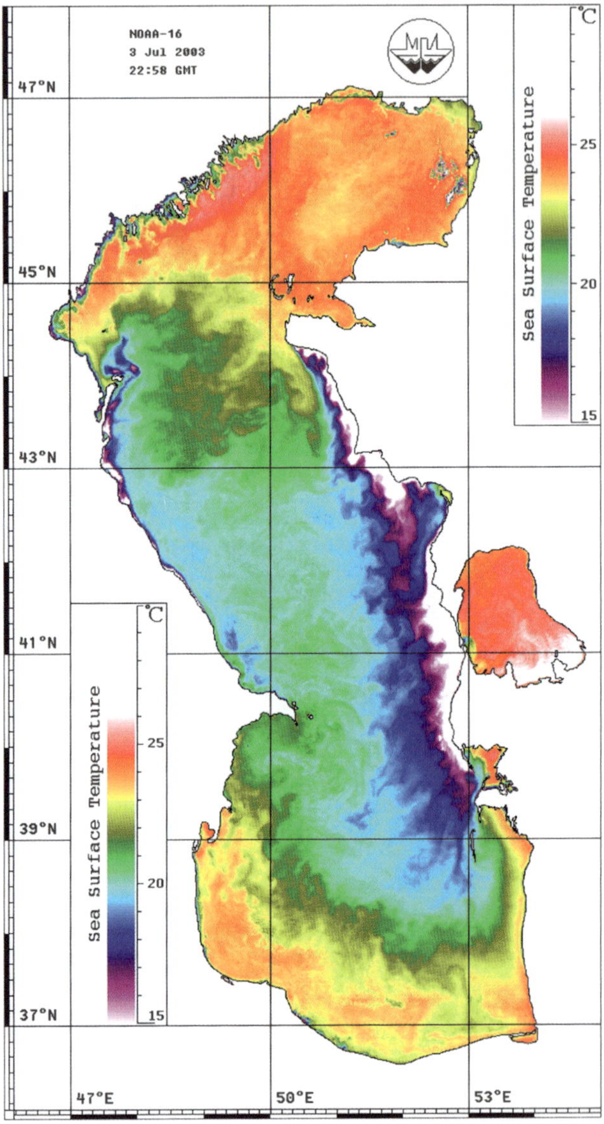

Fig. 12 SST field in the Caspian Sea (NOAA-16, 3 July 2003)

due to a decrease of the water temperature; and to investigate seasonal, interannual, and climate variability of the sea [32–34]. Figure 12 shows an upwelling event along the eastern coast of the Caspian Sea, which is stretching from Tyub-Karagan Peninsula (Kazakhstan) in the north to Cheleken Peninsula (Turkmenistan) in the south for 550 km. The width of the upwelling zone varies from 25 to 75 km. This is a yearly seasonal feature, observed from June to August. SST along the shores is less than 15°C (white color in Fig. 12), which is too cold for swimming in the summertime. There is an evident similarity between the upwelling zone in

Fig. 13 Satellite view on the ship loading in the port of Turkmenbashi. ©2013 DigitalGlobe (accessed on 26 March 2013 via Google Earth)

Portugal and in the eastern Caspian Sea, because their maximum development is observed only in July–August, which has a significant impact on the recreation and tourist areas, because very low SST prevents swimming in the sea during the vacation season. The tourist zone "Avaza" may be influenced by this natural phenomenon (see Fig. 12), which requires daily satellite monitoring of SST in the summertime and investigation of its seasonal and interannual variability.

In this chapter we will not discuss ice cover monitoring, because this is a very rare and local event in coastal waters of Turkmenistan, which is mainly observed in the Turkmenbashi Bay, for example, in winters 2007/2008 and 2011/2012. Finally, we can mention high resolution optical imagery provided by IKONOS, SPOT, QuickBird, FORMOSAT-2, and many other satellites. Spatial resolution of this imagery may be as high as 0.5–8 m which allows to monitor coastal and offshore infrastructure like ports (Fig. 13), oil terminals, oil/gas offshore platforms and pipelines, construction of buildings, and changes in the landscape and shoreline.

3 Kara-Bogaz-Gol Bay

Kara-Bogaz-Gol is a bay in the eastern part of the Caspian Sea which penetrates deeply into the mainland of Turkmenistan (Figs. 1 and 2) [35, 36]. This is the Caspian's largest salt-generating lagoon separated from the sea with two sandy spits extending meridionally for more than 90 km. These sandy spits form the Kara-Bogaz-Gol Strait 7–9 km long, 120–800 m wide, and 3–6 m deep. Due the difference of water levels in the Caspian Sea and the bay, waters from the sea rush at a speed of 50–100 cm/s along the strait to the bay where they completely evaporate (at a rate of 800–1000 mm/year, on the average). Therefore, with the average annual atmospheric precipitations in this region being no more than 110 mm, Kara-Bogaz-Gol represents an enormous natural evaporation basin of seawater [18, 35, 36].

Due to high evaporation the bay is filled with brine, the salinity of which reaches 270–300 psu. This brine is a concentrated solution of salts such as chlorides

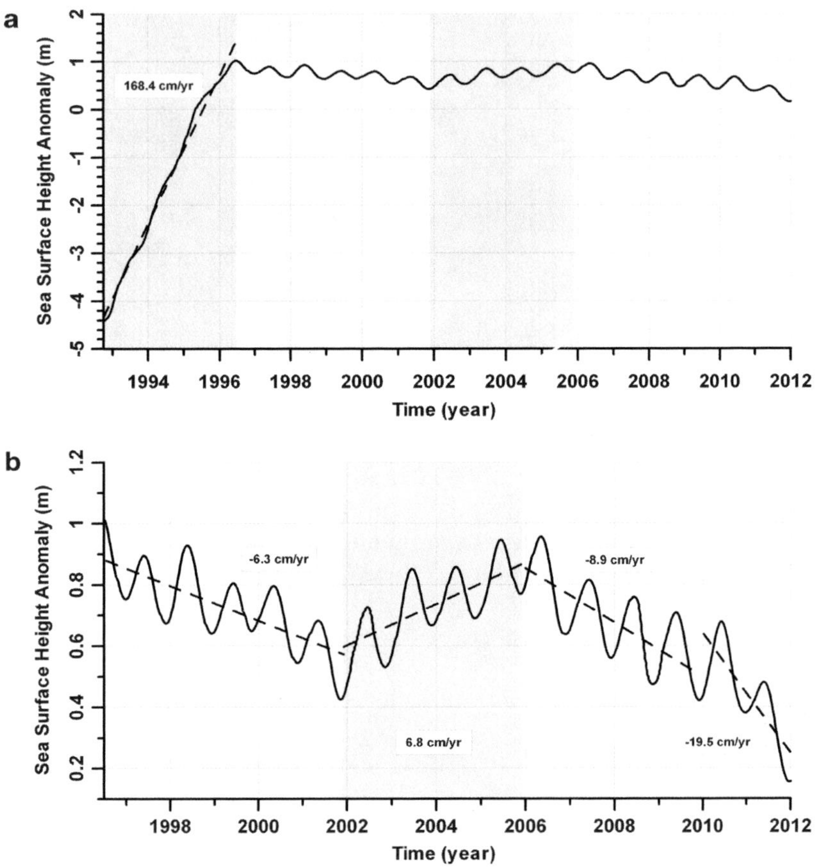

Fig. 14 Relative water level variability in the Kara-Bogaz-Gol Bay basing on the TOPEX/Poseidon, Jason-1, and Jason-2 satellite altimetry data: (**a**) for the time period September 1992–December 2011 and (**b**) for 1996–2011

of sodium, magnesium, potassium, magnesium sulfate, and small quantities of rare-earth elements. Kara-Bogaz-Gol Bay is the largest salt deposit where up to 20 salt minerals were found. The salt deposits of the bay accumulated dozen billion tons of various salts that make the most valuable raw material for the development of the chemical industry, agriculture, nonferrous metallurgy, medicine, and other branches of the economy of Turkmenistan [18, 35, 36].

In order to prevent further fall of the Caspian Sea level, which, in 1977 had been at its lowest for the past 400–500 years (−29 m), the Kara-Bogaz-Gol Strait was closed in March 1980 by a sandy dam. After the separation of the bay from the sea, it rapidly dried off. By the middle of 1984, the bay had become an almost completely dry salt lake. In order to revive, protect, and develop this unique bay and salt field on the Caspian Sea, it was decided to renew the water supply to Kara-Bogaz-Gol Bay. In September 1984, the Caspian Sea water was fed into

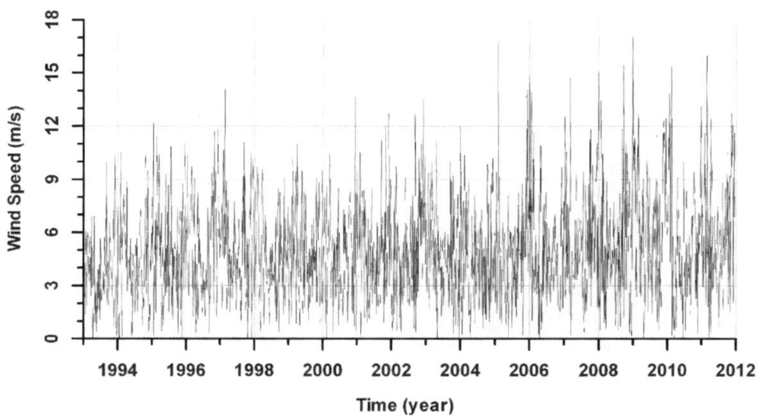

Fig. 15 Seasonal and interannual variability of the wind speed in Kara-Bogaz-Gol Bay at the crossover point of 31/168 tracks (see Fig. 2 for location) in 1993–2011 based on satellite altimetry data

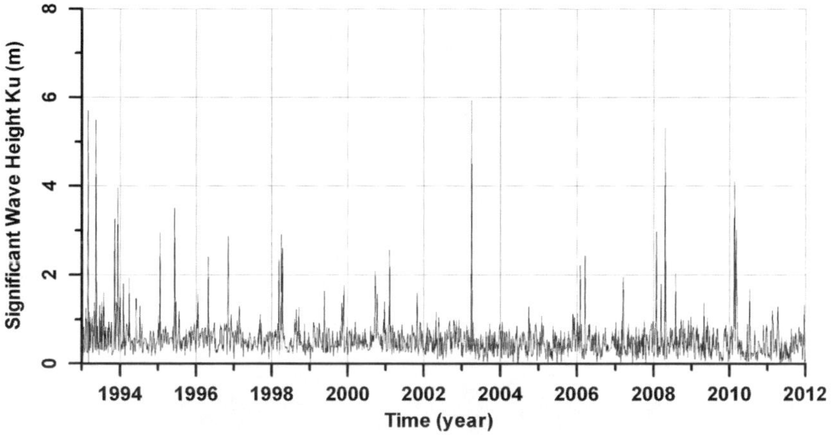

Fig. 16 Seasonal and interannual variability of the wave height in Kara-Bogaz-Gol Bay at the crossover point of 31/168 tracks (see Fig. 2 for location) in 1993–2011 based on satellite altimetry data

the bay at a rate of 1.5–1.6 km^3/year. This amount did not result in the restoration of the hydrological and hydrochemical conditions in the bay. In April 1992, the area of the bay reached 4,600 km^2, the absolute level mark was −33.71 m, and the depths varied from 0.2 to 1.4 m. In June 1992, the dam was destroyed and the natural seawater runoff to the bay resumed [18, 35, 36].

The process of refilling of Kara-Bogaz-Gol Bay coincided with the beginning of the TOPEX/Poseidon satellite altimetry mission, then followed by Jason-1 and Jason-2 missions (Fig. 14a) [7, 8, 18, 35, 36]. The area of the bay is crossed by two ground tracks of the above-mentioned satellites (Fig. 2), and in the

Fig. 17 Satellite view over Kara-Bogaz-Gol Bay on 20 March 2013 (MODIS-Terra, spatial resolution 250 m, true color with adjusted contrast and brightness)

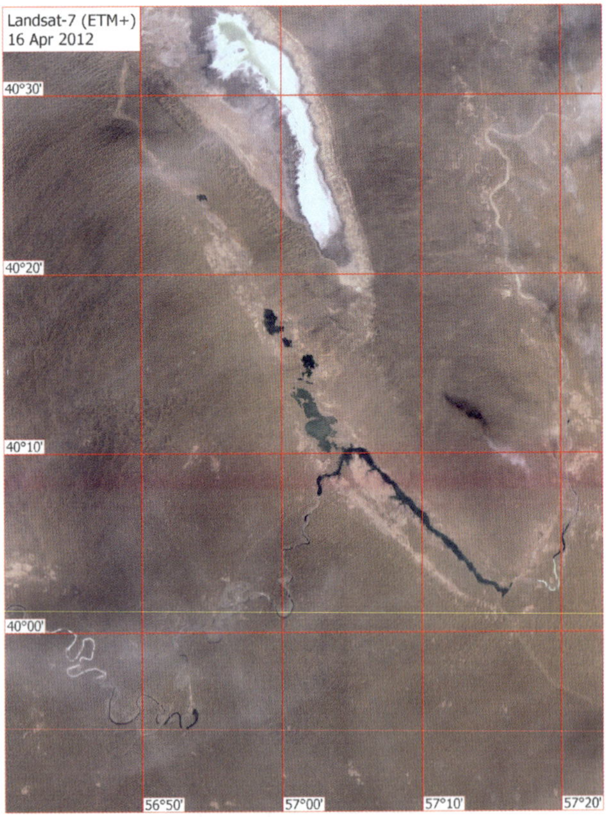

Fig. 18 Satellite image of the southern part of the Karashor Depression and the surrounding collectors acquired by Landsat-7 (ETM+) on 16 April 2012

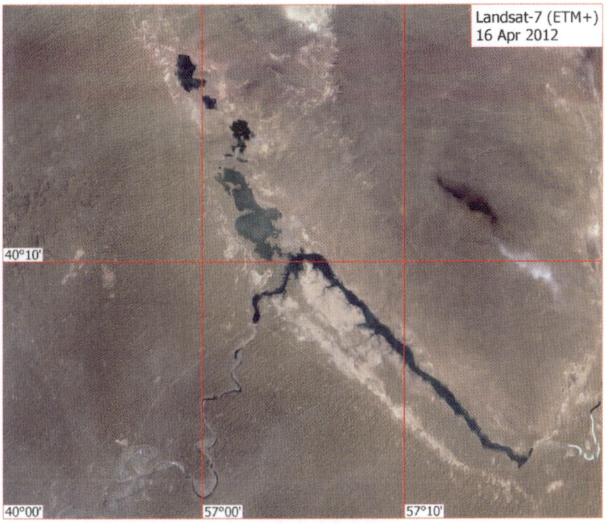

Fig. 19 Satellite image acquired by Landsat-7 (ETM+) on 16 April 2012 – a zoom on the collector southward of the Karashor Depression

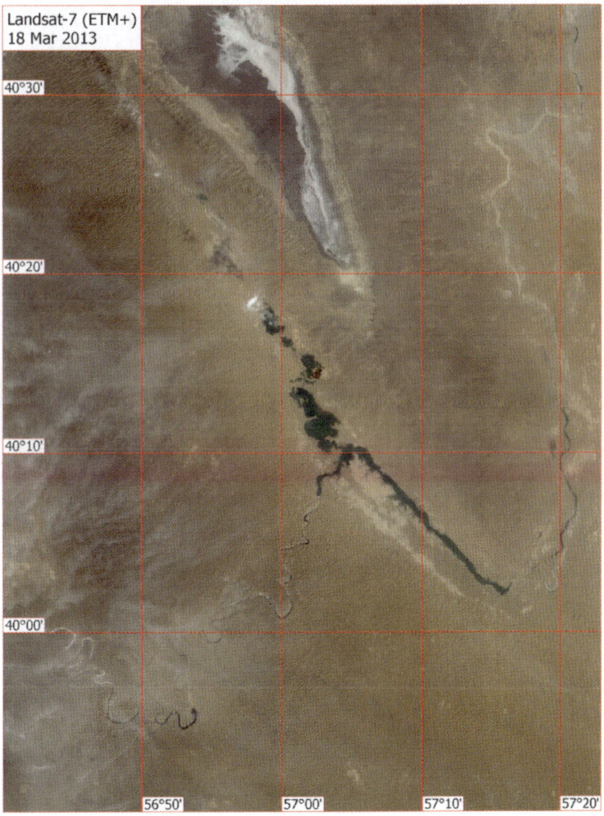

Fig. 20 Satellite image of the southern part of the Karashor Depression and the surrounding collectors acquired by Landsat-7 (ETM+) on 18 March 2013

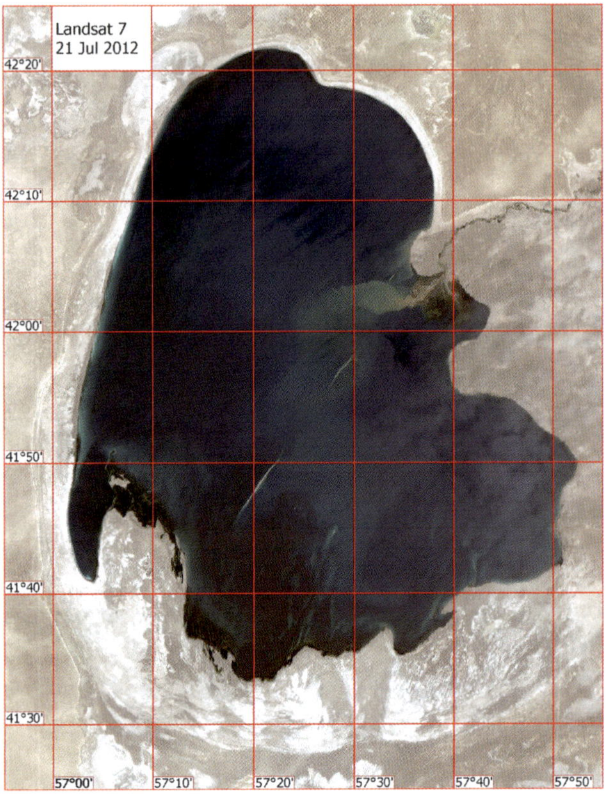

Fig. 21 Sarykamysh Lake on the Landsat-7 image acquired on 21 July 2012 (true color)

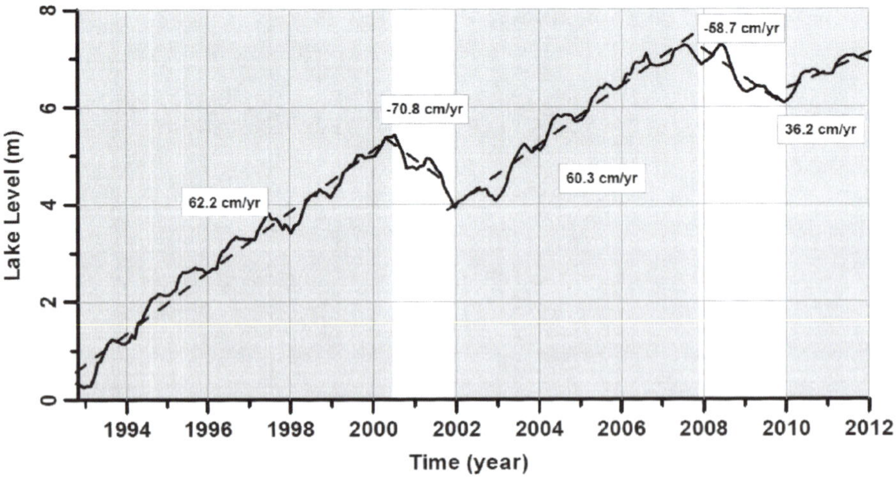

Fig. 22 Relative interannual and seasonal variability of the Sarykamysh Lake water level (m) in September 1992–December 2011. *Grey* areas show time periods when the lake level was rising

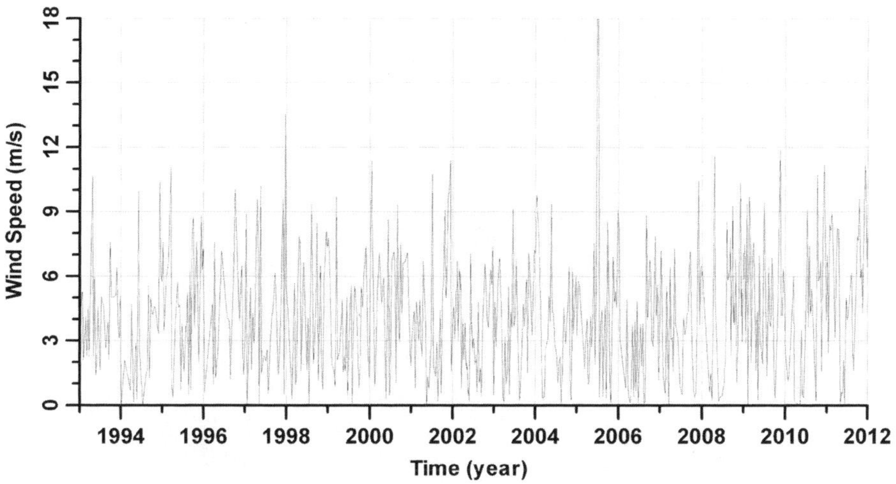

Fig. 23 Seasonal and interannual variability of the wind speed in Sarykamysh Lake in 1993–2011 based on satellite altimetry data

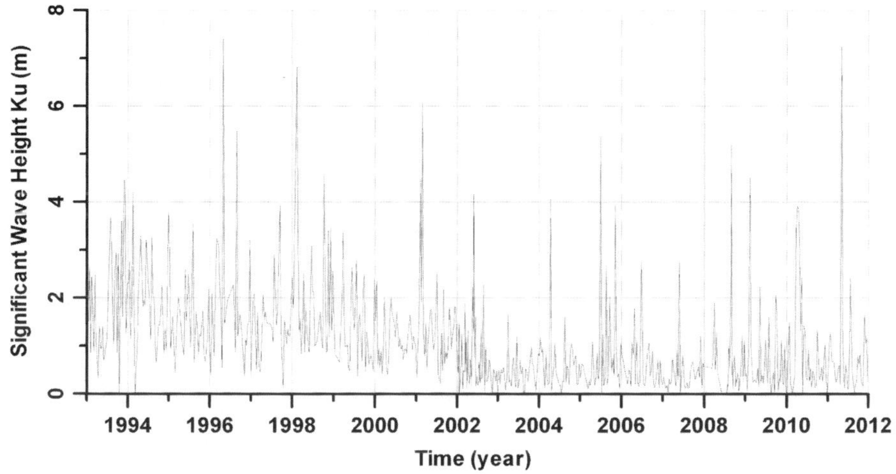

Fig. 24 Seasonal and interannual variability of the wave height in Sarykamysh Lake in 1993–2011 based on satellite altimetry data

crossover point we have water level measurements every 5 days. This allowed to monitor the filling of the bay, which occurred with a rate of 168 cm/year (Fig. 14a) [7, 8, 18, 35, 36]. Then, the level rise stopped in the bay and its variations started to reflect seasonal changes well correlated with the seasonal level changes in the Caspian Sea (Figs. 3 and 14) [18, 35, 36]. Thus, the rate of the level fall (until winter 2001/2002) in the bay was 6.3 vs. 5.9 cm/year in the Caspian Sea. For the time period 2002–2006 the bay level has been rising again with the rate +6.8 vs.

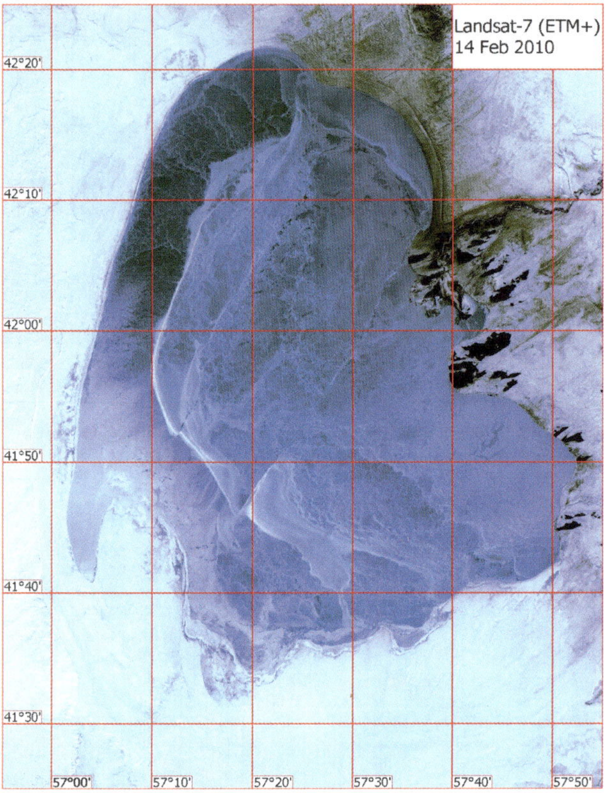

Fig. 25 Sarykamysh Lake covered by ice (*blue color*) on 14 February 2010 (Landsat-7, true color)

9.9 cm/year in the Caspian. Since 2006 till the end of 2009 the level of the bay was falling with a rate of 8.9 vs. 8.5 cm/year in the Caspian, and till the end of 2011 it accelerated to19.5 vs. 15.2 cm/year in the Caspian.

Satellite altimetry also allows to reconstruct values of the wind speed (Fig. 15) and wave height (Fig. 16) at the crossover point of 31/168 tracks right in the center of Kara-Bogaz-Gol Bay. This can be a valuable source of information for the coastal chemistry industry only, because there are no shipping activities and offshore oil/gas platform operations in the bay, as well as there are no tourist and recreation facilities. Figures 15 and 16 show variability of these parameters at the crossover point of 31/168 tracks between January 1993 and December 2011 with a time step of 5 days. It seems that the wind speed is quite reasonably reconstructed, which can be checked by a coastal meteorological station if there is any. The wave height of 0.5–1 m is in general fine, but several picks exceeding 2–3 m seem to be unrealistic.

We are not sure that SST, suspended matter, or chlorophyll concentration fields in Kara-Bogaz-Gol Bay are required for the chemical industry, but it is possible to obtain these parameters on a regular basis (see Figs. 8, 9, 11, and 12).

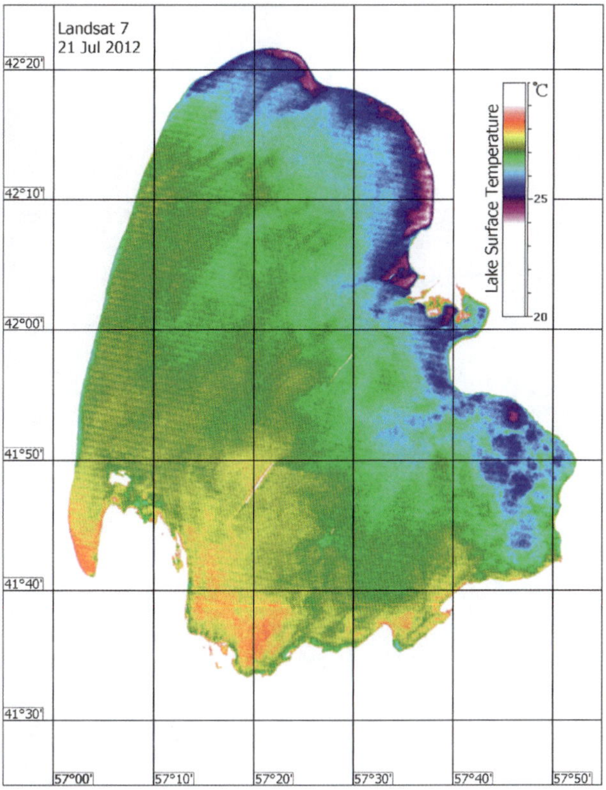

Fig. 26 SST distribution in Sarykamysh Lake on 21 July 2012 (Landsat-7)

For example, Fig. 17 shows peculiarities of suspended matter distribution in the bay on 20 March 2013. Wave-like areas with light colors represent waters with higher concentration of suspended matter.

4 Altyn Asyr Lake

The idea of Altyn Asyr Lake (Golden Age Lake) is to collect drainage waters from irrigated lands of Turkmenistan and forward them to the Karashor Depression by a system of canals. The project is described in details in [37, 38]. The Karashor Depression is located between Kara-Bogaz-Gol Bay of the Caspian Sea on the west and Sarykamysh Lake on the northeast (Figs. 1 and 2). Karashor is a deep dry depression. Its length along the long axis is about 110 km, it is 20 km wide in northwest, and it is gradually narrowing to the southeast. The bottom of the depression is located at a level of -28 m (28 m below the ocean level), which corresponds approximately to the present Caspian Sea level (−27.7 m by the end of 2012).

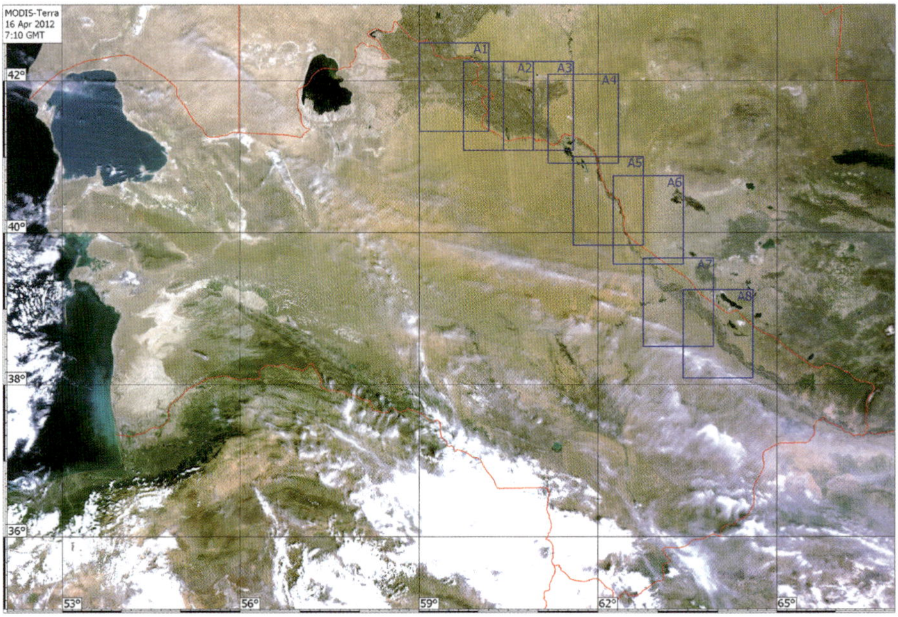

Fig. 27 Satellite view over Turkmenistan and Amu Darya River from MODIS-Terra (250 m resolution) on 16 April 2012. *Red lines* show borders between countries. *Blue rectangular frames* A1–A8 show the geographic location of a series of Landsat-7 (30 m resolution) images focused on the Amu Darya and surrounding areas (Figs. 28, 29, 30, 31, 32, 33, 34, and 35)

According to the Project, Altyn Asyr Lake will be 103 km long, 18.6 km wide, and 69 m deep. It will have a surface of about 1,915.8 km^2 and a volume of 132 km^3 [37, 38]. It is planned that the depression will receive annually 10 km^3 of CDW, thus it will be filled during about 15 years. When accumulation of water is completed, the lake will look similar to the modern western part of the Large Aral Sea [1]. Fortunately, the 107 ground track of Jason-2 satellite crosses the bottom of the northern part of the Karashor Depression (Fig. 2), and in the future it will be possible to follow water filling of the lake basing on satellite altimetry data as it has been done since 1993 for the Kara-Bogaz-Gol Bay case study (see Fig. 14).

On 15 July 2009 filling of collectors by water began in the vicinity of the Karashor Depression. Since that time, P.P. Shirshov Institute of Oceanology, Russian Academy of Sciences (Moscow, Russia), has combined efforts with Geophysical Center, Russian Academy of Sciences (Moscow, Russia), and Marine Hydrophysical Institute (Sevastopol, Ukraine) in order to monitor the Karashor Depression and surrounding collectors and canals by means of satellite remote sensing. For this task we have used optical scanners of medium spatial resolution (250 m) MODIS-Terra and -Aqua, as well as of high resolution (30 m) Landsat-5 (TM) and Landsat-7 (ETM+) [15–19, 21, 22, 38]. Satellite images of medium resolution are daily available and that of high resolution – about once a month, which is sufficient for the observed rate of construction.

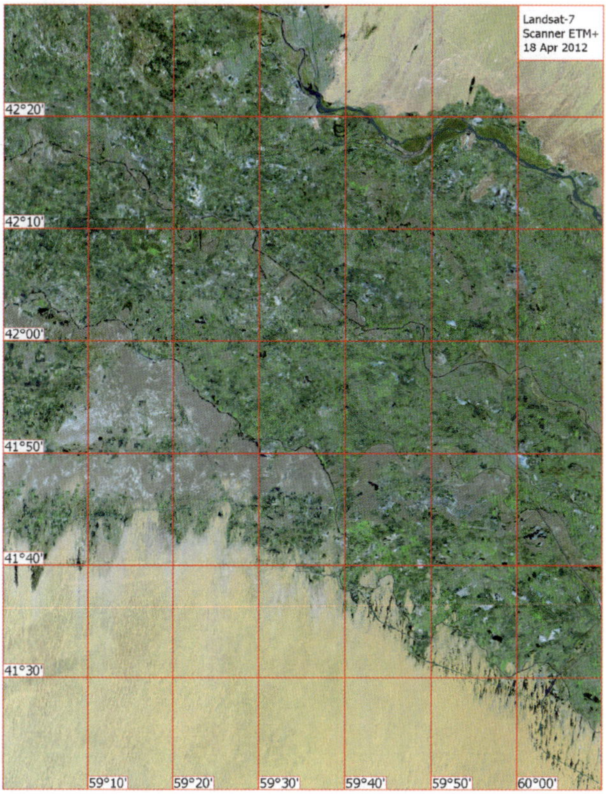

Fig. 28 High resolution satellite view on Amu Darya River and surrounding areas (frame A1, see Fig. 27 for location) from Landsat-7 (30 m resolution) on 18 April 2012

In the above-mentioned papers, as well as in this book, the chronology of the construction of the collectors in the vicinity of Altyn Asyr Lake is followed by means of satellite imagery. In this chapter we will not repeat this chronology, but will show recent satellite images of the area close to the Karashor Depression. Figures 18 and 19 show the state of water filling in the canals leading to the Karashor Depression on 16 April 2012. Due to the intensive deposition of salt (white color) and changes in the landscape we clearly detect the ancient meandering riverbed of Uzboy River as well as a system of canals leading to the Karashor Depression. The bottom of the southern part of the Karashor Depression is covered by salt (white area). In the southeastern corner of the satellite image we can see the Transturkmen (Main) collector which is dry. In the northwestern corner of the frame we can observe an 18-km section of the collector which joins the northernmost flooded area (dark colors) with the Karashor Depression. Almost 1 year later, on 18 March 2013, we did not detect significant changes in the working area close to the Karashor Depression as well as canals remain filled by the same amount of water (Fig. 20).

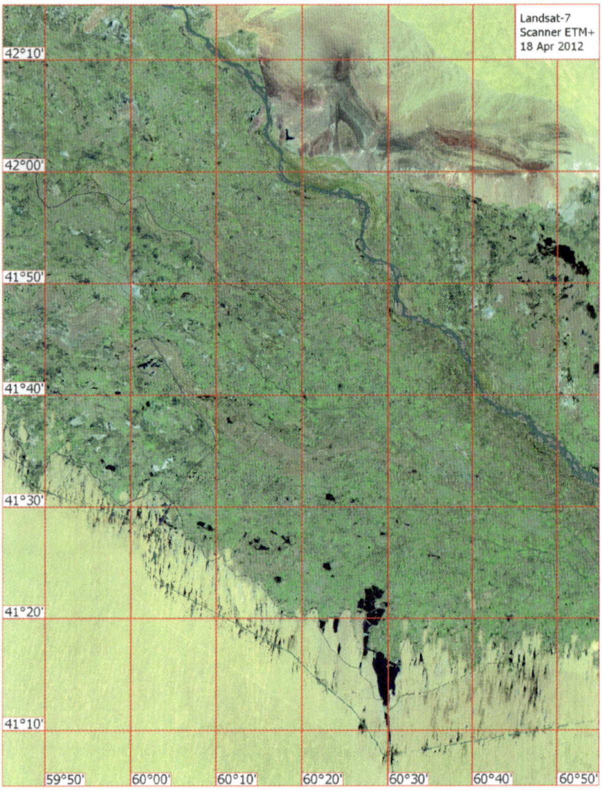

Fig. 29 High resolution satellite view on Amu Darya River and surrounding areas (frame A2, see Fig. 27 for location) from Landsat-7 (30 m resolution) on 18 April 2012

High resolution Landsat optical imagery (see a zoom in Fig. 19) showed its effectiveness in the monitoring of the Altyn Asyr Lake construction. Many details, including the wave-like structure of sandy dunes in the desert, canals, roads, salt deposits, small-scale features of the landscape, are clearly visible in the satellite images.

5 Sarykamysh Lake

Sarykamysh Lake was formed in 1971 as a result of flooding of a set of small Sarykamysh lakes located at the bottom of the natural Sarykamysh Depression located 200 km southwestward from the Aral Sea (Figs. 1 and 2). Its bottom lies at the −38 m absolute level (below the ocean level). This depression periodically was filled by the Amu Darya waters via the Daryalyk riverbed. Now Sarykamysh Lake is a large drainage water body, which has been used as a discharge collector of salty irrigation water from the fields (Fig. 21). Currently, the lake covers an area of about 3,900 km^2 [18]. Salinity of the lake waters has been continuously increasing: from 3 to 4 g/L in the early 1960s to 12–14 g/L in 1987 [9, 39, 40]. The maximum

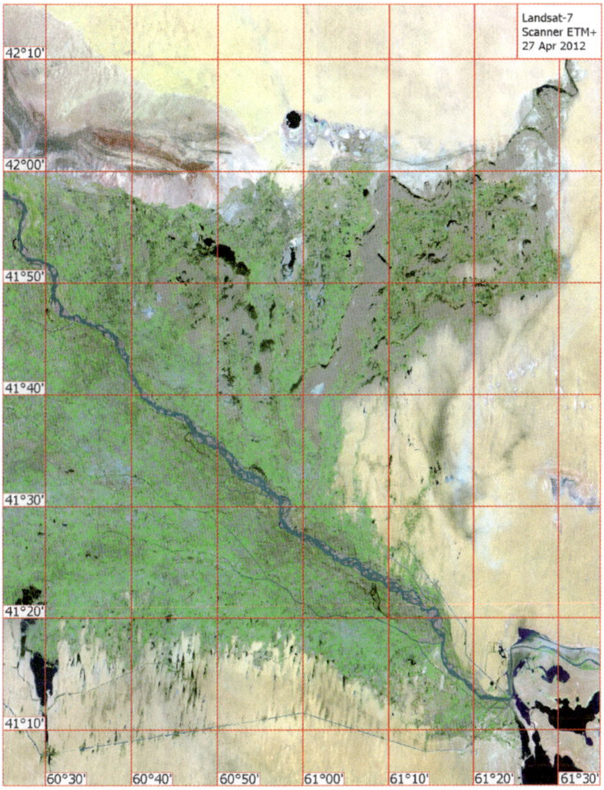

Fig. 30 High resolution satellite view on Amu Darya River and surrounding areas (frame A3, see Fig. 27 for location) from Landsat 7 (30 m resolution) on 27 April 2012

depth is about 40 m. Water of Sarykamysh Lake is salinized and contains biogenic matters, admixtures of pesticides, defoliants, and fertilizers, nevertheless there is a local fishery activity in the lake. Today Sarykamysh Lake is shared between Turkmenistan (the southern part) and Uzbekistan (the northwestern part).

Again, the same 107 ground track of TOPEX/Poseidon, Jason-1, and Jason-2 satellites crosses Sarykamysh Lake (Fig. 2). Since September 1992, Sarykamysh Lake has been progressively increasing in size, reaching its maximum level at the beginning of 2000 with an increase of almost 5 m at a rate of 62 cm/year as observed by the TOPEX/Poseidon and further by the Jason-1/-2 altimetry missions (Fig. 22). Then, during 1.5 years its level dropped by 1 m with a rate of about 71 cm/year, and then since 2002 its level was rising again with about the same rate (60 cm/year) till the end of 2007. During these 6 years the lake added about 3 m to its level. From the beginning of 2008 till the end of 2009 the level again dropped by 1 m with a rate of 59 cm/year. From the beginning of 2010 it has been rising again with a rate of 36 cm/year, and by the end of 2011 it almost reached the same level as it was in summer 2007 and 2008.

It seems that the wind speed (Fig. 23) is quite reasonably reconstructed from the altimetry data, which can be checked by a coastal meteorological station if

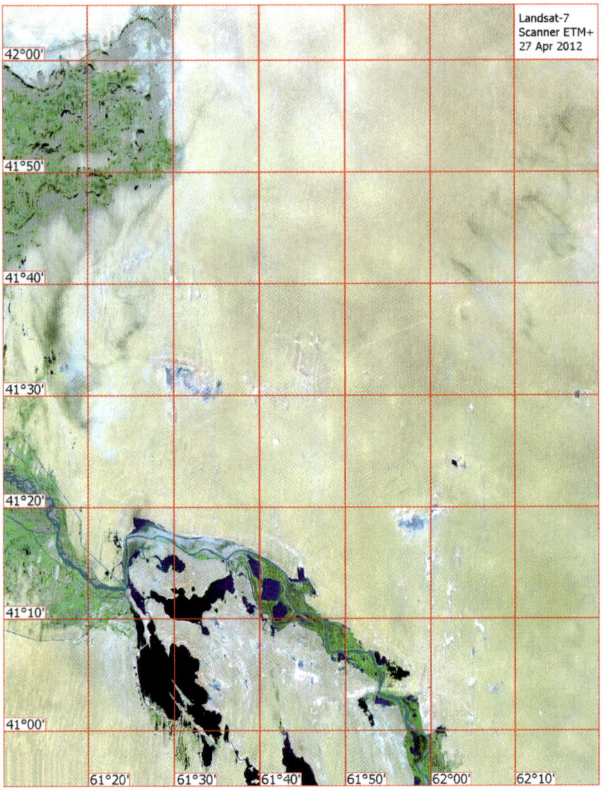

Fig. 31 High resolution satellite view on Amu Darya River and surrounding areas (frame A4, see Fig. 27 for location) from Landsat 7 (30 m resolution) on 27 April 2012

there is any. As concerns the wave height, there are too many picks exceeding 2–3 m, which are unrealistic for such a small lake (Fig. 24). Moreover, a notable shift of a signal to more logical lower values of the wave height has been observed since 2002. Less credible convertation of satellite altimetry data into the wave height is explained by a small size of the lake with known interference of the land with the signal.

High spatial resolution of Landsat imagery (30 m) (Fig. 21) allows to follow changes in the shape of the lake and its morphometric characteristics in details and to calculate with high accuracy the lake surface [18]. For example, in 2003 it was 3,782 km^2, in 2005 – 3,880 km^2, in 2007 – 3,956 km^2, and in 2009 – 3,874 km^2. The same optical images can very well show suspended matter distribution in the lake (Fig. 21), as well as the ice cover or distribution of floating ice in the lake (Fig. 25). The fine structure of the ice cover is visible thanks to high spatial resolution of Landsat imagery. Finally, Landsat sensors allow to measure SST in the lake, which is shown in Fig. 26. This SST map reveals a notable upwelling event along the northeastern coastline due to northeasterly winds. The upwelled water (blue colors) along the coast has SST, which is about 3°C less than waters in the opposite side of the lake (about 27.5°C).

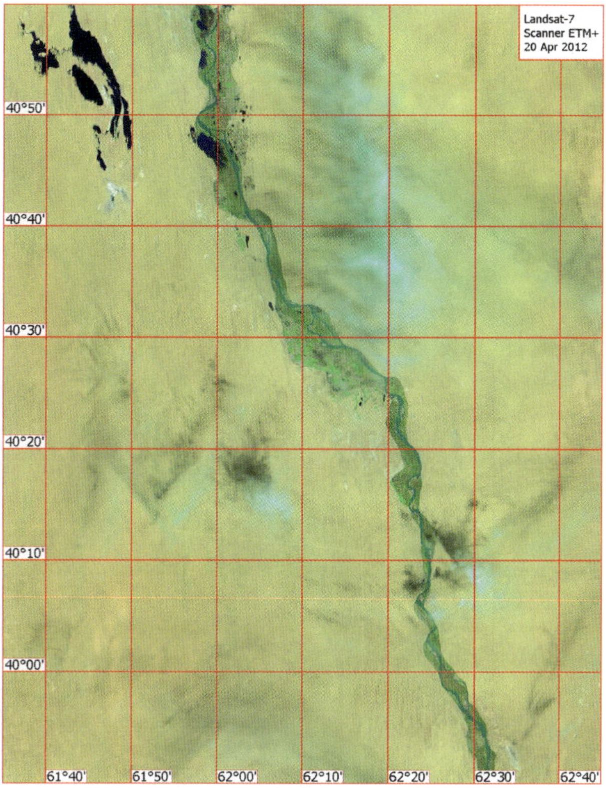

Fig. 32 High resolution satellite view on Amu Darya River and surrounding areas (frame A5, see Fig. 27 for location) from Landsat-7 (30 m resolution) on 20 April 2012

6 Amu Darya River

The Amu Darya is the largest river in Central Asia and it is formed as a result of confluence of Pyandzh (catchment area of 114,000 km^2) and Vakhsh (39,000 km^2) Rivers. The Amu Darya River length is about 1,450 km from the confluence of these rivers to the Aral Sea. The total catchment area is estimated as 465,000 km^2, the area of the effective drainage basin is 300,000 km^2 [41]. The river flows from the southeast to northwest partially in Turkmenistan, along the border with Uzbekistan, and in Uzbekistan before reaching the Aral Sea via a vast delta (Fig. 27).

Downstream of Atamurat town in the southeastern part of Turkmenistan (former name – Kerki), the Amu Darya water is consumed due to irrigation of areas, lost by evaporation in flooded alluvial plains, and transpiration by hydrofilous plant vegetation. Hydrometric works on Amu Darya River have been conducted since 1870s, but before the first quarter of the twentieth century the hydrological observations had not been systematic. The longest series of observations is available from the gauging stations of Kerki (1910–1920, 1925–1937, 1952–2006) and Chatly (new name – Samanbay, Uzbekistan) (1913–1917, 1931–1973) [41].

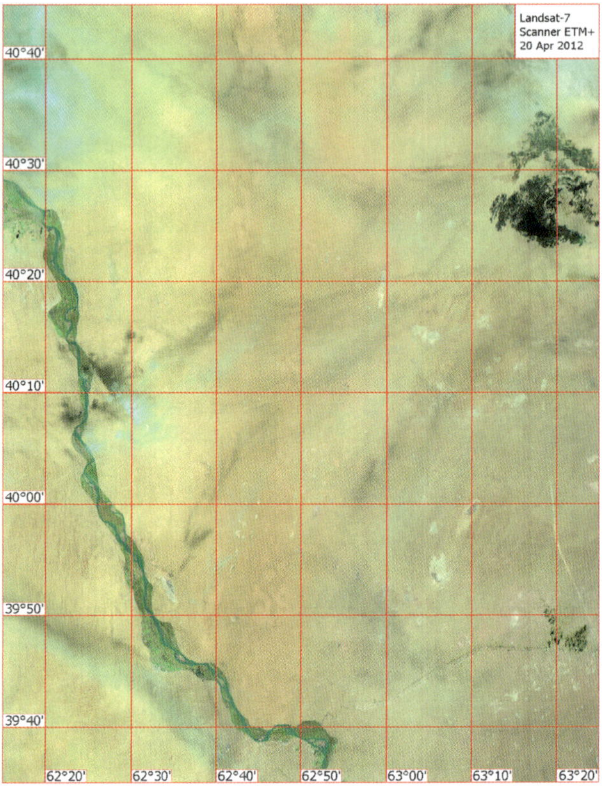

Fig. 33 High resolution satellite view on Amu Darya River and surrounding areas (frame A6, see Fig. 27 for location) from Landsat-7 (30 m resolution) on 20 April 2012

The range of annual river flow variation makes up to 51 km^3 (from 54 km^3 in 1986 to 105 km^3 in 1969). The 5-year period (1961–1965) was very dry with a mean volume of 62.2 km^3/year, i.e., 10 % less than long-time average, very abounding in water was the 8-year period from 1952 till 1959 with a mean annual volume of 74 km^3. Beginning from 1956 a considerable volume of the Amu Darya water is transferred by the Karakum Canal to arid regions of Turkmenistan. In last years the water delivery in the canal amounts to 10–11 km^3/year. According to the Basin Water Organization "Amudarya," in the 2001–2007 period, the total volume of water withdrawal from the Amu Darya varied from 26 to 44 km^3/year [41].

High resolution satellite imagery allows to monitor the areas around Amu Darya River, to control desertification processes, to calculate the normalized difference vegetation index (NDVI) and the normalized difference water index (NDWI) for further seasonal and interannual analysis and estimates of the state of vegetation and water capacity in the Amu Darya River area. This information is of vital importance for agriculture and water management in Turkmenistan and Uzbekistan. Figures 28, 29, 30, 31, 32, 33, 34, and 35 show the satellite image frames A1–A8, focused on Amu Darya River, which cover almost the whole length of the river in

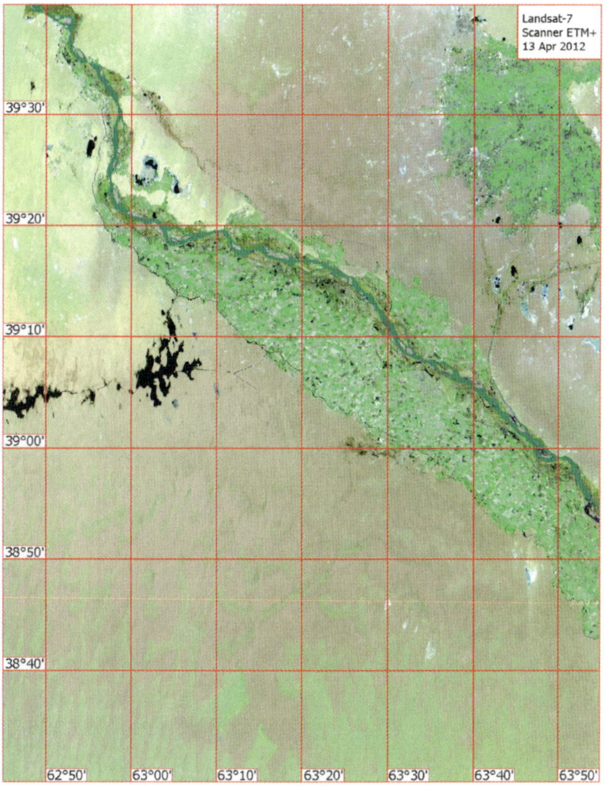

Fig. 34 High resolution satellite view on Amu Darya River and surrounding areas (frame A7, see Fig. 27 for location) from Landsat-7 (30 m resolution) on 13 April 2012

the eastern part of Turkmenistan at the border with Uzbekistan (Fig. 27). The images were acquired from Landsat 7 ETM+ scanner during 2 weeks in April 2012. Amu Darya River, a "green belt" of a different width along the river, a system of canals, lakes, and water reservoirs are clearly visible on the high resolution satellite imagery.

7 Conclusions

During the last two decades P.P. Shirshov Institute of Oceanology of the Russian Academy of Sciences (RAS) together with Geophysical Center RAS and Marine Hydrophysical Institute (Ukraine) performed a set of national and international scientific projects focused on the new satellite remote sensing technologies and developed complex satellite monitoring systems for the Caspian, Aral, Black, Azov, Eastern Mediterranean, and Baltic seas. Our experience in receiving, processing, and analysis of different multisensor satellite data was applied to monitoring of the Southeastern Caspian Sea, Kara-Bogaz-Gol Bay, Sarykamysh

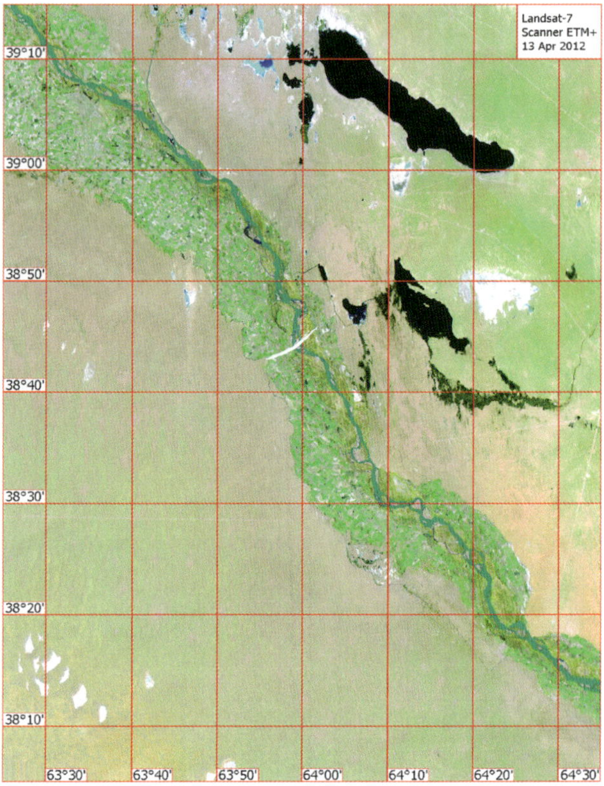

Fig. 35 High resolution satellite view on Amu Darya River and surrounding areas (frame A8, see Fig. 27 for location) from Landsat-7 (30 m resolution) on 13 April 2012

Lake, and artificial Altyn Asyr Lake, which since July 2009 has been under construction in Turkmenistan. It can be applied also to monitoring of reservoirs, lakes, rivers, and for agriculture by the analysis of the index of vegetation (NDVI) and vegetation liquid water content (NDWI). Satellite monitoring of water resources and land is of great importance for Turkmenistan and other countries located in arid zones especially now when significant changes in regional climate are observed.

In this chapter we demonstrated modern capabilities of satellite remote sensing technologies in environmental monitoring and showed examples of the use of satellite data and imagery for the analysis of morphometric characteristics, sea/lake level, temperature, water quality, wind speed, wave height in the main water bodies in Turkmenistan. Special attention was paid to the drainage lakes Sarykamysh and Altyn Asyr, which we started to monitor since the beginning of its construction in July 2009. High resolution satellite imagery is very effective in monitoring of the Amu Darya River area.

Further sea level, wind and waves, SST, suspended matter and chlorophyll concentration, and oil pollution monitoring at various points of the Caspian Sea and Kara-Bogaz-Gol Bay with the use of satellite remote sensing and other observations should allow us to follow the ecological state and future changes which are extremely important for designing, constructing, and operating industrial installations and infrastructure in the sea and on its coasts, and first of all for providing ecological security for economic activities in the Caspian Sea region.

In June 2011, we proposed to organize in Ashkhabad the National Center for Satellite Monitoring and Regional Climate Change in Turkmenistan with the support of the European Space Agency (ESA) and the Committee on Space Research (COSPAR) [17]. This Center could effectively monitor the ecological state, oil pollution, and sea level of the Caspian Sea waters of Turkmenistan, especially in such sensitive areas as the National tourist zone "Avaza," as well as in the areas of offshore oil and gas production, and in the areas of new offshore projects, such as the Trans-Caspian gas pipeline. The Center could monitor the state of Sarykamysh Lake and the filling of the new Altyn Asyr drainage lake, Amu Darya River, as well as other water bodies and land areas. The regional climate change in Turkmenistan and Central Asia should be an important task for the research in the Center. Satellite monitoring and climate research performed by the Center could provide very useful information for agriculture, water management, and science in Turkmenistan.

References

1. Kostianoy AG, Kosarev AN (eds) (2010) The Aral Sea environment, vol 7, The Handbook of Environmental Chemistry. Springer, Berlin
2. Stanchin I, Lerman Z (2008) Water in Turkmenistan. In: Spoor M, Arsel M (eds) The last drop. water, security, and sustainable development in Central Eurasia. Routledge, London
3. Lavrova OYu, Kostianoy AG, Lebedev SA, Mityagina MI, Ginzburg AI, Sheremet NA (2011) Complex satellite monitoring of the Russian Seas. IKI RAN, Moscow, 470 pp (in Russian)
4. Kostianoy AG, Lobkovskiy LI (2003) Satellite monitoring of the Caspian Sea. Vestnik Kaspiya N 6:56–65 (in Russian)
5. Kostianoy AG (2004) Satellite monitoring of the Caspian and Black seas. In: Proceedings of the "Caspian Ecology 2004" 2nd International Caspian and Black Sea Ecology Summit and Showcase, Istanbul, 25–26 November 2004, pp E29–E31 (in English), R131–R133 (in Russian), T167–T169 (in Turkish)
6. Litovchenko KTs, Kostianoy AG, Ivanov AYu (2004) Space radiolocation of oil pollution in the Caspian Sea. Vestnik Kaspiya N 1:96–106 (in Russian)
7. Lebedev SA, Kostianoy AG (2004) Satellite altimetry of the Caspian Sea. Vestnik Kaspiya N 3:82–101 (in Russian)
8. Lebedev SA, Kostianoy AG (2005) Satellite altimetry of the Caspian Sea. Moscow: Sea, 366 pp (in Russian)
9. Kostianoy AG, Zavialov PO, Lebedev SA (2004) What do we know about dead, dying and endangered lakes and seas? In: Nihoul JCJ, Zavialov PO, Micklin PP (eds) Dying and dead seas. Climatic versus anthropic causes, NATO ARW/ASI Series. Kluwer Academic Publishers, Dordrecht, pp 1–48
10. Kostianoy AG, Kosarev AN (eds) (2005) The Caspian Sea Environment, vol 5, Part P, The Handbook of Environmental Chemistry. Springer, Berlin, p 271

11. Ginzburg AI, Kostianoy AG, Sheremet NA, Kravtsova VI (2010) Satellite monitoring of the Aral Sea region. In: Kostianoy AG, Kosarev AN (eds) The Aral Sea Environment, vol 7, The Handbook of Environmental Chemistry. Springer, Berlin, pp 147–179
12. Ginzburg AI, Kostianoy AG, Sheremet NA, Kravtsova VI (2010) Satellite monitoring of the Aral Sea. In: Izrael YuA (ed) Problems of ecological monitoring and modelling of ecosystems, vol 23, pp 150–193 (in Russian)
13. Kostianoy AG (2011) Environmental problems of the Caspian Sea. Izvestiya Nat Acad Sci Azerbaijan (Proc Azerbaijan Acad Sci Earth Sci) N2:42–55
14. Kostianoy AG (2013) Inland seas and lakes: Central Asia case study. In: Jorgensen SE (ed) Encyclopedia of environmental management, vol 2. Taylor and Francis, New York, pp 1460–1467
15. Kostianoy AG, Lebedev SA, Solovyov DM (2010) Satellite monitoring of the Altyn Asyr Lake and water resources of Turkmenistan. In: Abstracts of the International Conference on "Science, Technique, and Innovation Technologies in the Epoch of Great Revival", Ashgabad, 12–14 June 2010, vol 1, pp 388–391 (in Russian, Turkmen, English)
16. Kostianoy AG, Lebedev SA, Solovyov DM (2011) Satellite monitoring of the Altyn Asyr Lake construction in 2009–2011. In: Abstracts of the international science conference on "Science, technique and innovation technologies in the Epoch of Great Revival", Ashgabat, 10–12 June 2011, Ylym, 2011 (in Russian)
17. Kostianoy AG, Lebedev SA, Zonn IS, Lavrova OYu, Solovyov DM (2011) Satellite monitoring of Turkmenistan. Moscow, Signal, 16 pp
18. Kostianoy AG, Lebedev SA, Solovyov DM (2011) Satellite monitoring of water resources in Turkmenistan. Int Water Tech J 1(1):4–15
19. Kostianoy AG, Lebedev SA, Solovyov DM (2012) Radar altimetry for monitoring of marine and inland waters in Turkmenistan. In: Abstracts of the ESA Symposium "20 Years of progress in radar altimetry", Venice, 24–29 September 2012, p 78
20. Kostianoy AG (2012) Satellite monitoring of ecological state of the Caspian Sea. In: Abstracts, The Caspian ecological forum, Turkmenbashi, 5–6 November 2012, pp 75–78 (in English, Russian and Turkmen)
21. Kostianoy AG, Zonn IS, Solovyov DM (2012) Construction of the Turkmen lake "Altyn Asyr" and the results of its satellite monitoring in 2009–2011. Int Water Tech J 2(1):59–66
22. Kostianoy AG, Zonn IS, Solovyov DM (2012) The Turkmen Lake "Altyn Asyr": satellite monitoring of its construction. Modern Probl Remote Sens Earth from Space 9(1):219–227 (in Russian)
23. Kosarev AN, Yablonskaya EA (1994) The Caspian Sea. SPB Academic Publishing, The Hague, 259 pp
24. Kostianoy A, Vignudelli S, Snaith HM, Cipollini P, Venuti F, Lyard F, Cretaux JF, Birol F, Bouffard J, Roblou L, Ginzburg A, Sheremet N, Kuzmina E, Lebedev S, Sirota A, Medvedev D, Khlebnikova S, Mamedov R, Ismatova Kh, Alyev A, Nabiyev T (2007) Value added satellite altimetry for coastal regions (ALTICORE). Vestnik Kaspiya (Caspian Sea Bulletin) N4:65–80
25. Lebedev SA, Kostianoy AG (2008) Integrated using of satellite altimetry in investigation of meteorological, hydrological and hydrodynamic regime of the Caspian Sea. J Terr Atmos Ocean Sci 19(1–2):71–82
26. Vignudelli S, Testut L, Calzas M, Mamedov R, Sultanov M, Alyev A, Cretaux JF, Kostianoy A (2008) Installation of pilot sea level monitoring station for satellite altimetry calibration/validation in the Caspian Sea at Absheron port, Baku, Azerbaijan (8–13 June 2008). Vestnik Kaspiya (Caspian Sea Bulletin) N4:40–55
27. Kouraev AV, Cretaux J-F, Lebedev SA, Kostianoy AG, Ginzburg AI, Sheremet NA, Mamedov R, Zakharova EA, Roblou L, Lyard F, Calmant S, Berge-Nguyen M (2011) Satellite altimetry applications in the Caspian Sea. In: Vignudelli S, Kostianoy AG, Cipollini P, Benveniste J (eds) Coastal altimetry. Springer, Berlin, pp 331–366. doi:10.1007/978-3-642-12796-0_13
28. Kostianoy AG, Lavrova OYu (eds) (2013) Oil pollution in the Baltic Sea. Springer, Berlin

29. Karabashev GS, Evdoshenko MA (2012) Influence of the wind field on the radiance of a marine shallow: evidence from the Caspian Sea. Oceanologia 54(4):655–673
30. Ginzburg AI, Kostianoy AG, Soloviev DM, Sheremet NA (2005) Structure of the upwelling off western coast of the Middle Caspian Sea (satellite observations). Issledovanie Zemli iz kosmosa 4:76–85 (in Russian)
31. Ginzburg AI, Kostianoy AG, Soloviev DM, Sheremet NA (2006) Frontal zone of upwelling near the eastern coast of the Caspian Sea as inferred from satellite data. Issledovanie Zemli iz kosmosa 4:3–12 (in Russian)
32. Ginzburg AI, Kostianoy AG, Sheremet NA (2004) Seasonal and interannual variability of the Caspian Sea surface temperature. Okeanologiya 44(5):645–659 (in Russian)
33. Ginzburg AI, Kostianoy AG, Sheremet NA (2005) Sea surface temperature. In: Kostianoy AG, Kosarev AN (eds) The Caspian Sea Environment. Springer, Berlin, pp 59–81
34. Ginzburg AI, Kostianoy AG, Sheremet NA (2012) Long-term variability of the Caspian Sea surface temperature (1982–2009). Modern Probl Remote Sens Earth from Space 9(2):262–269 (in Russian)
35. Kosarev AN, Kostianoy AG (2005) Kara-Bogaz-Gol Bay. In: Kostianoy AG, AN K (eds) The Caspian Sea Environment. Springer, Berlin, pp 211–221
36. Kosarev AN, Kostianoy AG, Zonn IS (2009) Kara-Bogaz-Gol Bay: physical and chemical evolution. Aquatic Geochemistry 15(1–2) Special Issue: Saline lakes and Global change: 223–236. doi:10.1007/s10498-008-9054-z
37. (2010) Turkmen Lake "Altyn Asyr", Ashgabat 104 pp (in Turkmen, Russian and English)
38. Zonn IS, Kostianoy AG (2013) The Turkmen Lake Altyn Asyr. In: Zonn IS, Kostianoy AG (eds) The Turkmen Lake Altyn Asyr and water resources in Turkmenistan. Springer, Berlin
39. Glazovsky NF (1995) The Aral Sea basin. In: Kasperson JX, Kasperson RE, Turner II BL (eds) Regions at risk: comparisons of threatened environments. United Nations University Press, Tokyo
40. Orlovsky L, Matsrafi O, Orlovsky N, Kouznetsov M (2013) Sarykamysh Lake—collector of drainage water: the past, the present and the future. In: Zonn IS, Kostianoy AG (eds) The Turkmen Lake Altyn Asyr and water resources in Turkmenistan. Springer, Berlin
41. Asarin AE, Kravtsova VI, Mikhailov VN (2010) Amudarya and Syrdarya rivers and their deltas. In: Kostianoy AG, Kosarev AN (eds) The Aral Sea Environment. Springer, Berlin, pp 101–121

Present-Day Condition of Ichthyofauna in Reservoirs of Turkmenistan

Firdauz M. Shakirova

Abstract The species composition of ichthyofauna in the inland water bodies of Turkmenistan was analyzed on the basis of many-year materials collected by the author and the data from numerous publications. Changes that occurred in the composition of fish population were analyzed. The anthropogenic factors that facilitated changes, mixing, and their wide dissemination were studied. This chapter contains the full list of fish inhabiting the studied water bodies with regard to the new taxonomic realities and status reports.

Keywords Acclimatization, Fish population formation, Fishery, Ichthyofauna, Irrigation construction, Rivers, Species composition, Water reservoirs

Contents

F.M. Shakirova (✉)
Tatar Department of Federal State Budgetary Research Institution "GosNIORH",
4 Tazi Gizzata Str., Kazan 420111, The Republic of Tatarstan (The Russian Federation)
e-mail: gosniiorh@gmail.com

I.S. Zonn and A.G. Kostianoy (eds.), *The Turkmen Lake Altyn Asyr and Water Resources in Turkmenistan*, Hdb Env Chem (2014) 28: 233–260, DOI 10.1007/698_2012_174,

1 Introduction

The present-day condition of the ichthyofauna in the water bodies of Turkmenistan was studied using materials collected by the author during many years on the Karakum Canal, the Murghab River, including water reservoirs, rivers flowing down the northern slope of the Kopetdag and data of publications ([1–5]; [6–14] and others). The species composition of fish in the studied water bodies (Table 1) is provided with regard to new taxonomic revisions and status reports ([15–19] and others).

Affected by numerous anthropogenic factors, including extensive irrigation construction, acclimatizing measures, intensive fishing, and others, the natural historic diversity of aqueous communities sustains deep changes. The irrigation construction conducted across the whole Central Asian region, including in Turkmenistan, since ancient times has become most intensive from the 1950s. The full or partial flow regulation of large rivers (Amudarya) as well as medium (Murghab, Tedjen) and small rivers flowing down the northern slope of the Kopetdag, the construction of water reservoirs and canals, lakes, and water bodies accumulating drainage waters (Turkmen Lake) brought about radical changes in the hydrography of these rivers which, quite naturally, affected their ichthyofauna. One time isolated river systems were united by means of hydraulic construction and redistribution of their flows to form a single water basin covering practically the whole irrigated territory of Turkmenistan, and this led to mixing of their faunas (Fig. 1).

No less significant for shaping the present-day composition of the fish population in inland water bodies was acclimatization of economically valuable Chinese species of pelagophil fish in them, including herbivorous, that were brought into the country still in the 1960s. In the course of acclimatization works the technique applied for separation of fish fry prevented from complete ridding of unrecorded species, hence, 19 fish species having no fishery and economic importance got into the water bodies of Turkmenistan together with grass carp (Chinese carp) (Fig. 2) and silver carp (Fig. 3). Among the "unplanned" invaders were goldfish (Amur population) (Fig. 4), black Amur (Fig. 5), white Amur bream (Fig. 6), Amur false gudgeon (Fig. 7), Korean sawbelly (Fig. 8), stone morocco (Fig. 9), Amur loach (Fig. 10), bitterling (Fig. 11), Chinese goby (Fig. 12), and others. Most of them adapted quite successfully just at once, formed individual populations, and began to disseminate actively in the inland water bodies of the republic. Acclimatization and appearance of self-replicating flocks of herbivorous fish in the Amudarya, Murghab, and Karakum Canal are of special interest. Their fraction in the total fish catch was 75–80% and sometimes even more than 90%.

Table 1 Composition of fish population in the middle Amudarya River, Sarykamysh Lake, Karakum Canal, Murghab River, and their reservoirs

	I					II		III	
Family and species	1	2	3	4	5	6	7	8	9
Acipenseridae family									
Acipenser nudiventris – bastard (spiny, fringe, barbell) sturgeon, ship	+	+	+	+	+	–	–	–	–
Pseudoscaphirhynchus hermanni – little Amu-dar shovelnose	+	– ?	+	–	–	–	–	–	–
P. kaufmanni – big Amu-dar shovelnose	+	– ?	+	+	–	–	–	–	–
Cyprinidae family									
Abramis brama orientalis – breams (eastern)	+	+	+	+	+	–	–	–	–
A. sapa – (southern, Aral) white-eye	+	+	–	–	–	–	–	–	–
Alburnoides bipunctatus eichwaldi – bystranka, rifle minnow	+	–	+	+	–	+	+	+	+
A. taeniatus – striped bystranka	+	+	+	+	+	–	–	+	+
Aristichthys nobilis[a] – bighead, spotted silver carp	+	+	+	+	+	–	–	–	–
Aspius aspius iblioides – Aral asp	+	+	+	+	+	–	–	–	–
Aspiolucius esocinus – pice asp	+	–	+	–	–	–	–	–	–
Barbus brachycephalus – Aral (Caspian) barbell	+	+	+	+	+	–	–	–	–
B. capito conocephalus – barbell bulat mai	+	+	+	+	+	–	–	–	–
Capoetobrama kuschakewitschi – ostroluchka	+	+ Only in header Daryalyk	+	+	–	–	–	–	–
Carassius auratus gibelio – goldfish, golden (Chinese) carp	+	+	+	+	+	–	–	+	+

(continued)

Table 1 (continued)

	I					II		III	
Family and species	1	2	3	4	5	6	7	8	9
Chalcalburnus chalcoides aralensis – Aral shemaya	+	+	+	+	+	–	–	–	–
Ctenopharyngodon idella[a] – grass (Chinese) carp	+	+	+	+	+	–	–	+	+
Cyprinus carpio – (European, mirror) carp	+	+	+	+	+	+	+	+	+
Garra rossica – disc-jaw	–	–	–	–	–	+	–	+	–
Gobio gobio lepidolaemus – gudgeon	+	+	+	+	–	+	+	+	+
Hemiculter leucisculus[a] – Korean sawbelly	+	+	+	+	+	–	–	+	+
Hypophthalmichthys molitrixa[a] – silver carp	+	+	+	+	+	–	–	+	+
Leucisus idus oxianus – orfe, ide	+	+	–	–	–	–	–	–	–
L. latus – Murghab dase	–	–	–	+	–	+	–	+	–
Mylopharyngodon piceus[a] – black carp, black Amur	+	+	+	+	+	–	–	–	–
Parabramis pekinensis[a] – white Amur bream	+	+	+	+	+	–	–	+	+
Pelecus cultratus – sabrefish, rasorfish	+	+	+	+	+	–	–	–	–
Abbotina rivularis[a] – Amur false gudgeon	?	–	+	+	–	–	–	–	–
Rhodeus ocellatus[a] – bitterling	+	+	+	+	–	–	–	–	–
Pseudorasbora parva[a] – stone morocco	+	+	+	+	+	–	–	+	+
Rutilus rutilus – Aral roach	+	+	+	+	+	–	–	–	–
Scardinius erythrophthalmus – rudd, redeye	+	+	–	–	–	–	–	–	–
Schizothorax intermedius – marinka	+	–	–	–	–	–	–	–	–
Sch. pelzami – transcaspian marinka	–	–	–	–	–	+	–	+	–

Table 1 (continued)

Family and species	I					II		III	
	1	2	3	4	5	6	7	8	9
Varicorhinus capoeta heratensis – transcaspian khramulya	+	–	+	+	+	+	+	+	+
Balitoridae family									
Dzihunia amudarjensis – Bukhar (Khivin) stone loach	+	?	–	–	–	–	–	–	–
Nemacheilus oxianus – Amu-dar stone loach	+	?	–	–	–	–	–	–	–
Paracobitis longicauda	+	?	+	–	–	+	–	+	+
Schistura cristata – crested stone	–	–	–	–	–	+	–	+	–
Sch. sargadensis – Turkmenian stone loach	–	–	–	–	–	–	–	+	–
Cobitidae family									
Sabanejewia auratas aralensis – golden spiny loach	+	+	+	+	+	+	–	+	+
Misgurnus anguillicaudatus[a] – Amur loach, oriental weatherfish	–	?	+	+	+	–	–	–	–
Family Siluridae									
Silurus glanis – (European) wells, (European) catfish, sheatfish	+	+	+	+	+	+	+	+	+
Esocidae family									
Esox Lucius – (northern) pike	+	+	–	–	–	–	–	–	–
Salmonidae family									
Salmo trutta trutta – brook trout	+	–	–	–	–	–	–	–	–
Gasterosteidae family									
Pungitius platygaster aralensis – Aral stockleback	+	+	–	–	–	–	–	–	–

(continued)

Table 1 (continued)

	I					II		III	
Family and species	1	2	3	4	5	6	7	8	9
Percidae family									
Sander lucioperca – Sander, zander, European pike-perch	+	+	+	+	+	–	–	+ Got in 1994–1996	–
Perca fluviatilis – river perch	+	?	–	–	–	–	–	–	–
Poecilidae family									
Gambusia holbrooki[a] – mosquito fish, topminnow	+	+	+	+	+	+	+	+	+
Adrianichthyidae family									
Oryzias sinensi[a] – Japanese medaka	–	–	+	–	–	–	–	–	–
Odontobutidae family									
Micropercops cinctus[a]	–	–	+	–	–	–	–	–	–
Gobiidae family									
Rhinogobius cheni[a] – (Amur) Chinese goby	+	+	+	+	+	–	–	+	+
Channidae family									
Channa argus[a] – snakehead	+	+	+	+	+	–	–	+ Got in 1994–1996	–
Total number of families	12	10	11	8	8	5	3	8	6
Total number of species	43	32	37	33	26	12	6	23	16
Total number of invader species and (in %)	11 (25.6)	11 (34.4)	15 (40.5)	13 (39.4)	11 (42.3)	–	–	9 (37.5)	7 (43.8)

(I) *1* – Amudarya River (middle reaches); *2* – Sarykamysh Lake; *3* – Karakum Canal; *4* – Khauzkhan Reservoir; *5* – Kopetdag Reservoir
(II) 1942 (according to [25, 36–38]). *6* – Murghab River basin; *7* – Tashkeprin Reservoir
(III) 1978–2002 (our data). *8* – Murghab River basin; *9* – Saryyazyn Reservoir
Notes: (+) species met in a water body; (–) species not found in a water body; (?) species assumed to inhabit a water body, but not found by us
[a]Invaders

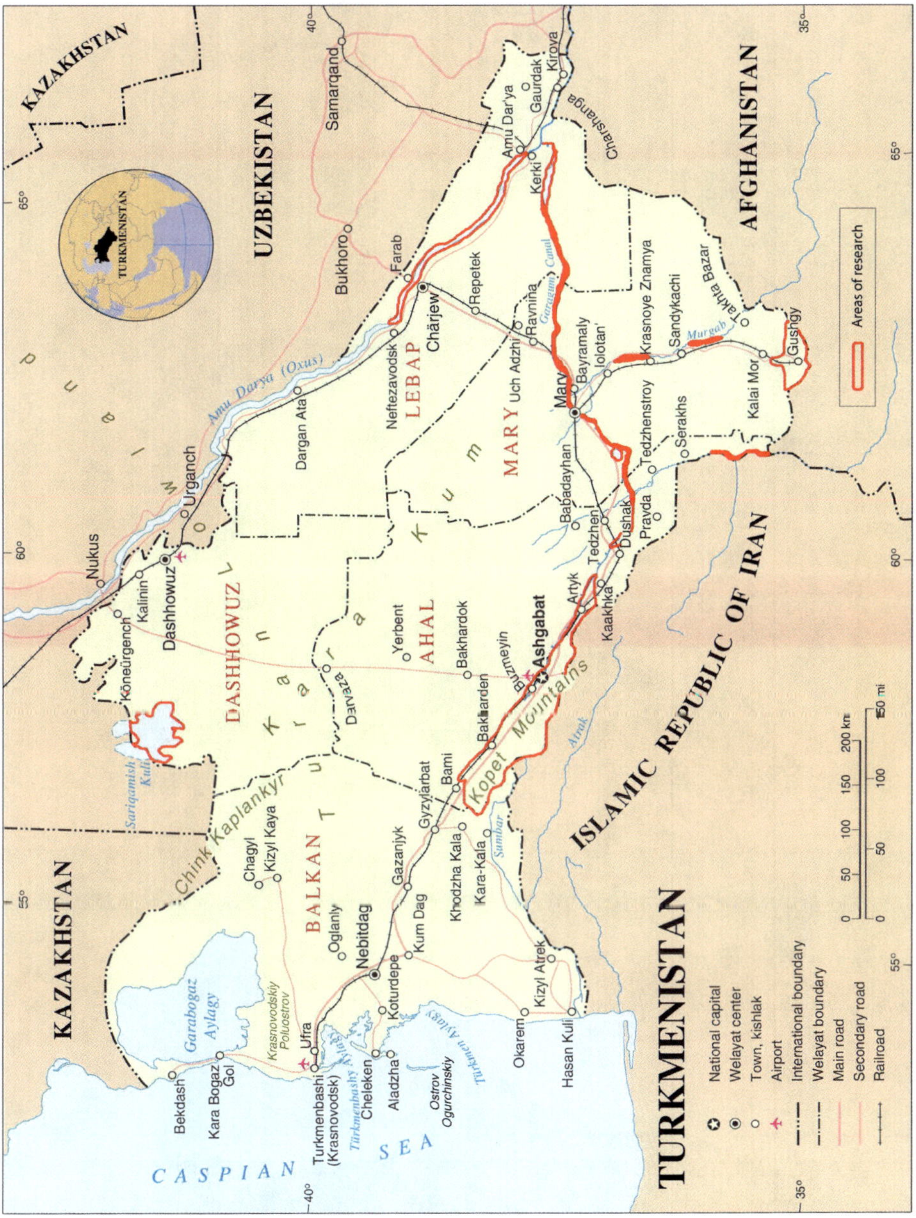

Fig. 1 Map of areas covered by studies

Fig. 2 *Ctenopharyngodon idella* – grass (Chinese) carp

Fig. 3 *Hypophthalmichthys molitrix* – silver carp

Fig. 4 *Carassius auratus gibelio* – goldfish, golden (Chinese) carp

Fig. 5 *Mylopharyngodon piceus* – black carp, black Amur

Fig. 6 *Parabramis pekinensis* – white Amur bream

Fig. 7 *Abbotina rivularis* – Amur false gudgeon

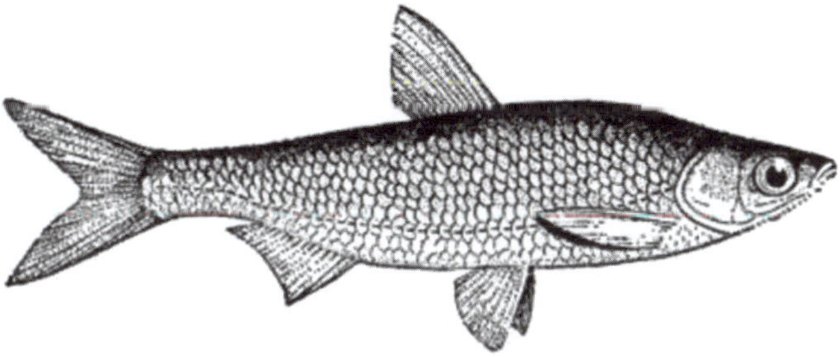

Fig. 8 *Hemiculter leucisculus* – Korean sawbelly

Fig. 9 *Pseudorasbora parva* – stone morocco

Fig. 10 *Misgurnus anguillicaudatus* – Amur loach, oriental weatherfish

Fig. 11 *Rhodeus ocellatus* – bitterling

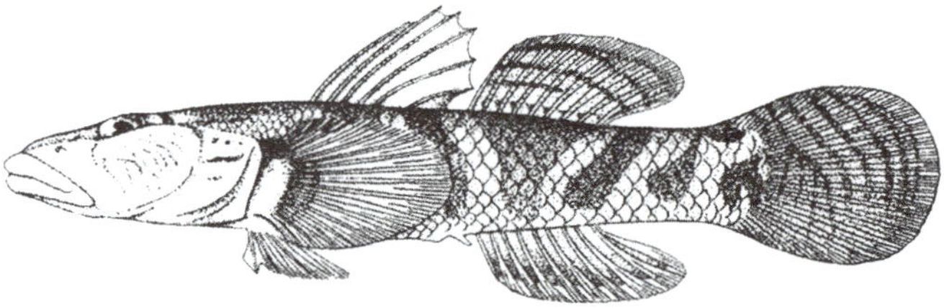

Fig. 12 *Rhinogobius cheni* – (Amur) Chinese goby

2 The Amudarya River

The ichthyofauna in the middle reaches of the Amudarya numbers presently over 40 species, the quarter of which (11 species) is represented by invaders (Table 1). Some of them, such as European pike-perch (Fig. 13), breams (eastern) (Fig. 14), Aral shemaya (Fig. 15), in the mid-1970s got into the middle stream of the river and settled down there as a result of regulation of the Amudarya flow and sea regression.

Fig. 13 *Sander lucioperca* – European pike-perch, sander, zander

Fig. 14 *Abramis brama orientalis* – breams (eastern)

Fig. 15 *Chalcalburnus chalcoides aralensis* – Aral shemaya

3 The Sarykamysh Lake

The ichthyofauna of the Sarykamysh Lake, the largest water body accumulating drainage waters, was formed from the species invaded from the Amudarya River, and bodies being a part of a land reclamation system ([20]). The fish population consists of the aboriginal species of the Aral–Amudarya basin, and Chinese invaders penetrated spontaneously and purposefully resettled here for fishery development in 1969–1974. For the time of existence of the modern lake about 36 fish species were found in it [20]. In 1980–1987 there were registered 27 species, while at present – 32 species, of which 11 species or 34.4% were invaders (see Table 1). The comparison of the fish species composition in the Sarykamysh Lake and other water bodies in the Amudarya River basin showed that until the 1980s (the years of its significant salinization) the composition of the aboriginal species of the lake's ichthyofauna was similar to that of the Aral Sea and also the lower reaches of the Amudarya and related water bodies. It was found that compared to the modern composition of ichthyofauna in the plain part of the Amudarya the fish population demonstrates less diversity as it lacks rheofil and stenogaline species typical of the river. The main limiting factors for their distribution in the Sarykamysh Lake are lacustrine hydrologic regime and rather high water salinity

(12–15 g/l). Out of rheofil species that inhabited the plain part of the Amudarya River in the 1970s, the big Amu-dar shovelnose (Fig. 16), little Amu-dar shovelnose (Fig. 17), and ostroluchka (Fig. 18) were found in the Sarykamysh Lake. At present only ostroluchka may be found in the Daryalyk header [20].

Fig. 16 *Pseudoscaphirhynchus kaufmanni* – big Amudarya shovelnose

Fig. 17 *Pseudoscaphirhynchus hermanni* – little Amudarya shovelnose

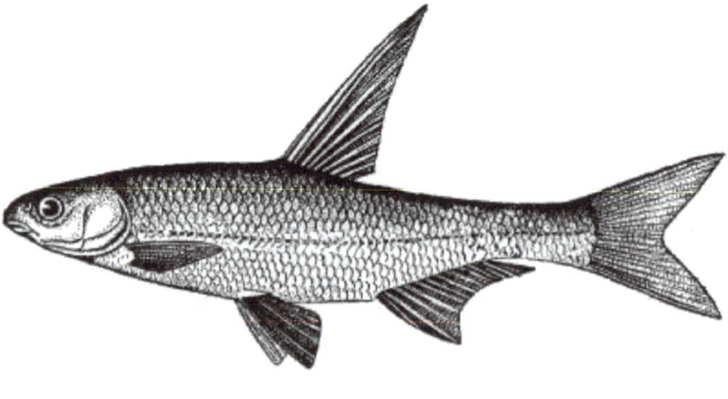

Fig. 18 *Capoetobrama kuschakewitschi* – Ostroluchka

4 The Turkmen Lake "Altyn Asyr"

It should be also mentioned here that the Central Asia's largest water body accumulating collector and drainage waters that is presently formed (since 2009) is as a result of implementation of the "Turkmen Lake" Project in the north of Turkmenistan. Its capacity will be 4–5 times greater than that of the Sarykamysh Lake [21, 22]. The results of the Sarykamysh Lake studies showed that the permanently growing water salinity distinguishing the dynamics of the water-salt regime of similar water bodies was the dominating factor of their development and evolution [13, 20, 21]. It is known that the fish population of the newly created lake with the formed from the species getting into it via the collector and drainage systems from original water bodies which include the lower and middle reaches of the Amudarya River, the Karakum Canal, the Tedjen and Murghab river basins and irrigation and drainage network of the Dashoguz velajat located in the north of Turkmenistan. Considering the correlation of the volumes of diverted collector and drainage waters and degree of their salinity, it may be assumed that the ichthyofauna of the lake will consist of the semi-anadromous fish, aboriginal fish of the Aral–Amudarya fauna, and Chinese introducents adapted in the region. The researchers anticipated to find 25–30 fish species in the lake, including carp (Fig. 19), breams (eastern), sabrefish (Fig. 20), Aral roach (Fig. 21), Aral shemaya, Aral asp (Fig. 22), Aral (Caspian) barbell (Fig. 23), khramulya (Fig. 24),

Fig. 19 *Cyprinus carpio* – carp (European, mirror)

Fig. 20 *Pelecus cultratus* – sabrefish, rasorfish

Fig. 21 *Rutilus rutilus aralensis* – Aral roach

Fig. 22 *Aspius aspius iblioides* – Aral asp

Fig. 23 *Barbus brachycephalus* – Aral (Caspian) barbell

Fig. 24 *Varicorhinus capoeta heratensis* – transcaspian khramulya

(European) catfish (Fig. 25), European pike-perch and under favorable conditions also goldfish, silver carp and bighead, grass (Chinese) carp, white Amur bream, (northern) pike (Fig. 26), and snakehead (Fig. 27) being of commercial significance. The fish

Fig. 25 *Silurus glanis* – (European) catfish, (European) wells, sheatfish

Fig. 26 *Esox lucius* – (Northern) pike

Fig. 27 *Channa argus* – snakehead

species not very demanding to external factors may also reveal mass development, such as stripped bystranka (Fig. 28), Korean sawbelly, stone morocco, Aral stickleback (Fig. 29), Chinese goby, and others being of no fishery significance [21]. The water with the salinity level of 4–6 g/l will be favorable for reproduction and

Fig. 28 *Alburnoides taeniatus* – striped bystryanka

Fig. 29 *Pungitius platygaster aralensis* – Aral stickleback

fattening of many of the mentioned fish species. The researchers noted that in the spawning period the fish sperm cells and roe, the fish embryos, and larvae are less stable compared to adult fish. For many species living there, except Aral stickleback, the salinity level should be no more than 10–12 g/l. After attaining such salinity the fish productivity of the lake will be about 10–15 kg/ha and may be even more if the measures on fishery regulation, protection of natural reproduction, fishery improvement, and fish farming are applied [20, 21]. Further development of KDV will depend on the purposes of water use – irrigation, technological and domestic needs, recreation, fishery development, etc., with regard to the water quality priority considerations. In each case the requirements for them will be different [13].

5 The Karakum Canal

In the Karakum Canal the ichthyofauna was formed from the fish of the Amudarya River so it differs only slightly from the original fish stock. Here we can find 36 species, of which 15 species or over 40% are invaders (see Table 1).

Beginning from 1984 the migration of roe and larvae of the Chinese pelagophillic fish was observed in the stretch of phase II of the Karakum Canal construction nearby the city of Mary (400th km of the Karakum Canal). However, the effect of their multiplication in this stretch was insignificant due to the lack of water bodies possessing forage base for the fish fries at their early stages of development. The subsequent research confirmed the important role of spawning grounds located in the head of the canal. It was found that nearly one-third of the migrants (26.5%) occurred in the upstream spawning grounds, nearly the half (43.5%) migrated down from the spawning places located at the 275–325th km of the river stretch, including 14.9% from the 175th to 200th km stretch [4]. The results of recent research [4] showed that the spawning grounds for pelagophillic fish were found on the 270–420th km, 440–599th km, and 630–635th km from where the fish roe and larvae migrate downstream.

6 The Khauzkhan Reservoir

The ichthyofauna of the Khauzkhan reservoir located at the 456th km of the Karakum Canal is inhabited at present by 33 fish species, of which 13 species or 39.4% are invaders. The reservoir was populated by way of self-settlement of fish from the Amudarya and as a result of fishery-acclimatizing works conducted in the basin. Here you can also find the representatives of ichthyofauna typical of the lower Amudarya and the Aral Sea that lived here until the 1980s until its considerable salinization, in particular, bastard (Fig. 30), Aral roach, Aral asp, Aral shemaya, golden spiny loach (Fig. 31) and also breams (eastern), sabrefish, goldfish, carp, catfish, and European pike-perch got into the reservoir due to the sea recession. At present such rheofil fish species as big and little Amu-dar shovelnose, pike asp (Fig. 32), Amu-dar stone loach (Fig. 33), transcaspian khramulya (Fig. 34), and others are met [3, 6, 23, 24] (see Table 1).

The ichthyofauna of the Khauzkhan Reservoir demonstrates a rather high population of the Chinese pelagophillic fish species naturalized in the past in the Karakum Canal and later on in the Amudarya River (1987). The changes witnessed presently in the structure of the fish population in the reservoir are connected with the decreased rate of natural multiplication of the species acclimatized in the head of the Karakum Canal [3, 4] and the long-time effect of irrational fishing.

Fig. 30 *Acipenser nudiventris* – ship, bastard (spiny, fringe, barbell) sturgeon

Fig. 31 *Sabanejewia auratas* – golden spiny loach

Fig. 32 *Aspiolucius esocinus* – pice asp

Fig. 33 *Nemacheilus oxianus* – Amudarya stone loach

Fig. 34 *Varicorhinus capoeta heratensis* – transcaspian khramulya

7 The Saryyazyn Reservoir and the Murghab River

The ichthyofauna of the Saryyazyn reservoir was formed as a result of settling here of the fish from the Murghab River – the drainless river system with its extremely specific fish population and purposeful acclimatizants let out into the water body to create productive commercial fish shoals [3, 5, 23].

The fish population was formed parallel to regulation of the Murghab River flow and construction of reservoirs started in the late nineteenth century, such as second and third Gindukush reservoirs – in 1895, Iolotan – in 1910, Tashkeprinsky – in 1940, and Kolkhozbentsky – in 1941. For nearly two decades the Tashkeprinsky reservoir was the head one. It was supplemented with the Saryyazyn reservoir constructed in 1958 and later on the "new" Saryyazyn reservoir. As all three reservoirs may be considered as development of one water body it will be quite appropriate to compare the biological indicators of fish and fishing of the Saryyazyn reservoir with that of the Tashkeprinsky reservoir.

The historically established fish fauna in the Murghab River basin before man's interference numbered ten species (see Table 1). The first acclimatizing works were conducted here in 1895. They helped to naturalize here such species as European mirror carp and European catfish, while the efforts on acclimatizing Aral (Caspian) barbell and Aral asp failed. In the 1930s the mosquito fish (Fig. 35) were let into the Murghab basin to control anopheles mosquitoes. They adapted well and spread widely here. The fish population of the river in this period reached 13 species [3, 5, 25]. However, the Tashkeprinsky reservoir was inhabited only by six species – gudgeon (Fig. 36), transcaspian marinka (Fig. 37), bystranka (Fig. 38), (European mirror) carp, European catfish, and mosquito fish, three of which were of commercial significance.

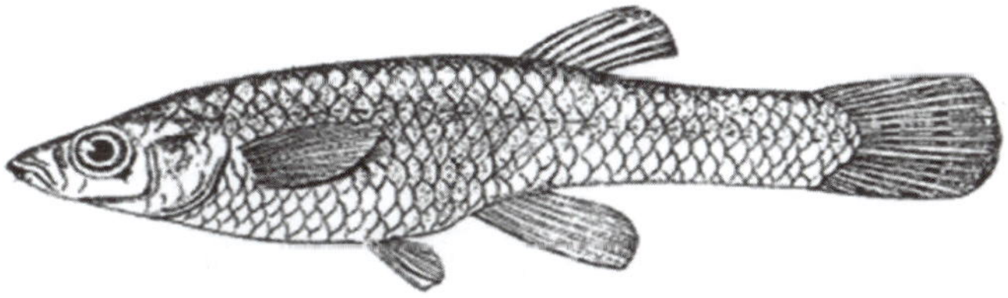

Fig. 35 *Gambusia holbrooki* – mosquito fish, topminnow

Fig. 36 *Gobio gobio lepidolaemus* – Gudgeon

Fig. 37 *Schizothorax pelzami* – transcaspian marinka

Fig. 38 *Alburnoides bipunctatus eichwaldi* – bystryanka, riffle minnow

In the Saryyazyn reservoir the aboriginal ichthyofauna is similar to that of the Tashkeprinsky reservoir, and this is quite understandable as the first stage of its biota formation occurred here. The follow-on broadening of the fish species composition in the reservoir was observed in the mid-1970s after completion of the second phase of the Karakum Canal construction (1966) that connected the Amudarya and Murghab basins via the operating pumped canal through which some species penetrated into the water body, in particular, striped bystranka. The species composition was also extended due to acclimatizing measures in the Murghab River basin in 1973–1974 that started from release of the fries of grass (Chinese) carp and silver carp into the water body, the natural multiplication of which was observed in the river in 1976. Together with the commercial herbivorous fish let into the water basin five species of casual acclimatizants got there quite by chance among which there were stone morocco, Korean sawbelly, Chinese goby, goldfish, and white Amur bream, the natural reproduction of which was first registered in the Murghab in 1985.

In 1994–1996 the first mature species of European pike-perch and snakehead were found here, although in the past they were never registered in this water body. The snakehead got into the water bodies of Central Asia in the period of acclimatizing the Chinese pelagophil fish. In the 1960s it accidentally got into the ponds of Uzbekistan and also into the Syrdarya, while by 1965 it already populated the river as far as the Aral Sea [26]. In the late 1960s the snakehead was met in the lower reaches of the Amudarya, while in the early 1970s – in the middle reaches from where it migrated to the Karakum Canal and its reservoirs.

In our opinion the snakehead and European pike-perch got into the Murghab via the pumping canal from the Karakum Canal where they were widespread. In the time when these species were found in the Murghab they broadly occurred in the Gindukush and Iolotan reservoirs and in the next 5 years they reached the Saryyazyn reservoir. So far (until 2010) the snakehead and European pike-perch were not found in the reservoir. Their invasion into the Saryyazyn reservoir may have negative consequences as it is here that some valuable commercial fish species populating the basin are fattening.

We should mention here about the accidental invader in the Murghab basin – the Korean sawbelly that is found both in the river proper and in the reservoirs. The materials collected in 1996 showed that the Korean sawbelly was one of the most abundant species in the Saryyazyn reservoir and thanks to its large sizes (its length – to 30.5 cm and weight – to 250 g) and it was actively caught by the local population. Perhaps it reveals high potential growth capacity. In the Murghab basin this sawbelly actively devours noncommercial fish species (8 or 15% of species having no commercial significance) and, therefore, at present its population is quickly growing due to abundance of food. And we can also judge about sufficiency of the forage base for this fish by indirect indicators – high Fulton's condition factor that reached 2.0 and maximum body weight index [27].

Therefore, as a result of the fishery measures taken in the Murghab basin we have at present 16 fish species in the Saryyazyn reservoir of which 7 are invaders. Undoubtedly, the second stage of the ichthyofauna formation in the reservoir characterized by broadening of the species composition is ongoing. Thus, before acclimatizing the herbivorous fish in the Murghab basin the fishing of transcaspian khramulya, carp, and (European) catfish were of commercial scale, but their shares in the catches were unequal. The catches of carp were 80–85%, (European) catfish 10–15%, transcaspian khramulya 5–10%, while the fish productivity of carp in the then leading Tashkeprinsky reservoir was 11.1–17.4 kg/ha, of others – 0.2–1.4 kg/ha.

Nowadays the fishing is concentrated mainly in the Saryyazyn reservoir in which the fish catches are greater than in all water bodies in the Murghab basin in the 1950s. The reserves of the main commercial fish species, such as carp, (European) catfish, transcaspian khramulya, dropped significantly due to their unwise use and deterioration of their propagation conditions caused by great fluctuations of the water level in the spawning period. It was found that the time of the first mass spawning of carp coincided with the water drawdown from the reservoir for presowing irrigation which led to drying out of spawning grounds and death of laid eggs. And also considering the fact that the first portion of eggs laid by the carp takes no less than 71.4% of the whole spawn it can be assumed that the population of this fish replenishes not due to its reproduction in the reservoir, but due to spawning during floods when water inundates large territories.

8 Conclusions

Therefore, the studies of ichthyofauna in two major reservoirs of Turkmenistan have shown that the formation of their fish populations has both common and specific features. In the Saryyazyn reservoir being the head one in the Murghab basin, the main factor controlling the biomass and population of the fish is silting. In the Khauzkhan reservoir the silting in the head Kelif reservoir also affects significantly the population of the pelagophillic fish species, both local Aral (Caspian) barbell and acclimatized silver carp, bighead, Chinese carp, black carp, and white Amur bream. But this process produces its effect on the ichthyofauna of the Khauzkhan reservoir with some delay.

Today the fish population in both reservoirs is distinguished by a relatively high number of the Far Eastern acclimatized fish species. Successful naturalization of herbivorous fish improved drastically the fish productivity of the water bodies without damaging the aboriginal ichthyofauna. The acclimatizants being herbivorous pelagophils are not rivals to the local species. Propagating successfully the invaders created rather quickly the self-reproducing communities. Despite the decreased catches of local fish species due to depletion of their reserves caused by irrational development, we presently face the growing use (and not always optimal) of the Far Eastern invaders which escalating catches may also affect their populations.

The conducted researches have shown that the potential fish productivity of reservoirs evaluated by the primary products is not high, although the fish productivity and fish growth rates in reservoirs are rather high and exceed fish growth even in the natural areas. In our opinion this may be explained by specifics of their feeding in reservoirs where, for example, the main food of the silver carp is detritus which takes 45.4–97.2% or 90% on the average of its ration. It is detritus and not phytoplankton that supports the fish productivity in reservoirs [4, 5, 28]. Presently the trophic role of detritus and its nutrition value for aqueous animals is the proved fact ([29, 30] and others). It is found that the detritus is fixed by the aqueous animals for 65% on the average and its nutritional value depends on its origin and, to a great extent, on its age, i.e., the degree of the organic matter transformation. At some stages of detritus existence its feed value may even exceed the nutritive quality of phytoplankton and other kinds of live food. But in all cases when detritus is used as food the average daily relative increment of aqueous animals was higher than for other feeds [30].

The practical results of works on acclimatizing the herbivorous fish and their application in the fishery practices without damage to the local fauna have demonstrated the vital role of this fish in rehabilitation and creation of productive ecosystems in inland water bodies [1, 31, 32]. And this is still more important as they may be applied effectively for reduction of the organic and biogenous pollution of water bodies in conditions of integrated use of water resources.

At present the acclimatizants may be found in once isolated mountain rivers in the Paropamiz and Kopetdag. Investigations of 15 rivers showed that out of 17 fish species living there 12 species were aboriginal and 5 species or 29.4% were invaders, such as Korean sawbelly, stone morocco, carp, mosquito fish (Table 2). In the Kashan and Kushka rivers there were found for the first time the goldfish, Korean sawbelly, and stone morocco; in the Tedjen River – the goldfish and Korean sawbelly; in the Kazganchai – the stone morocco, while in the Yanbash – the stone morocco and mosquito fish. The researchers found out that the crested stone (*Schistura cristata*) (Fig. 39) was detected in the Murghab River basin not only in the Kushka and Kashan rivers, but also in the Tedjen River. The area of occurrence of the Turkmenian stone loach (*Schistura sargadensis*) (Fig. 40) put on the Red Books of Turkmenistan [33, 34] is much broader and it lives not only in the Keltechinar and Chyrlak rivers (former Sharlavuk), but in the Meanachai, Dushak, and Kushka rivers, too. The Kessler's stone loach (*Nemacheilus kessleri*) is not obviously present in the ichthyofauna of Turkmenistan. Despite our energetic efforts we failed to trace this fish. Instead of this, as it was mentioned above, the Turkmenian stone loach was caught for the first time in the Kushka River. The Kessler's stone loach was first described by G.V. Nikolsky (1940) [25] by the four specimens fished in the Kushka River. Three of them are kept in the Zoological Museum of the Moscow Lomonosov University. Our morphological analysis revealed complete identity of the features of these specimens and Turkmenian stone loach from the Kushka River. In other words, the fish specimens caught by G.V. Nikolsky are the Turkmenian stone loach. The habitat of the transcaspian marinka and bystranka, riffle minnow that disappeared today from some rivers is narrowing [6]. In the 1930s–1940s, (eastern) chub (*Leuciscus cephalus orientalis* Nordmann) (Fig. 41) was rather widespread in the Atrek River basin (Sumbar, Chandyr rivers), but in 2003 and 2007 it was not found here. The Kura barbell (*Barbus lacerta cyri* Filippi) (Fig. 42) that was first detected in 1925 in the Sumbar River and being rather common that time was not found in 2003 and 2007 during investigations in the river [3, 9].

Therefore, today the ichthyofauna in the water bodies of Turkmenistan has changed a lot [35]. Some aboriginal species are no longer found here, but at the same it is replenished with some new invaders represented by 1 order (*Cyprinodontiformes* – Cyprinodontoids), 4 families (*Poeciliidae* – (livebearers) mosquitofishes; *Oryziatidae* – medakas (ricefishes); *Odontobutidae* – (slippers) loachgobies; *Channidae* – snakeheads) and 13 genus (*Ctenopharyngodon* – grass (Chinese) carp; *Hemiculter* – sawbellies; *Hypophthalmichthys* – silver carps; *Mylopharyngodon* – black Amurs; *Parabramis* – white Amur breams; *Abbottina* – false gudgeon; *Pseudorasbora* – stone moroccos; *Misgurnus* – (mud) loaches; *Oryzias* – medakas; *Gambusia* – mosquito fishes; *Micropercops; Rhinogobius; Channa* – snakeheads). Some of them play an important role in the country's fishery industry.

Table 2 Species composition of fish in the rivers of Paropamiz and Kopetdag

Species	Rivers 1	2	3	4	5	6	7	8	9	10	11	12	13	14
Schistura cristata – crested stone	■□	■□	■□	■□	■□	■■		■□	■□		■□	■□	■□	
Schistura sargadensis – Turkmenian stone loach		□□				■□	■□	□□	□□					In Turkmenistan Red Book
Narrowing habitat														
Schizothorax pelzami – transcaspian marinka		■□	■□	■■	■■	■■		■■			■■	■□	■□	In Turkmenistan Red Book
Alburnoides bipunctatus eichwaldi – bystranka, riffle minnow	■□	■□	■□		■□	■■					■■	■■	■□	In Turkmenistan Red Book
Leuciscus cephalus orientalis – (Eastern) chub, skelly										■■ Common in Atrek basin				In Turkmenistan Red Book
Barbers lacerta cyri – kura barbell										■■ Occurred in Sumbar				In Turkmenistan Red Book
Nemacheilus kessleri – Kessler's stone loach		■■												
Acclimatizants														
Carassius auratus gibelio – goldfish, golden (Chinese) carp	□□	□□	□□											
Hemiculter leucisculus – Korean sawbelly	□□	□□	□□											
Pseudorasbora parva – Amur stone morocco	□□	□□		□□	□□									
Cyprinus carpio – (European mirror) carp	■□	■□	■□		■■									
Gambusia holbrooki – mosquito fish, topminnow	■□	■□	■□		□□									

1 – Kashan; *2* – Kushka; *3* – Tedjen; *4* – Kazganchai; *5* – Yanbash; *6* – Keltechinar; *7* – Cyrlak (former Sharlavuk); *8* – Meanachai; *9* – Dushak; *10* – Atrek basin (Sumbar, Chandyr rivers); *11* – Archabil (former Firyuzinka); *12* – Altyyab (former Chulinka); *13* – Sekizyab; *14* – Species put on the Turkmenistan Red Book

(■) Literary data

(□) Our research data

Fig. 39 *Schistura cristata* – crested stone

Fig. 40 *Schistura sargadensis* – Turkmen stone loach

Fig. 41 *Leuciscus cephalus orientalis* – (Eastern) chub, skelly

Fig. 42 *Barbers lacerta cyri* – Kura barbell

References

1. Aliev DS (1976) Role of herbivorous fish in reconstruction of fishery ichthyofauna and biological improvement of water bodies. Problems Ichthyol 16(2):247–262 (in Russian)
2. Aliev DS, Verigina IA, Svetovidova AA (1963) Composition of fish species brought with white grass carp and bighead from China. Problems of fishery use of herbivorous fish in the USSR water bodies. AN TSSR Publishers, Ashgabad, pp 178–180 (in Russian)
3. Aliev DS, Sukhanova AI, Shakirova FM (1988) Fish of inland water bodies in Turkmenistan. Ylym, Ashgabad, p 156 (in Russian)
4. Aliev DS, Sukhanova AI, Shakirova FM, Malakhova TV (1994) Herbivorous fish in Turkmenistan. Ylym, Ashgabad, p 326 (in Russian)
5. Pavlov D.S. et. al (1994). The Book summarizes the results of ichthyological Research on the "new" Saryyazyn reservoir. The material is important to create a Database and monitoring in the reservoir as a model of yhe arid zone
6. Salnikov V.B. et. al. (1998). This thesis devoled to the ichthyofauna of rivers Paropamiz and Kopetdag, based on our expedition on the river to explore the diversity and status of fish populations
7. Salnikov VB (2006) Ichthyofauna of water bodies of Kugitang. J Problems Desert Dev Ashgabad 3:18–23 (in Russian)
8. Salnikov VB (2006) Prospects of artificial rearing of Amudarya shovel-nosed fish in Turkmenistan. J Problems Desert Dev Ashgabad 4:44–51 (in Russian)
9. Salnikov VB (2008) Ichthyofauna of the Sumbar and Chandyr rivers in the Etrek River basin: reduction of species diversity and protection measures. In: Proceedings of the research-practical conference on 75th anniversary of the Khazar State nature reserve, Ashgabad-Turkmenbashi, pp 118–122 (in Russian)
10. Salnikov VB (2008) Modern state of ichthyofauna in water bodies of Koytendag and its protection. In: Proceedings of the research-practical conference on 75th anniversary of the Khazar State nature reserve, Ashgabad-Turkmenbashi, pp 123–126 (in Russian)
11. Charyev R (1984) On some consequences of importation and acclimatization of white amur (*Ctenopharyngodon idella* (Val.) *Cyprinidae*) in the Karakum Canal. Problems Ichthyol 24(3):385–392 (in Russian)
12. Shakirova FM (2007) Change in the ichthyofauna composition in the inland water bodies of Turkmenistan after acclimatizing the Chinese herbivorous fish. Theses of the reports at the international scientific conference "natural and invasion processes of formation of biodiversity of aqueous and land ecosystems", Rostov-on-Don, 5–8 June 2007, pp 325–327 (in Russian)
13. Shakirova FM (2010) Perspectives of use of saline collector-drainage waters in the zone of the Turkmen Lake for fishery development. In: Proceedings of international scientific conference, Ashgabad, 24–25 March 2010, pp 233–235 (in Russian)
14. Shakirova FM, Sukhanova AI (1993) Ichthyofauna of Turkmenistan (composition and distribution). Newslett Turkmenistan AN Ser Biol Sci 3:35–45 (in Russian)
15. Vasilieva ED, Kozlova MS (1989) Morphological characteristic of common sawbelly (*Hemiculter leucisculus*) (*Pisces, Cyprtnidae*). Bull Moscow Soc Nat Res Div Biol 94(3):36–46 (in Russian)
16. Vasilieva ED, Makeeva AP (1988) Morphological peculiarities and taxonomy of Pisces, Eleotridae of the USSR fauna. Zool J XVII:1194–1204 (in Russian)
17. Reshetnikov YuS, Bogutskaya NG, Vasilieva ED et al. (1997) Annotated listing of pisciform and fish of freshwater basins in Russia. Problems Ichthyol 37(6):723–771 (in Russian)
18. Reshetnikova Y (ed) (2003) Atlas of freshwater fish of Russia, vols. 1 and 2. Nauka, Moscow, p 379 and p 253 (in Russian)
19. Reshetnikov YS (ed) (2010) Fish in nature reserves of Russia (in 2 vols.). Partnership of scientific publications, vol. 1. KMK, Moscow. Freshwater Fish, p 627 (in Russian)
20. Sanin MV, Kostyukovsky VI, Shaponenko SI et al (1991) The Sarykamysh Lake and water bodies accumulating collector and drainage waters. Nauka, Moscow, p 150 (in Russian)

21. Salnikov VB (2010) Perspectives of fishery use of the Turkmen Lake. In: Proceedings of the scientific conference, Ashgabad, 24–25 March 2010, pp 217–219 (in Russian)
22. Esenov PE, Aganov SE (2010) Ecological and economic aspects of CDW use in the zone of the Turkmen Lake "Altyn Asyr". In: Proceedings of international scientific conference, Ashgabad, 24–25 March 2010, pp 239–242 (in Russian)
23. Reshetnikov YuS, Shakirova FM (1993) Zoogeographical analysis of Central Asian ichthyofauna by the list of freshwater fish. Problems Ichthyol 33(1):34–45 (in Russian)
24. Salnikov VB, Reshetnikov YuS (1991) Formation of fish population in artificial water bodies of Turkmenistan. Problems Ichthyol 31(4):565–575 (in Russian)
25. Nikolsky GV, Tsentilovich FF (1951) Fish in the Murghab basin. In: Collection of proceedings of the MGU Zoological Museum, vol. 7, pp 105–112 (in Russian)
26. Dukravets GM, Machulin AI (1978) Morphology and ecology of Amur snakehead (*Ophiocephalus argus warpachowskii* Berg) acclimatized in the Syrdarya basin. Problems Ichthyol 18(2 (109)):222–228 (in Russian)
27. Shakirova FM, Nikolaev AA (1991) Peculiarities of the biology of sawbelly (*Hemiculter leucisculus* (Basilewsky)) in the Murghab basin. Newslett Turkmenistan AN Ser Biol Sci 1:2–9 (in Russian)
28. Shakirova FM (1985) Growth of white bighead and white amur in the Saryyazyn reservoir. Newslett Turkmenistan AN Ser Biol Sci 1:22–26 (in Russian)
29. Panov DA, Sorokin YuI, Motenkova LG (1969) Experimental study of nutrition of silver carp fries. Problems Ichthyol 9(1 (54)):38–152 (in Russian)
30. Pavlyutin A.N. (1974). Radioisotope method proved conclusively that the detritus is assimilated with high efficiency water animals. And the food value of detritus is largely dependent on its age
31. Nikolsky GV, Aliev DD (1974) Role of Far Eastern herbivorous fish in the ecosystems of natural water bodies of invasion. Problems Ichthyol 14(6):974–978 (in Russian)
32. Nikolsky GV, Aliev DS, Milanovsky YuE (1979) Reclamation fish. Znanie, Moscow, pp 3–64 (in Russian)
33. Red Book (1999) Povidovye essays of rare and endangered species of Turkmenistan included in the Red Book (1999, 2011) described by F.M. Shakirova. So called them, analyzing the ichthyofauna of rivers Paropamiz and Kopetdag
34. Red Book of Turkmenistan (2011) Invertebrate and vertebrate animals, vol.2, 3nd edn. Ashgabad, Turkmenistan pp 125–159 (in Russian)
35. Shakirova FM (2011) Present state and changes in the ichthyofauna composition in the water bodies of Turkmenistan. In: Proceedings of the 1st All-Russia Conference with International Participation "Modern State of Bioresources in the Inland Water Bodies", vol. 2, Borok, 12–16 September 2011, ACWAROS, Moscow, pp 817–824 (in Russian)
36. Nikolsky GV, Tsentilovich FF (1955) Notes on hydrobiology of the Murghab River basin. Proc Murghab Hydrobiol Station 3:40–53 (in Russian)
37. Nikolsky GV (1953) Biological peculiarities of fauna complexes and importance of their analysis for zoogeography. In: Essays on general problems of ichthyology. AN SSSR, Moscow-Leningrad, pp 65–76 (in Russian)
38. Berg LS (1948–1949) Freshwater fish of the USSR and nearby countries. Parts 1–3. AN SSSR, Moscow-Leningrad, p 1381 (in Russian)

Wetland Birds of the Hydrographic Network of Altyn Asyr

Eldar A. Rustamov and Anna V. Belousova

Abstract The term "hydrographic network of Altyn Asyr" was introduced, the spatial–temporal dynamics of wetlands and wetland birds' habitats during migration and in wintering areas was described, data on the composition, type of staying, and monitoring of wetland birds on the water bodies of the hydrographic network of Altyn Asyr was shown, a forecast for the growth of the avifauna and wetland birds' population on the Altyn Asyr Lake was given.

Keywords Ecosystems, Hydrographic Network, IBAs, Karakum Darya River, Lake Altyn Asyr, Turkmenistan, Waterbirds, Waterfowl, Wetlands, Wintering of birds

Contents

E.A. Rustamov (✉)
Expert in Wetlands and Waterbirds, Ministry for Nature Conservation of Turkmenistan, Ashgabat, Turkmenistan
e-mail: elldaru@mail.ru

A.V. Belousova
All-Russian Research Institute for Nature Protection, Moscow, Russia

I.S. Zonn and A.G. Kostianoy (eds.), *The Turkmen Lake Altyn Asyr and Water Resources in Turkmenistan*, Hdb Env Chem (2014) 28: 261–290, DOI 10.1007/698_2012_217,

1 Introduction

In Turkmenistan, due to positive climatic and natural factors, there are environmental conditions for the wintering of many species of wild avifauna, particularly for wetland birds, which is mainly due to the absence of a period of deep winter depression, which is present in the northern regions of Eurasia. Favorable temperature conditions on the plains, and, as a consequence, good food supply are the main prerequisites for the formation of wintering clusters of wetland birds in inland water bodies of the country, which are connected, in one way or another, with the Turkmen Lake Altyn Asyr. Indeed, as observations have shown, the emergence of new, suitable for birds, wetlands on artificial reservoirs, as well as expansion of the irrigated land has a positive impact on the number of wintering birds, contributing to their concentration near water bodies and to the formation of new wintering areas. So far there has been no summarizing work on the inventory and the cross-spectrum analysis of the wetland birds' wintering, including all wetlands of the plain Turkmenistan, despite the fact that the beginning of the study of bird fauna of the country started already in the second half of the nineteenth century and the research was the most intensive in the twentieth century when artificial wetlands were appearing and when they were expanding their network.

The first survey about wintering of waterbirds in Turkmenistan was conducted by Laptev in the South-Eastern Caspian [1] and was continued in the same area by Isakov and Vorobiev [2, 3]. In inland water bodies regular winter surveys have been held since 1967 in the areas of the Amu Darya River, Karakum Darya (the Karakum Canal), Murgab, Tejen and Kopetdag foothills, and later also in the north of the country by: Khakyev in 1970–1976; Rustamov in 1977–1984 and 1999–2012; Poslavsky in 1985–1989 and in 1994, and by other ornithologists [4–10]. In 1970–1994 accounts were conducted by aircraft (airplane An-2, helicopters Mi-1, Mi-2, Mi-4, and Mi-8), and in other years – by ground methods from the shore by car. As a rule, as one account was taken one survey of a water body (lake, reservoir). There were used 8–10 times zoom binoculars, and since 2005 – 40–60 times zoom telescopic tubes [5, 7, 11]. Color photographs of birds are taken from the book of Rustamov "Wildlife of Turkmenistan and its protection" [12].

2 Habitat Transformation of Wetland Birds in Connection with the Development of the Hydrographic Network of Altyn Asyr

The main purpose of the design and construction of the Turkmen Lake Altyn Asyr, as it is known, is to collect drainage water from the huge surfaces of irrigated agriculture lands that emerged on the territory of Turkmenistan as a result of wild lands development and transformation of ancient oasis lands. The large-scale construction of irrigation systems that took place in the twentieth century and is continuing nowadays highly influenced the structure and spatial dynamics of the biotic components of desert and anthropogenic landscapes. Therefore, regarding the diversity of wetlands ecosystems it is not sufficient to consider the Turkmen Lake Altyn Asyr isolated from the whole basin of the artificial drainage, it will not reveal the whole picture of diversity change of such an important group of birds as wetland birds. Almost the entire hydrographic network of Turkmenistan (with the exception of the south-west), both natural and artificial, is directly or indirectly related to the Turkmen Lake Altyn Asyr. The authors found it useful to consider global (in the secular context) changes in the waterbirds, including all water bodies of this system. Based on the foregoing, under the term "hydrographic network of Altyn Asyr" we deem reservoirs, lakes, flooded areas, and canals of all types, connected directly or indirectly by waterways with the Turkmen Lake Altyn Asyr.

Turkmenistan, as well as the whole south-western part of Central Asia, is one of the areas of transit flights and wintering areas for birds that gather and migrate to the south from areas of Northern Eurasia, mainly from Western Siberia, Kazakhstan, and Eastern Europe. In desert conditions water as an environmental factor and wetlands as habitats are important and crucial especially for water limnophilous birds. In autumn and winter on the plains one can observe vegetation of some species of herbaceous plants, development of quite a variety of invertebrates that lead an active life, even despite the short-term cooling periods. Favorable environmental (climate: mild winters) and ecological conditions of the region determine the relevant feed. As far as waterbirds are concerned – it is also the lack of freeze-up and the availability of submarine and surface feed.

However, at the beginning and in the middle of the last century, overall scarcity of water resources in flat parts of Turkmenistan determined also a low concentration of waterfowl, especially in winter, despite the fact that the population of these birds in Eurasia as a whole, compared to the present time, was ten times bigger. In Turkmenistan, with the exception of shallow waters of the South-Eastern Caspian, there were no large concentrations of them [13]. Naturally, on the plains of the country waterfowl tended to stay in wetland habitats in the valleys of the Amu Darya, Murgab, and Tejen Rivers. However, a small number of birds stayed for wintering because of small surfaces and the capacity of wetlands, and they were used mainly for stops during seasonal migrations. In floodplains of the Amu Darya River birds tended to stick to coastal and island stretches and small floodplain lakes and oxbow lakes, as the river was not regulated, reservoirs were not built on it. There were no large concentrations of birds either on the Murgab or Tejen Rivers, despite the fact that in the valleys of these rivers there was conducted construction of reservoirs, though small, but still, they did build reservoirs there in order to regulate water collection and reduce discharges of flood water in the desert.

Habitat transformation of wetland birds' habitats on the plains of Turkmenistan was uneven for 100–120 years. Of course, in the first and even in the second half of the last century it was impossible to assume that such great crucial transformation processes and development of wetland ecosystems would take place and that, ultimately, there would be an integrated hydrographic network of Altyn Asyr. In the twentieth century wildlands and oasis lands were gradually developed, which was accompanied by irrigation and drainage construction. Specific changes in the hydrographic network in the southern half of the country were associated mainly with the complex and gradual process of construction of the Karakum Canal (now Karakum Darya), as referred to in an article by I.S. Zonn which is included in this book.

To avoid repetitions, we will omit the spatial–temporal characteristic of wetland birds' habitat transformation in the area of Karakum Darya. However, it should be emphasized that in the expansion process of constructing, strengthening Karakum Darya's influence and stabilizing developed areas in its zone as well as in adjacent to the old oases, there was transformation of arid ecosystems and their replacement by wetlands with all the components of biodiversity, especially in case of wetland birds. It is sufficient to consider the example of the lake system Kelif Uzboy that once consisted of fluvial reservoirs: on its large (Lake Karashor and Chaskak) and small (Kargaly, Petdeli, Swan, Turkmen, Twenty) stilling and filtration lakes there emerged conditions not only for nesting but also for wintering of wetland birds. The total area of all the Kelif Lakes by the mid-1950s amounted to 93 km^2 (82 km^2 – the water table), and even already then there appeared a wintering area not only for the *Anseriformes* but also for other wetland birds. The lakes turned out to be so significant that in 1970 there was formed the Kelif Ornithological Reserve [14]. However later, in the process of silting and overgrowing this lake system began to lose its value for mass wintering of waterfowl and its area decreased by ten times and now totals 8.5 km^2. However as a result of the creation of the new reservoir Zeid in the mid-1990s (in 2005 the surface area was 365 km^2), the role of Uzboy Kelif as a place of concentration of waterbirds, especially in winter, started to increase again.

Thus, one of the main causes of the emergence of the new wintering area in southern and south-eastern Turkmenistan (or the southern part of the hydrographic network of Altyn Asyr) turned out to be the anthropogenic formation of wetlands that has caused since the 1960s underflooding of lower parts of the desert along the route of Karakum Darya and near-delta areas of the Murgab and Tejen Rivers, during the withdrawal and discharge of drainage waters from agricultural arrays of the river valleys in the Central Karakum.

The development of Karakum Darya as an artificial river led to the unification of the basins of the Amu Darya, Murgab, Tejen and small rivers of the Kopetdag macroslope, causing the growth of surfaces of already existing and the emergence of new anthropogenic landscapes and wetland ecosystems and, as a consequence, biotope redistribution of birds during migration, their nesting and wintering. It is vital to add that this has led to the restructure of the other components of the wildlife in the zoogeographical context; for example, there was infiltration of elements of ichthyofauna from the Amu Darya basin into the Murgab and Tejen, as well as the partial union of their theriofauna.

The hydrological situation was changing not only in the southern half but also in the north and in the far east of Turkmenistan. Thus, in the early 1960s there was an increase in cultivated areas for cotton and rice in the lower lands of the Amu Darya, which increased the flow of irrigation water, and, therefore, the volume of drainage water, which began to be diverted (since 1961) to the Sarykamysh basin, 150 km to the west of the so-called Khorezm – Kunyaurgench oasis, and that has been dumped there to this day. As a result, since the mid-1960s Lake Sarykamysh has been rapidly increasing its area (in 1975 – about 1,500 km^2, in 2000 – more than 2,500 km^2, and now – about 3,900 km^2), and by now it has become one of the largest artificial lakes of the Central Asian region. Uncontained growth of the volume of drainage water has led to an increase not only of the water table but also of the capacity of the entire Sarykamysh basin as a territory suitable for the habitat of wetland birds. One could observe the emergence of a new area of birds' concentration during migration and partly wintering (in the first half of winter) in Central Asia, and new nesting sites for the *Anseriformes* and colonial nesting species such as pelicans, cormorants, gulls, and terns. Due to the drop of the Aral Sea level and drying of the Amu Darya delta there happened a "relocation" of some nesting places of colonial nesting waterbird species from Southern Sub-Aral area to Sarykamysh.

At the top of the Amu Darya delta at the border with Uzbekistan since 1977 there was carried out the construction of new reservoirs – Tuyamuyun or Dueboyun (according to the project – more than 600 km^2, however, today the water table is about 130 km^2) and Soltansanjar (350 km^2), from which in 1982 there was built the Tuyamuyun left bank water supply canal (20 km), which joined the Malyab Canal, and together they are now called Turkmen Darya (180 km). The expansion of the delta of the Amu Darya to the west and the emergence of the large transboundary Lake Sarykamysh in northern Turkmenistan, as well as large drainage lakes in the east – in depressions of the right-bank of Amu Darya – in the middle of its course (see below) – this is the second large-scale transformation of the surface hydrographic network of the country.

Speaking more about Sarykamysh, it should be noted that there with the formation of a vast wetland already in 1975–1976 there were registered 39 species of waterbirds, of which five are *Anseriformes* [15], and today their number, respectively, increased to 90 and 24 species [16]; in the middle of the last century there were nesting only Eurasian Bittern, Common Shelduck, Mallard, Eurasian Coot and Black-winged Stilt [17]. It is natural that over the past decades the number of wetland birds' species has grown many times. Approximately 75% of all species of waterbirds stop during migration, about 25% stay to nest. Large wintering and migratory clusters of waterbirds in Sarykamysh before its large-scale filling did not exist due to the shortage of water – the lake's surface at the end of the nineteenth century and in the middle of the twentieth century was not more than 100 km^2. Now wintering exists there, but usually in the first half of winter (till mid-December), until the temperature falls to the limits pessimal for wintering wetland birds (complete freezing).

It must be said that in the early 1980s there were formed new reservoirs also at the right bank of the Amu Darya in the Sundukli Desert at the border with Uzbekistan. This is a system of lakes Soltandag (103 km^2), Gyzylburun (11 km^2) and Tailak (3 km^2), emerged as a result of filling depressions of the same name by large volumes of drainage water diverted along collectors and the river bed of the Mehejan from the Karshyn Region of neighboring Uzbekistan. In these wetlands there were also observed concentrations of waterbirds that settled for wintering, moving during migrations from the north-east, that is, from the valley of the Zeravshan River and plain lakes of neighboring Uzbekistan. At first, these drainage lakes existed in isolation, but later they were connected to the Amu Darya, and thus, with the whole hydrographic network of Altyn Asyr.

Finally, a large-scale transformation of the hydrography of Turkmenistan as a habitat for wetland birds started recently as a result of the implementation of the following project since 2000 – construction of the Transturkmen (Main) Collector of 720 km length and construction of the Turkmen Lake Altyn Asyr. Aspects of this construction are thoroughly described in the chapter by I.S. Zonn and A.G. Kostianoy, included in this book, as well as in the book "Turkmen Lake Altyn Asyr" [18], and that is why this information was not included in this chapter.

Thus, long-term and large-scale construction covering almost the entire territory of the country (by the way, not only of Turkmenistan but partially of neighboring Uzbekistan) has led to redistribution of huge volumes of water and to the emergence of the artificial, along with the natural one, hydrographic network consisting of different types of canals, the main ones of which are irrigation canals (main and secondary irrigation) and drainage canals – tertiary, secondary, and main. The network of canals as a system of complex hydraulic constructions with numerous branches is a zone of underflooding on the vast, usually low areas in different parts of the Karakum Desert. However, irrigated and developed were not only desert areas but also territories inside oasis, previously not cultivated in the end deltas of the Murgab and the Tejen, on the piedmont plains of Kopetdag, as well as newly developed lands (wildlands), such as arrays of Khauzkhan and Shasenem. Irrigation construction on the plains of Turkmenistan and the corresponding transformation of ecosystems has been continuing till now.

In the first half of the last century in the valleys of the Amu Darya, the Murgab and the Tejen wintering areas witnessed a relatively small number of waterbirds; however due to a large number of water bodies that appeared (reservoirs, large and small lakes, drainage floods) in the Karakum were formed wetlands' habitats, which serve as places of concentration of waterfowl during migration and wintering and during the nesting period. This is especially true for the southern half of the country. In addition there emerged huge territories of agricultural lands, where in the second half of the last century there were mainly cotton fields, but during last 20 years it has changed towards winter cereals and forage crops. As a result, there emerged places with good food supply not only for White-fronted and Greylag Geese and Ruddy Shelducks, but also Common Cranes; when these fields are filled with irrigation water or meltwater, then some dabbling ducks also concentrate there. Finally, as a result of climate warming, an essential role plays continuation of vegetation of native grasses and weed plants, in free, as well as in flooded or underflooded sandy areas or winter fields.

In the hydrographic network of Altyn Asyr (Fig. 1) we identified 67 wetlands (sites), combined into 9 districts, of which 13 water bodies (19.4%) were not surveyed.

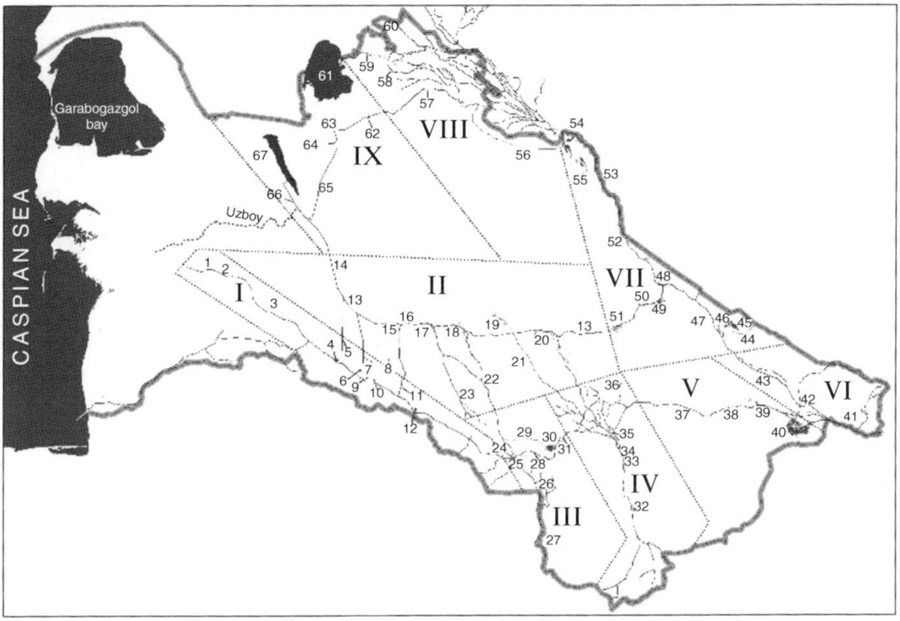

Fig. 1 Wetlands of the hydrographic network of Altyn Asyr: I. Near-Kopetdag Region: 1 – Chokrak Lake, 2 – Donuzaji Lake, 3 – Uzynshor flooded area, 4 – Kopetdag Reservoir, 5 – Geokdepe Collector (Akhal flooded area), 6 – Rukhabat Collector (Ovadandepe flooded area), 7 – Ashgabat Collector (Northern Ashgabat flooded area, Djaparlytakyr Collector and Lake 37th km), 8 – Akbugday or Gyaur Collector (Lake Bozkel, Chukurtakyr and Deryatakyr flooded areas), 9 – Kurtli Water Reservoir, 10 – Ashkhabad Water Reservoir, 11 – Lake Mergen (the former Lake Kulankyrlan), 12 – Gyaurs Fishery; II. Central Karakum Region: 13 – Main Turkmen Collector, Altyn Asyr, and flooded areas: 14 – Atabeg, 15 – Mollakurban, 16 – Kelili, 17 – Garajaovlak, 18 – Ayrakly, 19 – Khangui and Gushliburun, 20 – Djarsay flooded area and Djarsay Collector, 21 – Murghab or Shikhmansur Collector, 22 – Tejen or Garavekil Collector, 23 – Kaka Collector or former Soltandesht flooded area; III. Tejen-Khauzkhan Region: 24 – Tejen Fishery, 25 – Tejen Water Reservoir, 26 – Khorkhor Water Reservoir, 27 – Dostluk Water Reervoir, 28 – Main Canal, 29 – Karakum Darya in the limits of Khauzkhan oasis, 30 – Khauzkhan Water Reservoir, 31 – Karakum Darya between settlement Energetik and Khauzkhan Water Reservoir, IV. Murgab Region: 32 – Saryazy Water Reservoir, 33 – Soltanbent Water Reservoir, 34 – Yelotan Water Resrvoir, 35 – Gindikush Water Reservoir, 36 – Seyrab flooded area (5th water discharge); V. Kelif Region: 37 – Karakum Darya between settlements Zakhmet and Nichka, 38 – Karakum Darya between settlements Nichka and Karametniyaz; 39 – Kelif Lakes; 40 – Zeyit Water Reservoir; VI. Upper Amu Darya Region: 41 – Amu Darya between settlements Kelif and Mukry, 42 – Amu Darya between settlements Mukry and Atamurat (former Kerki), 43 – Amu Darya between settlements Atamurat and Garabekaul; VII. Middle Amu Darya Region: 44 – Mekhejan Collector and flooded area, 45 – Lake Soltandag, 46 – Lakes Kyzylburun, Taylak and Turangyldyz, 47 – Amu Darya between settlement Garabekaul and Turkmenabat Town, 48 – Amu Darya between Turkmenabat and Seidi (former Neftezavodsk) towns, 49 – Lake Ulyshor (former Kattashor), 50 – Lake Rakhmankel (former Ramankeldogajik), 51 – Lake Eraji, 52 – Amu Darya between settlements Seidi and Birata (former Darganata), 53 – Amu Darya Valley between settlement Birata and Tuyamuyun Water Reservoir, 54 – Tuyamuyun Water Reservoir, 55 – Soltansanjar Water Reservoir; VIII. Lower Amu Darya Region: 56 – Turkmen Darya (former Ilyaly Canal), 57 – Malyab Canal, 58 – Dostluk or Kolli Collector, 59 – Daryalyk Collector, 60 – Lake Kernay or Aybovur; IX. Sarykamysh Region: 61 – Lake Sarykamysh, 62 – Lake Zengibaba or Goyungyrlan, 63 – Lake Uzynshor, 64 – Lake Atabayshor, 65 – Dashoguz Collector, 66 – Akyaila flooded area, 67 – Turkmen Lake Altyn Asyr (project)

By the mid-1980s irrigation and drainage lakes of the future hydrographic network of Altyn Asyr became a prominent, if not a dominant type of wetlands of Turkmenistan. Drainage lakes turned out to be so to say ecological oases – areas of biological diversity maintenance, on the other hand, they have become involved in the socioeconomic sphere and are used for recreation, fishing, hunting, and cutting reed, etc. Preservation or loss of their social and environmental significance depends largely on their current state, that is on the stage of ecological succession, speed and direction of the main processes of succession, as well as on possible technical reclamation activities and solutions, which might predetermine environmental progress or ecological regression of these ecosystems and, therefore, maintenance or loss of their biosphere and socially useful functions.

3 Composition and Status of Wetland Birds

Of the 135 species of wetland birds of the avifauna of Turkmenistan for water bodies of the hydrographical network of Altyn Asyr 120 species are considered here, of which at the time of special accounts (2005–2011) authors registered in total 81 species, 27 species were met during the examination of water bodies during other periods, and 12 species were included based on the published data. The greatest number of species was found in Waders – 40 species (33.3%), and also in *Anseriformes* – 27 (22.5%), among which – dabbling, diving ducks, and shelducks (respectively, 10, 9, and 2 species), Geese (4 species) and Swans (2), and the remaining groups – Grebes (5 species), Pelicans and Cormorants (2 species each), Herons (9), Ibises (2), Storks (1), Flamingo (1), Cranes (3), Rails (8), Skuas (2), Gulls (10), and Terns (8).

Diversity and status of birds according to the nature of travel and their relative abundance are shown in Table 1, which demonstrates that 16 species may be met in water bodies throughout the year, that is, both during the nesting time and during seasonal flights and wintering. Nesting birds have 41 species, or one third of the registered species; migratory and wintering birds – 27 species. It should be noted that the birds are distributed in wetlands of the hydrographic network of Altyn Asyr unevenly, which depends both on the ecological characteristics of a particular bird species and on the location of the corresponding water body in relation to the transit routes of birds and its biotope conditions, which determines, in its turn, its capacity as a wetland in a given season. Figures 2, 3, 4, 5, and 6 show 30 different wetland birds.

4 Population Dynamics

Long-term changes in the population of waterfowl during wintering can be examined based on the example of Kelif Lakes as a model area of the southern half of the hydrographic network of Altyn Asyr, where the bird census has been held since

Table 1 Species diversity and status of waterbirds at water bodies of the network of Altyn Asyr

Character of stay species (shown in figures 2–31 with the same numbers)	All water bodies (except of Altyn Asyr Lake)				Altyn Asyr Lake (after filling by water)				
	Summer staging	Migrating–breeding	Migrating	Wintering	First stage – Ak-Yaila flooded areas, present state	Summer staging	Migrating–breeding	Migrating	Wintering
Little Grebe[2] – *Tachybaptus ruficollis*	X	XXX	XXX	XX	+	X	XX	XXX	X
Great Crested Grebe – *Podiceps cristatus*		XX	XXX	XXX	+		X	XXX	X
Red-necked Grebe – *Podiceps grisegena*			X	X	+			X	X
Horned Grebe – *Podiceps auritus*			X	X				X	X
Black-necked Grebe – *Podiceps nigricollis*	X	X	XXX	XX	+	X	X	XX	X
Great White Pelican[3] – *Pelecanus onocrotalus*		X	XX	X				X	
Dalmatian Pelican – *Pelecanus crispus*		X	XX	XX				X	
Great Cormorant[3] – *Phalacrocorax carbo*	XXX	XXX	XXXX	XXXX	+	XX	XXX	XXX	XX
Pygmy Cormorant – *Phalacrocorax pygmaeus*		X	XX	XX			X	X	
Eurasian Bittern[4] – *Botaurus stellaris*		XX	XXXX	XXX			XX	XXXX	XXX
Little Bittern – *Ixobrychus minutus*		XX	XXXX	X	+			XX	
Black-crowned Night Heron – *Nycticorax nycticorax*	X	XX	XXX		+	X	X	XX	

(continued)

Table 1 (continued)

Character of stay species (shown in figures 2–31 with the same numbers)	All water bodies (except of Altyn Asyr Lake)				Altyn Asyr Lake (after filling by water)				
	Summer staging	Migrating–breeding	Migrating	Wintering	First stage – Ak-Yaila flooded areas, present state	Summer staging	Migrating–breeding	Migrating	Wintering
Squacco Heron – *Ardeola ralloides*	X	X	X						
Cattle Egret – *Bubulcus ibis*	X	X	X						
Great Egret[5] – *Ardea alba*	XXX	XX	XXXX	XXX	+	XX	XX	XXX	X
Little Egret – *Egretta garzetta*	X	XXX	XXX	X			XX	XXX	
Grey Heron – *Ardea cinerea*	XX	XXX	XXX	XX	+	XX	XX	XXX	X
Purple Heron[6] – *Ardea purpurea*	X	XX	XX			X	X	XX	
Common Spoonbill – *Platalea leucorodia*			XX					X	
Glossy Ibis[7] – *Plegadis falcinellus*		X	XXX				X	XX	
Black Stork – *Ciconia nigra*			XX	XX					
Greater Flamingo[8] – *Phoenicopterus roseus*				XX		X	?	XXX	
Mute Swan – *Cygnus olor*	XX	XX	XXX	XXX	?	X	X	X	XX
Whooper Swan – *Cygnus cygnus*			XX	XXX	?			XX	XX
Greater White-fronted Goose[9] – *Anser albifrons*			XXXX	XXXXX				XXX	XXX
Lesser White-fronted Goose – *Anser erythropus*			XX	XX				XX	X
Greylag Goose[10] – *Anser anser*			XXXXX	XXXXX				XXXX	XXX

Ruddy Shelduck – *Tadorna ferruginea*	XX	XXX	XXX	XXXX	+	XX	XXX	XXX	XXX
Common Shelduck – *Tadorna tadorna*	XX	XXX	XXX	XXX	+	XX	XXX	XXX	XXX
Eurasian Wigeon[11] – *Anas penelope*	X		XXXX	XXXX	+			XXX	XX
Gadwall – *Anas strepera*	XX	X	XXXX	XXX	+	X	X	XXX	XX
Eurasian Teal – *Anas crecca*	X		XXXXX	XXXXX	+	X		XXXX	XXX
Mallard – *Anas platyrhynchos*	XXX	XXX	XXXXX	XXXXX	+	XXX	XXX	XXXXX	XXX
Northern Pintail[12] – *Anas acuta*			XXXX	XX				XXXX	X
Garganey – *Anas querquedula*			XX	X				XX	
Northern Shoveler – *Anas clypeata*			XXX	XX				XXX	X
Marbled duck – *Marmaronetta angustirostris*		X	XX	XX			X	X	
Red-crested Pochard[13] – *Netta rufina*	X	XX	XXXXX	XXXXX	+	X	XX	XXXX	XXXX
Common Pochard[14] – *Aythya ferina*			XXXXX	XXXXX	+			XXXX	XXXX
Ferruginous Duck[15] – *Aythya nyroca*	X	XX	XX	XXX		X	XX	XX	XX
Tufted Duck[16]– *Aythya fuligula*			XXXX	XXXXX	+			XXXX	XXXX
Greater Scaup – *Aythya marila*			XX	XX				XX	XX
Common Goldeneye – *Bucephala clangula*			XXX	XXX				XXX	XX
Smew – *Mergellus albellus*			XXX	XXX				XXX	XX
Red-breasted Merganser – *Mergus serrator*			XX	XXX				XX	XX

(continued)

Table 1 (continued)

Character of stay species (shown in figures 2–31 with the same numbers)	All water bodies (except of Altyn Asyr Lake)				Altyn Asyr Lake (after filling by water)				
	Summer staging	Migrating–breeding	Migrating	Wintering	First stage – Ak-Yaila flooded areas, present state	Summer staging	Migrating–breeding	Migrating	Wintering
Common Merganser – *Mergus merganser*			XXX	XXX	+			XXX	XX
White-headed Duck[17] – *Oxyura leucocephala*			XX	XX				XX	X
Common Crane – *Grus grus*			XXXXX	XXX				XXXX	
Water Rail – *Rallus aquaticus*			XXXX	XXX	+			XXXX	XX
Little Crake – *Porzana parva*	X	X	XX			X	X	XX	
Baillon's Crake – *Porzana pusilla*	X	X	X			X	X	X	
Common Crake – *Porzana porzana*	X		XX			X		XX	
Corn Crake – *Crex crex*			XX					XX	
Common Moorhen[18] – *Gallinula chloropus*	X	XXX	XXXX	XXX	+	X	XXX	XXXX	XX
Purple Swamphen – *Porphyrio porphyrio*		XXX	XXX	XX			XX	X	
Eurasian Coot[19,20] – *Fulica atra*	XX	XXX	XXXXX	XXXXX	+	XX	XXX	XXXXX	XXXX
Stone Curlew[21] – *Burhinus oedicnemus*		XXX	XXXX		+		XXX	XXXX	
Little Ringed Plover[22] – *Charadrius dubius*	XX	XX	XXX		+	XX	XX	XXX	
Ringed Plover – *Charadrius hiaticula*		XXX	XXX				XXX	XXX	

Kentish Plover – *Charadrius alexandrinus*	XX	XXX	XXXX	XXX	+	XX	XXX	XXXX	XXX
Greater Sand Plove – *Charadrius leschenaultia*	XX	X	XX		+	XX	X	XX	
Caspian Plover – *Charadrius asiaticus*	X		XX			X		XX	
Eurasian Golden Plover – *Pluvialis apricaria*			X					X	
Grey Plover – *Pluvialis squatarola*			XX	XX				XX	XX
White-tailed Lapwing[23] – *Vanellus leucurus*			XXXX	X	+			XXXX	
Red-wattled Lapwing – *Vanellus indicus*		XX							
Northern Lapwing – *Vanellus vanellus*			XXXX	XXX				XXX	X
Ruddy Turnstone – *Arenaria interpres*		X	XX				X	XX	
Black-winged Stilt[24] – *Himantopus himantopus*		XXXX	XXXX		+		XXX	XXXX	
Pied Avocet – *Recurvirostra avosetta*			XXX					XXX	
Eurasian Oystercatcher – *Haematopus ostralegus*		XX	XXX				XX	XXX	
Little Stint[25] – *Calidris minuta*			XX					XX	
Temminck's Stint – *Calidris temminckii*			XX					XX	
Curlew Sandpiper – *Calidris ferruginea*			XXX					XX	
Dunlin – *Calidris alpina*		XXXX	XXXX		+		XXX	XXXX	

(continued)

Table 1 (continued)

Character of stay species (shown in figures 2–31 with the same numbers)	All water bodies (except of Altyn Asyr Lake)				Altyn Asyr Lake (after filling by water)				
	Summer staging	Migrating–breeding	Migrating	Wintering	First stage – Ak-Yaila flooded areas, present state	Summer staging	Migrating–breeding	Migrating	Wintering
Broad-billed Sandpiper – *Limicola falcinellus*		X					X		
Ruff – *Philomachus pugnax*	XX		XXXX		+	XX		XXXX	
Jack Snipe – *Lymnocryptes minimus*			XXX					XXX	
Common Snipe – *Gallinago gallinago*			XXX	XX				XX	
Eurasian Woodcock – *Scolopax rusticola*			XX	XXX				XX	
Black-tailed Godwit – *Limosa limosa*			XXX	XX				XXX	X
Whimbrel – *Numenius phaeopus*			XXX	XX				XXX	
Eurasian Curlew – *Numenius arquata*			XX	XX				XX	
Spotted Redshank – *Tringa erythropus*			XX	X				XX	X
Common Redshank[26] – *Tringa totanus*		XX	XXXX	XXXX	+		XX	XXXX	XX
Marsh Sandpiper – *Tringa stagnatilis*			XXX	XX				XXX	X
Common Greenshank[27] – *Tringa nebularia*			XX	X				XX	X
Green Sandpiper – *Tringa ochropus*			XXX	XXX				XXX	XX

Wood Sandpiper – *Tringa glareola*			XXXX		+			XXXX	
Terek Sandpiper – *Xenus cinereus*			XX					XX	
Common Sandpiper – *Tringa hypoleucos*			XX	X				XX	
Red-necked Phalarope[28] – *Phalaropus lobatus*			XXXXX		+			XXXXX	
Collared Pratincole – *Glareola pratincola*		XX	XXX				X	XXX	
Black-winged Pratincole – *Glareola nordmanni*			X					X	
Pallas's Gull[29] – *Larus ichthyaetus*	X	XX	XXX	XX		X	XX	XXX	
Little Gull – *Larus minutus*			XX	XX				XX	
Black-headed Gull[30] – *Larus ridibundus*	XX		XXXXX	XXXX	+	XX		XXXXX	XX
Slender-billed Gull – *Larus genei*		XX	XXX	X			XX	XXX	
Mew Gull – *Larus canus*			XXX	XXX				XXX	XX
Caspian Gull – *Larus cachinnans*		XXX	XXXX	XXXX	+		XXX	XXXX	XXX
Gull-billed Tern – *Sterna nilotica*		XX	XX				X	XX	
Sandwich Tern – *Sterna sandvicensis*		XXX	XXX				XXX	XXX	
Caspian Tern[31] – *Sterna caspia*		XX	XX				XX	XX	
Common Tern – *Sterna hirundo*		XXXX	XXXXX		+		XXXX	XXXX	

(continued)

Table 1 (continued)

Character of stay species (shown in figures 2–31 with the same numbers)	All water bodies (except of Altyn Asyr Lake)				Altyn Asyr Lake (after filling by water)				
	Summer staging	Migrating–breeding	Migrating	Wintering	First stage – Ak-Yaila flooded areas, present state	Summer staging	Migrating–breeding	Migrating	Wintering
Little Tern – *Sterna albifrons*		XXX	XXX				XXX	XX	
Whiskered Tern – *Chlidonias hybridus*	XX		XXX			XX		XX	
Black Tern – *Chlidonias niger*	X		X			X		X	
White-winged Tern – *Chlidonias leucopterus*	XX		XX			XX		XX	

Population estimation: X – singular encounter; XX – rare; XXX – not numerous; XXXX – common; XXXXX – numerous)
Comments: 2–31 – number of figures with the corresponding bird species. Vagrant birds that are not registered at the time of the field work and made listed by the authors from the literature: Swan goose – *Anser cygnoides*, Siberian Crane – *Grus leucogeranus*, Demoiselle Crane – *Grus virgo*, Mongolian plover – *Charadrius monqolus*, Eurasian Dotterel – *Ch. morinellus*, Sanderling – *Calidris alba*, Great Snipe – *Gallinago media*, Bar-tailed Godwit – *Limosa lapponica*, Red Phalarope – *Phalaropus fulicarius*, Pomarine Jaeger – *Stercorarius pomarinus*, Parasitic Jaeger – *St. parasiticus*, Ross's Gull – *Rhodostethia rosea*

Fig. 2 (**a**) Little Grebe (photo – Suleiman Kankul). (**b**) Great White Pelicans and Great Cormorants (photo – Arazmyrat Amanov). (**c**) Eurasian Bittern (photo – Suleiman Kankul). (**d**) Great Egret (photo – Suleiman Kankul). (**e**) Purple Heron (photo – Suleiman Kankul). (**f**) Glossy Ibises (photo – Evgeny Agryzkov)

1967. From the very first it should be noted that in the 45-year period wetlands of the hydrographic network of Altyn Asyr have witnessed changes, whose main feature has been the reduction in the total number of birds, annual changes in the ratio of wintering birds species.

The total number of birds on Kelif Lakes is characterized by sharp fluctuations in different years (Fig. 7). For example, in the winter of 1973/1974 more than 450,000 birds of all species of *Anseriformes* and other wetland birds were counted, in 1983/1984 – only 2,000 birds. In all years of counts, on average, there were 53,800 birds counted. Significant fluctuations are characteristic also for the number of species (in average 14 species were observed): in the unusually cold winters of 1968/1969, 1971/1972, and 1976/1977, when these lakes and other water bodies were almost

Fig. 3 (**a**) Greater Flamingo (photo – Suleiman Kankul). (**b**) Greater White-fronted Goose (photo – Nicky Petkov). (**c**) Greylag Goose (photo – Lars Lachmann). (**d**) Eurasian Wigeon (photo – Nicky Petkov). (**e**) Northern Pintail (photo – Nicky Petkov). (**f**) Red-crested Pochard (photo – Nicky Petkov)

completely frozen, there were recorded from 5 to 9 species, in the warmer, more favorable winters of 1973/1974, 1975/1976, 1977/1978, and 2001/2002 – from 21 to 25 species.

The dominant species on the lakes are Eurasian Coot, from *Anseriformes* – Mallard and Red-crested Pochard; their share, based on the average data, is, respectively, 30.7%, 21.9% and 14.9%. Codominants (1–10%) are Eurasian Wigeon, Gadwall, Eurasian Teal, Northern Pintail, Northern Shoveler, Common Pochard, Tufted Duck. In the category of rare species (<1%) – Pelicans, Cormorants, from *Anseriformes* – Mute Swan and Whooper Swan, Greylag Goose, Ruddy Shelduck, Common Goldeneye, Common and Red-breasted Mergansers, Smew, and Lesser White-fronted Goose, Ferruginous, White-headed and Marbled Ducks.

Fig. 4 (**a**) Common Pochard (photo – Nicky Petkov). (**b**) Ferruginous Duck (photo – Nicky Petkov). (**c**) Tufted Duck (photo – Nicky Petkov). (**d**) White-headed Duck (photo – Nicky Petkov). (**e**) Common Moorhen (photo – Suleiman Kankul). (**f**) Eurasian Coot (photo – Vladislav Vasiliev)

Dynamics of the population correlation for the major groups of waterfowl has always depended on the degree of development and transformation of these lakes. For instance, in the late 1960s and the beginning of 1970s along with the stable maintenance of their level, set and controlled from the late 1950s, there was rapid development of surface and especially underwater vegetation (by the way, that's exactly the time when the issue was resolved regarding the release and growth of herbivorous fish – grass carp [*Ctenopharyngodon idella* (Valnciennes)] and silver carp (*Aristichthys nobilis* Rich) into Kelif Lakes, and, as a consequence, phyto- and zoobenthos. The result is good food supply for wintering of the birds of the wetland complex that eat both plant and animal food. It is no coincidence that the population of birds started to increase exactly in that period. The percentage of Eurasian Coot

Fig. 5 (**a**) Flying Coots (photo – Robert Kozubov). (**b**) Stone Curlew (photo – Yashin Atajanov). (**c**) Little Ringed Plover (photo – Suleiman Kankul). (**d**) White-tailed Lapwing (photo – Suleiman Kankul). (**e**) Black-winged Stilt (photo – Suleiman Kankul). (**f**) Little Stint (photo – Suleiman Kankul)

was the highest in 1968–1972 and in the average totaled 56%, which reduced the percentage of Diving ducks (29%, of which 27% was Red-crested Pochard) and Dabbling ducks (13%), whose populations were also high. One can assume that such a pattern existed earlier, since the beginning of the 1950s when surveys were not conducted.

The mid-1980s were characterized by strong siltation of lakes, which led to their almost complete degradation and loss of their feeding value both for Eurasian Coot and for *Anseriformes*. In 1984–1988 the percentage of dabbling ducks (16%, of which 10% Mallard) and diving ducks (39%, including Red-crested Pochard – 26%) increased compared with the previous period; the percentage of Eurasian Coot decreased (to 27%). These changes occurred along with the sharp fall in the total

Fig. 6 (**a**) Common Redshank (photo – Suleiman Kankul). (**b**) Common Greenshank (photo – Suleiman Kankul). (**c**) Red-necked Phalarope (photo – Suleiman Kankul). (**d**) Pallas's Gulls (photo – Yashin Atajanov). (**e**) Black-headed Gull in winter plumage (photo – Suleiman Kankul). (**f**) Caspian Tern Colony (photo – Yashin Atajanov)

number, so we can talk only about the decrease in the percentage of Eurasian Coot and not about the actual increase in the number of Diving ducks. For the majority of wetland birds, except for Waders and, to some extent, *Ciconiiformes*, wintering conditions gradually worsened. On the other hand, the slow sedimentation led to a habitat patches, which apparently caused an increase in a variety of other birds (up to 18%).

In 2002–2006, when Kelif Lakes in the form in which they were three decades ago almost ceased to exist, the process of construction and expansion of the reservoir Zeyit was started in the southeast of Kelif Uzboy and wintering of birds had been "transferring" to it. The emerged water body is under the filling process; the banks, especially the southern ones, are being blurred, so the strip of coastal

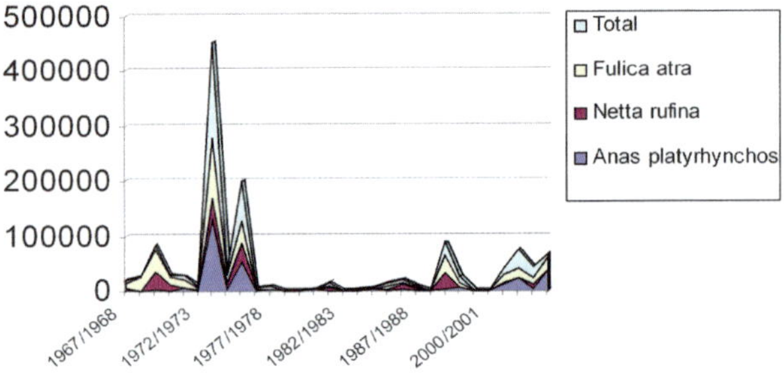

Fig. 7 Long-term dynamics of the population of common species of wetland birds on Kelif Lakes

vegetation in many places has not been formed yet. In recent years, the total number of birds here began to increase significantly, despite the disturbing factor due to randomly developing fisheries. The ratio of groups of species has also changed. The percentage of Eurasian Coot does not exceed, on average, 34%; the percentage of dabbling ducks (41%) has increased as never before, especially that of Mallard (36%), however the percentage of diving ducks has fallen to the minimum (20%); among diving ducks the percentage of Red-crested Pochard and that of Common Pochard total 5% and 10%, respectively. The water body still doesn't have sufficient underwater plant matter, phyto- and zoobenthos. The process of wintering formation at Zeit Water Reservoir is most likely to be similar to the process of filling the depressions of Kelif Uzboy and wintering formation there in the 1950s, but then there was a higher number of birds of the wetland complex.

The dynamics of birds' population in these lakes, apparently, is typical also of other wintering areas of the southern half of the hydrographic network of Altyn Asyr, and to a large extent correlates with the cycles of population changes (of course, with different quantitative figures) of waterbirds in the nesting habitats along with the decadal climate change in Central Asia. Thus, according to Krivenko [19], at the end of the warm and dry period of the 1960s compared to the previous decade, everywhere, including in Kazakhstan (and in the south of Western Siberia), a relatively small population of wetland birds was determined by a general increase in aridity. In addition, at the same time a disturbing factor greatly increased, which led, among other negative factors, to the dramatic decline in the abundance of birds. However, in the next cool and moist period, with significant abundance of water, although brief (1970–1972), the number of breeding pairs in these regions increased by 50%. Apparently, therefore, their abundance in the wintering areas in Turkmenistan, in particular, on Kelif wetlands, also increased dramatically (on an average, from 22,183 to 119,213 birds). In 1973–1979 in this vast region a warm and dry period was again developing, which, coupled with anthropogenic pressure, caused in Kazakhstan and Southern Siberia sharp regression of lake systems and a reduction in the population of nesting birds from 1.8 to 0.9 million pairs [19].

This also coincides with a double reduction in their abundance also on Kelif Lakes (on an average, from 119,213 to 62,182 birds). Then – again a cool-wet phase of 1980–2005, which lasted up to the present time [20]. Along with the increasing irrigation there was a noticeable increase in the population of a number of species in the nesting field, such as Greylag Goose, Eurasian Coot, Ruddy Shelduck, Common Shelduck, and Mute Swan [19]. However, the growing impact of the anthropogenic factor suppressed the natural tendency of the population growth [21, 22], which had a negative effect on the population of waterbirds, particularly *Ciconiiformes,* and it continued to decrease. Apparently it is no coincidence that in Kelif wintering areas their population in the second half of the 1980s fell by more than six times (on an average, from 62,182 to 9,528 birds). However, the number of birds on these lakes later stabilized and started to increase, by also six times – on an average, from 9,528 to 61,325 birds. Still, based on the average performance, this figure is twice less than it used to be in the first half of the 1970s.

5 The Role of the Lake as a Promising Wetland of the Hydrographic Network of Altyn Asyr and as One of the Important Birds' Areas of Turkmenistan

The Karashor Depression, reserved for the Turkmen Lake Altyn Asyr, more specifically, its northern and eastern parts, is not yet a wetland in the truest sense of the word. However, it is considered one of the Important Birds' Areas (IBAs) not only of Turkmenistan but also of Central Asia [23]. Among 50 IBAs of Turkmenistan wetlands total 32 (64%), of which 9 (18%) are coast-marine and 23 (46%) are part of the hydrographical network of Altyn Asyr. Three wetland IBAs (Fig. 1) are located most closely to Altyn Asyr, because they are situated in the area of a single flyway of wetland birds. This is an IBA, which covers the ancient riverbed Uzboy to the southwest of Karashor (see Fig. 1) and two IBAs to the northeast – Sarykamysh and Zengibaba (or Goyungyrlan).

Currently, in the IBA Karashor we have found: resident birds – 10 species, migrating–breeding – 20, migrating–wintering – 8, migrating – 85. Since the Karashor Depression has not been filled yet, there prevails a complex of plain-chink species, of the so-called desert type of avifauna. Among resident birds, in the nesting areas there can be observed: Saker Falcon, Golden Eagle, Long-legged Buzzard, Chukar, Rock Pigeon, Eurasian Eagle-Owl, Little Owl, Brown-necked Raven, Streaked Scrub-Warbler, etc.; among migratory-nesting species – Egyptian Vulture, Short-toed Snake-Eagle, Common Kestrel, Greater Sand Plover, Alpine Swift, Finsch's Wheatear, Eurasian Hoopoe, etc. All of these are associated with chinks and ravines of the eastern and northern parts of Karashor, except for Greater Sand Plover, which is found on the outskirts of the solonchak and on the takyrs at the western border of the depression. After filling of the depression a complex of waterbirds should be formed (see below).

On Sarykamysh there is already a formed complex of birds, numbering in total more than 250 species. Among them there are more than 100 species of wetland birds, including 24 species of *Anseriformes* [16], and their total number currently is more than 20,000 in all seasons, except for unusually cold winters when Sarykamysh gets frozen. Naturally, the proportion of the number of different species changes. Among endangered species – Dalmatian Pelican nests and stops during migration, and Ferruginous Duck also migrates. Moreover, the number of some species under IUCN/Birdlife International criteria exceeds the 1% level of their biogeographic populations: Dalmatian Pelican and Great White Pelican, Great Cormorant and Pygmy Cormorant, Common Pochard and Tufted Duck and Gull-billed Tern. Also there nest the following birds: Pallas's Gull and Caspian Gull. During migration the following ones dominate: Eurasian Coot, Mallard, Red-crested Pochard, Common Pochard and Tufted Duck. Among the species of codominants – Eurasian Wigeon, Gadwall, Eurasian Teal, Northern Pintail, Northern Shoveler, etc. On the neighboring in the east and west chinks of South Ustyurt following predatory birds nest: Golden Eagle, Common Kestrel and, probably, Lesser Kestrel; Saker Falcon and others; during migration and wintering there can be observed White-tailed Eagle, Steppe Eagle, and Eastern Imperial Eagle.

Lake Zengibaba (Goyungyrlan), as a wetland and IBA, supports waterbirds in the warm season, because in winter the lake usually freezes. Properties of the complex of wetland species that stay here during seasonal migration are characteristic for the lake, but their population is not as big as on Sarykamysh. Basically, these are: ducks, coots, waders, gulls and some terns, of which Gull-billed Tern can nest. However, one should also pay attention to the species that normally are not characteristic for aquatic biotopes, such as Saker Falcon and Lesser Kestrel, individual pairs may nest at the southern coast of the IBA on the remnant hill of Goyungyrlan.

After filling of the Ak-Yaila Depression (from the Ak-Yaila well till the Gumsepshen well), that is, upon the final completion of the 1st phase of the project, the water will go into the Karashor Depression, and the Turkmen Lake Altyn Asyr can be regarded then as a new intrazonal habitat of birds, including not only solonchaks and gullies-chink biotopes but also wetland biotopes. The fact is that in the Karashor Depression a solonchak there was no vegetation originally, except for rare halophyte in small shallow saline lakes in the lowest places of the most northern part of the depression. On the eastern and north-eastern borders of the solonchak and along the elevated sections of the depression, where water in the near future will not come, in particular, along the ravines Uch-agyzchay coming down from the north – from the Kaplankyr Plateau, grow sparse Halocnemum associations (*Halocnemum strobilaceum*). Along the western border of Karashor there is a strip of takyrs with communities of blue-green algae and lichens with insignificant amount of ephemera and 1-year thistle. In the higher parts of this strip grow such half-shrubs as *Anabasis salsa* and various Russian thistle (*Salsola arbuscula, Salsola rigida, Salsola gemmascens*), sometimes even shrubs of black saxaul (*Haloxylon aphyllum*). Further and higher, already in the adjacent sand edge of Uchtagankum, vegetation cover consists of psammophyte-shrub groups. It is clear that such a habitat, which Karashor is now, in terms of wetland birds is not suitable for their existence and their permanent residence. Here one can briefly observe only small groups of shorebirds during their migrations.

In the long term the ecological situation and stability of the Turkmen Lake Altyn Asyr as a wetland ecosystem and habitat of wetland birds, of course, will largely depend on its weediness by underwater and coastal hydrophilous plants (Fig. 8). It is fairly safe to say that the development of hydrophilous vegetation and associated fauna, and the dynamics of waterfowl on the lake will correspond to the "Sarykamysh type". At depths up to 0.5 m, first grow various Pondweeds (*Potamogeton* sp.), at depths up to 1.5-m-thick reeds of Common reed (*Phragmites australis*), which will take most of the coastal strip, especially along the western shore. In addition to reed, in shallow water somewhere will grow Coon's tail (*Ceratophyllum demersum*), cane species of Club-rush (*Scirpus* sp.) and Cattail (*Typha* sp.). Depending on the extent and dynamics of water salinity one may predict the development of the algae: Stoneworts (*Charophyta*), as well as Spike Watermilfoil (*Myriophyllum spicatum*), Bladderwort (*Utricularia* sp.) and Duckweed (*Lemna* sp.).

The abundance of vegetation will contribute to the great diversity and biomass of aquatic invertebrates, which will serve as a corresponding food base for vertebrates and, above all, fish. Further, after aquatic organisms enter the lake, including fish, under corresponding conditions, self-replicating, growing in numbers populations should be forming, which over a short time will be able to provide food resources for such fish-eating birds as Gulls, Terns, Pelicans, Cormorants, to some extent, for Herons, and among prey species – for White-tailed Eagle, Eurasian Marsh Harrier and Osprey.

Currently wetland birds are seen only at the site of the 1st stage of Lake Altyn Asyr, that is, at the spills of Ak-Yaila. The number of species here is not more than 36, which is much less than in the same Sarykamysh and other water bodies of the hydrographic network of Altyn Asyr. This is despite the fact that the 1st stage is in the area of a historically formed flyway along the ancient riverbed of the West Uzboy. The water has been let flow through Ak-Yaila since the summer of 2009, but banks have not yet been covered with hydrophilous vegetation of the required thickness, which will create protection and feeding conditions for wetland birds, that is, conditions under which their diversity and population will be increasing.

Of course, it all depends on the actual volume of waste water inflow, based on which we can predict the water-salt balance of Lake Altyn Asyr. This is, in general, will determine the development of the flora and fauna of the lake and the surrounding areas, particularly vegetation, and animal populations, that depend on the vegetation, as already noted, fish, fish-eating and other wetland birds that are at the top of the emerging environmental (food) chain.

In the fauna of Turkmenistan there were as a whole revealed: 418 species of birds, of which 135 are wetland birds, including 33 *Anseriformes* [25], 27 of them are listed for the entire hydrographic network of Altyn Asyr. In fact, the total number of the birds at Lake Altyn Asyr, after its stabilization, can exceed 230 species; 116 will comprise the complex of wetland birds, including 46 nesting species (Table 1). According to our forecasts avifauna will be formed in the first place, on the basis of the birds species listed below. Mallard currently inhabits all the lakes of northern Turkmenistan, and it is more frequently observed during surveys at Lake Sarykamysh, so it will be one of the first to inhabit Lake Altyn

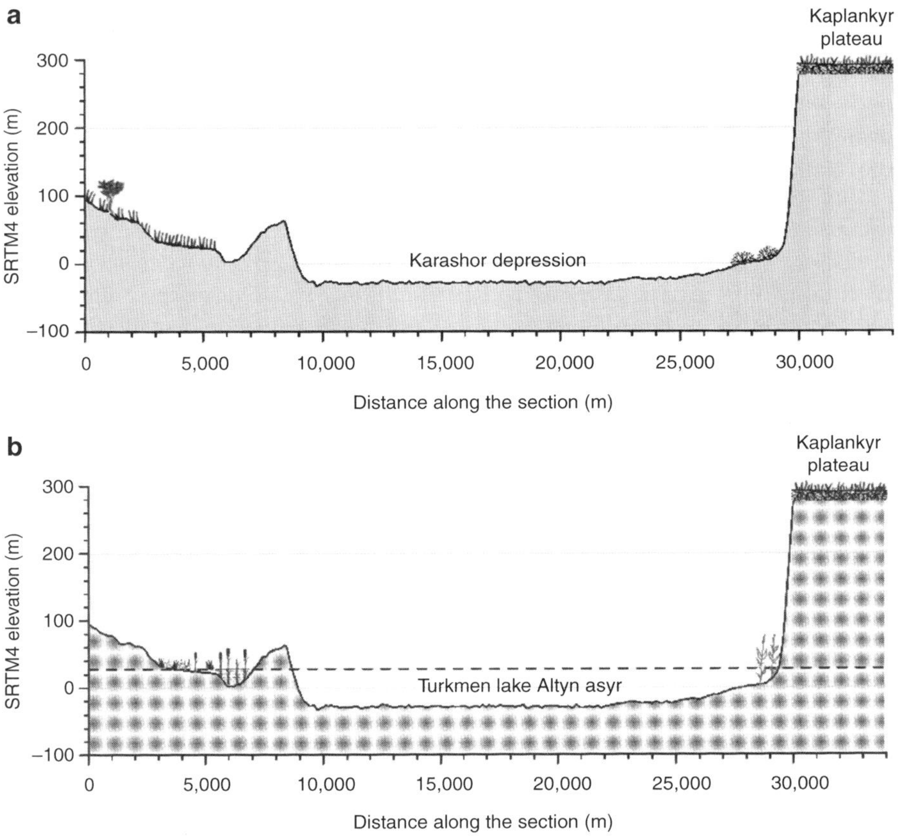

Fig. 8 The scheme of the profile of the Karashor Depression for the width of the Goklengui well before (**a**) and after (**b**) its filling

Asyr. But the number of mallards at nesting sites will be small due to the inability of the appearance of islands while flooding, except for one, 2 km east of the present Geklengui well as the rest of the bottom Karashor is flat; such islands could serve as a safe nesting place not only for Mallard, but also for other species of wetland birds. Eurasian Teal will be a migratory and wintering bird. It seems that Gadwall will become not a migratory, but also a nesting and even wintering bird. The future status of Eurasian Wigeon – just a migratory and wintering bird; Northern Pintail will be quite rare during migration, and Northern Shoveler will be a migratory and wintering bird. Red-crested Pochard in future will be a migratory and nesting bird and numerous in winter. Common Pochard will be seen both during migration and in winter. Ferruginous Duck in northern Turkmenistan is already now a rare migratory species; Tufted Duck is a migratory bird and stay for winter here, and will certainly be more numerous than Ferruginous Duck. Common Goldeneye, as well as on Sarykamysh, will be rare migratory and wintering bird. Smew in northern

Turkmenistan, as well as Common Merganser, migrates through and stays for winter regularly, but in small numbers, Red-breasted Merganser – even more rarely, only during migration. Ruddy Shelduck and Common Shelduck will no doubt nest, or maybe they already nest there; there are quite suitable nesting biotopes at Uzboy and at Karashor. Greylag Goose will not be numerous in winter and will be seen only during migration, as in the vicinity of the lake there will not be, at least in the near future, arable land for grain crops and fallow. Mute Swan will be a wintering species, and with appropriate development of the sloughs and the density of the reed will be able to nest. Whooper Swan will spend the winter with mute swans, but in small numbers. In addition, more or less common in winter and during migration will be Grebes, Pygmy and Great Cormorants, Grey Heron and Great Egret; among waders – Common Redshank, Black-winged Stilt and Avocet, Gulls and Terns. On the underflooded shallow areas there will be likely to be met Greater Flamingo, it is possible that this species will try to nest, as well as Great White Pelican and Dalmatian Pelican.

Of the birds that do not form the wetland complex, there will remain species inhabiting chinks (see above), and besides from other biotopes there will come birds that are not common for the wetland complex, particularly Common Pheasant, and synanthropic ones – Indian Myna, Eurasian Magpie, Eurasian Tree Sparrow, Laughing Dove and Eurasian Collared Dove and others.

Thus, in the future, the water surface of Lake Altyn Asyr, and most importantly – its dense riparian vegetation, should become a shelter for a lot of migratory and nesting birds and in warm winters – wintering wetland birds. This optimal picture can exist then development of vegetation will occur on the coastal stretches where the water line will be above the salt brine; it means that it would reach solid soil areas on the north-western edge of the depression, for example, in the well Geklengui, or water will flood at least low-lying parts of gullies and ravines of the northern and north-eastern chinks. The water level in the Lake Altyn Asyr will reach a point of about +28 m (absolute sea level), thus the maximum depth will reach 58 m.

It will not be crucial how much, in this case, the surface of the water body will be, because for a concentration of 20,000 wetland birds (minimum figure according to the IUCN/Birdlife International criteria that is needed for recognition of the international significance of a wetland as IBA) during their seasonal migrations (the water reservoir is located on one of the most important migration routes) you need not necessarily a certain surface of the water body (although it should not be $<40\ km^2$), but its food resources and resistance to adverse weather conditions. The water table and the growth of the reservoir, of course, will depend on evaporation, direct precipitation and on the associated amount of temporary drains (for example, from the Kaplankyr Plateau along the dry riverbed Uch-agyzchay) into the basin with heavy rains, as well as on water filtration and its absorption by salt brine etc. The authors were not supposed to predict the actual water balance, that is why it is still difficult to say how the ecosystem of this wetland in terms of time will be developing. Mineralization of the accumulated in the lake waters should not exceed 10–12 g/l, the excess will have a negative impact on fish production [24] and, in general, on biological productivity of Altyn Asyr, for example, as it is the case in Lake Sarykamysh, where in May 2009 we observed concentration of fish-eating and other wetland birds mainly in the near-mouth area and at the very mouth of Daryalyk at the eastern bank of Sarykamysh, but not at the opposite south-western banks, which indicates that large fish capacity remains in places where relatively "fresh", not strongly mineralized waters from Daryalyk penetrate.

After putting in operation Lake Altyn Asyr in northern Turkmenistan, along with wetlands Sarykamysh and Zengibaba (Goyungyrlan), in the next 10 years one more vast wetland will appear on the migratory route of the general direction from south-west to north-east, that connects the south-eastern Caspian Region through Uzboy corridor with delta of the Amu Darya and the Aral (with what is left of it). It should be noted that with the laying of the northern Dashoguz Collector two more small water bodies were formed – Uzynshor and Atabayshor (Fig. 1), the auxiliary role of which on the migratory route will also be positive. The branched location of the channels with water bodies in the desert not only in northern Turkmenistan, but also in its whole plain part has a positive effect on the migration situation and distribution of both wetland birds and terrestrial (with rare exceptions) birds. It seems essential to constantly monitor the environment and birds' population at the appropriate stages of flooding and the development of the Turkmen Lake Altyn Asyr.

6 Conclusions

1. The main purpose and objective of the Turkmen Main Collector of the Turkmen Lake Altyn Asyr is to collect and use, if possible, the whole collector-drainage water of Turkmenistan.
2. Favorable climate conditions on the plains of Turkmenistan and, as a consequence, good food supply, are the main prerequisites for the formation of wintering clusters of waterbirds on inland wetlands. However, the general lack of water resources on the plains of Turkmenistan in the beginning and in the middle of the twentieth century conditioned a small concentration of waterbirds on wetlands, especially in the winter period.
3. Large-scale construction, covering almost the entire flat part of the country, has led to redistribution of huge volumes of water and the formation of an artificial hydrographic network, one of the final stages of which is the Turkmen Lake Altyn Asyr. The authors propose the term of the "hydrographic network of Altyn Asyr", which unites 67 wetlands of 9 areas (basins).
4. Of 135 species of waterbirds of Turkmenistan, for wetlands of the hydrographic network of Altyn Asyr 120 species have been cited. The greatest number of species of waders – 40 species (33.3%), *Anseriformes* – 27 (22.5%), Grebes (5 species), Pelicans and Cormorants (2 species each), Herons (9), Ibises (2) and Storks (1), Flamingo (1), Cranes (3), Rails (8), Skuas (2), Gulls (10), and Terns (8).

On Lake Altyn Asyr after its filling and stabilization the total number of birds can exceed 230 species, of which 116 will be waterbirds, 46 of which may be nesting birds.

5. The dynamics of the population of waterbirds of the "hydrographic network of Altyn Asyr" to a large extent will be correlated with cycles of their population changes (of course, with different quantitative figures) in the nesting area along with interdecadal climate change in North and Central Asia.
6. The development of Lake Altyn Asyr as a wetland ecosystem will be determined by the volume of waste water inflow, based on which the water-salt balance of the lake will be forming, which in its turn will cause the biodiversity development of Lake Altyn Asyr, in the first place, vegetation and hydrobionts, on which will depend the composition and abundance of waterbirds that are at the top of the emerging food chain of the lake ecosystem.
7. Further forecasts of the development and stabilization of the Turkmen Lake Altyn Asyr as an ecosystem will depend on the timely and high-quality space and ground-based monitoring and control of the natural environment of this region, its biological component, in particular birds and their population dynamics.

References

1. Laptev MK, Sulima VL, Freiberg LR (1934) All-Union wildfowl reserve in Gazan-Kuli, Turkmenistan. Izvestiya of the Turkmen interministerial committee on nature protection and development of natural resources, #1. Ashgabat, pp 3–119
2. Isakov YA (1940) Ecology of the wintering of waterbirds in the South Caspian region. In: Proceedings of the All-Union wildfowl reserve Hasan-Kuli, issue 1, Moscow, pp 160–238
3. Isakov YA, Vorobiev KA (1940) Review of wintering sites and migration of birds in the South Caspian region. In: Proceedings of the All-Union wildfowl reserve Hasan-Kuli, issue 1, Moscow, pp 3–159
4. Mikheev AV (1970) On wintering of waterbirds on artificial water bodies of Central Asia. Scientific records of the V.I. Lenin Moscow Pedagogical Institute, #272, pp 168–174
5. Rustamov EA (1992) Wintering of waterbirds in Turkmenistan: sites of concentration, zone division, composition and structure of the population. Biol Sci (Vysshaia shkola, Moscow) 3(339):50–62
6. Rustamov EA (2004) The current state, perspectives and principles of the organization of wintering counting of wetland birds in Central Asia (on the example of Turkmenistan). Biosphere territory "Ysyk-Kyl", issue 2, Bishkek, pp 25–30
7. Rustamov EA, Poslavsky AN, Karavaev AA, Myatlev VD, Chernov VY (1990) Geography, ecology and protection of wintering areas of waterbirds of Turkmenistan. In: Nature protection in Turkmenistan, issue 8. Ashgabat, pp 56–100
8. Rustamov EA, Khakyev A (1978) On wintering of waterbirds on Kelif Lakes. Izvestiya of the Turkmen Academy of Sciences. Ser Biol Sci 4:96–99
9. Khakyev A (1975) State of the population of wintering waterbirds in the area of the V.I. Lenin Karakum Canal. In: Proceedings of the All-Union conference on migration of birds, Part 1, Moscow, pp 98–99
10. Eminov A (1972) Population and location of waterbirds wintering at the Karakum Canal. In: Resources of waterbirds in the USSR, their reproduction and usage, Part 2. Moscow State University, Moscow, pp 85–86

11. Vasiliev VI, Karavaev AA, Rustamov EA (1984) Distribution and population of waterbirds in wintering areas of Turkmenistan. In: Ecological and fauna-related aspects of studying animals of Turkmenistan. Ashgabat, pp 92–132
12. Rustamov AK (2011) Wildlife of Turkmenistan and its protection (on the example of vertebrates). Ylym, Ashgabat, pp 1–246
13. Dementiev GP (1952) Birds of Turkmenistan, vol 1. Turkmen Academy of Sciences, Ashgabat, pp 3–546
14. Rustamov AK (1975) Migrations of birds due to anthropogenic impact on arid lands. In: Proceedings of the All-Union conference on migration of birds, Part 1, pp 87–89
15. Velikanov VP, Khokhlov AN (1979) On avifauna and particularities of the biology of waterbirds and near-water birds of Lake Sarykamysh. Environment and birds of the coasts of the Caspian Sea and adjacent lowlands. In: Proceedings of the Kyzyl-Agach state nature reserve, issue 1, Baku, Azgosizdat, pp 236–240
16. Amanov A (2012) Anseriformes Birds of the North-Eastern zone of the Caspian sea. Caspian Ecological Forum. Türkmen Döwlet Neşirýat Gullugy, Aşgabat, p 15
17. Rustamov AK (1948) On the current state of the Sarykamysh depression. Rep Acad Sci USSR 60(8):1449–1451
18. Turkmen Lake "Altyn asyr" (2010) Türkmen Döwlet Neşirýat Gullugy, Aşgabat, pp 7–99
19. Krivenko VG (1991) Waterbirds and their protection. Agropromizdat, Moscow, pp 1–271
20. Krivenko VG, Vinogradov VG (2008) Birds of the aquatic environment and rhythms of the climate of Northern Eurasia. Nauka, Moscow, p 588
21. Azarov VI (1984) The population of waterbirds in nesting sites and its changes on the lakes of Tobol-Ishim forest-steppe. In: The current state of resources of waterbirds, Moscow, pp 119–121
22. Auezov EM, Vinogradov VG (1986) Summer counting of waterbirds on the water bodies of Turgay depression. In: Studying of birds of the USSR, their protection and rational usage – Abstracts, IX All-Union ornithology conference, Part 1, Leningrad, pp 46–47
23. Rustamov EA, Welch GR, Brombacher M (eds) (2009) Important bird areas in Turkmenistan. Priority sites for conservation. MNPT, Ashgabat, pp 9–198
24. Salnikov VB (2010) Perspectives of the usage of the Turkmen Lake for fishery. Role of the Turkmen Lake Altyn Asyr in the improvement of the ecological situation of the region. In: Proceedings of the international scientific conference, TDNG, Ashgabat, 24–25 March 2010, pp 217–219
25. Rustamov EA (ed) (2013) Birds of Turkmenistan: field illustrated indicator. UNDP, Ashgabat, pp 3–680 (to appear)

International Cooperation of Turkmenistan in the Water Sector

Jon Marco Church

Abstract The aim of this chapter is to provide an overview of cooperation of Turkmenistan with neighboring countries, donor countries, as well as international organizations, including financial institutions. Its core is an analysis of the major drivers of cooperation and an overview about the different types of interactions and relations between Turkmenistan and its international partners. This is not an attempt to evaluate the quality or quantity of Turkmen initiatives or actions and no recommendation was produced. This is an effort to systematize information that is available to the public and to reflect on the experience of the author working in the country and region on water issues.

Keywords Bilateral, International, Multilateral, Turkmenistan, Water

Contents

J.M. Church (✉)
University of Reims, International Research Center on Sustainability, 57 bis rue Pierre Taittinger, 51096 Reims Cedex, France
e-mail: jon-marco.church@univ-reims.fr

I.S. Zonn and A.G. Kostianoy (eds.), *The Turkmen Lake Altyn Asyr and Water Resources in Turkmenistan*, Hdb Env Chem (2014) 28: 291–310, DOI 10.1007/698_2012_173,

1 Introduction

"Putting forward new proposals and initiatives on environmental issues, Turkmenistan stands ready for intensifying and promoting positive cooperation on global scale." These were the words that closed President Berdimuhamedov's message to the participants in the conference "Environmental Cooperation of Turkmenistan with Major International Organizations: Achievements and Success," held in Ashgabat on November 21–22, 2011. One of the objectives of that conference was to discuss the proposal to establish a Caspian Environment Council and to create a Regional Center for Climate Change in Ashgabat. I do not know how far these initiatives went, but what I know for sure is that the meeting promoted positive cooperation for this book, particularly for this chapter. My presentation at the conference focused more on cooperation between the United Nations Economic Commission for Europe (UNECE) and Turkmenistan in the water sector. This contribution will however go beyond this, starting from a short description of the situation in the four major transboundary water bodies, followed by a brief historical perusal of the last twenty years.

The aim of this chapter is to provide an overview of cooperation of Turkmenistan with neighboring countries, donor countries, as well as international organizations, including financial institutions. Its core will therefore be an analysis of the major drivers of cooperation, what Peter Haas called "influencing factors" [1], together with an overview about the different types of interactions and relations between Turkmenistan and its international partners. It goes without saying that this is not an attempt to evaluate the quality or quantity of Turkmen initiatives or actions and that no recommendation will be produced. This is an effort to systematize information that is available to the public. Moreover, this has nothing to do with the chapter dedicated to international cooperation in the Environmental Performance Review of Turkmenistan and its final recommendations that have recently been adopted by the UNECE Committee on Environmental Policy [2].

Few scientific papers have been published specifically on this topic in English [3–6], which is one of the reasons that persuaded me to accept the invitation of the editors of this book. So far, most scholarly research and development cooperation has focused on the regional scale, especially on the Aral Sea [7–10]. This is due to the relative difficulty in obtaining firsthand information about the situation in the country, particularly about strategic issues such as this. The general feeling is however that the situation is improving and this text is a demonstration thereof. The sources used for this analysis are, first of all, official documents and publications by the Turkmen government [11–14] and by international organizations [15–21], some of which are available online. Their interpretation relies heavily on the author's familiarity with the issue and the country, having specialized in regional environmental cooperation and having served the UN in Turkmenistan.

2 Transboundary Waters

As it can be observed from the map below, there are four transboundary water bodies in Turkmenistan, shared with a total of eight countries, based on hydrological boundaries:

1. Amu Darya–Sarygamysh Lake–Aral Sea basin (Afghanistan–Tajikistan–Uzbekistan–Kyrgyzstan–Kazakhstan)
2. Murgab river basin (Afghanistan)
3. Tejen river basin (Afghanistan–Iran)
4. Atrek–Caspian Sea basin (Azerbaijan–Iran–Kazakhstan–Russia)

Source: UNECE (2011) Second assessment of transboundary rivers, lakes and groundwaters

The Aral Sea basin is the largest catchment area in Central Asia and one of the largest closed water systems in the world [7–9]. Its main effluents are on one hand the Amu Darya, which is the most important river in Central Asia and flows from Afghanistan and Tajikistan to Turkmenistan and Uzbekistan, and on the other hand the Syr Darya, which flows from Kyrgyzstan to Uzbekistan, Tajikistan, and

Kazakhstan. With regard to the Amu Darya, a small but significant amount of water originates from Afghan territory, but its exploitation is currently very low because of the war and of the socioeconomic situation in the country. Most of the water is generated instead on Tajik territory, but the country uses only a small proportion of it. Hoping to achieve energy and food security, Tajikistan, which is the poorest among former Soviet republics and has recently experienced a civil war, is investing heavily in the development of hydropower production and of irrigated land. This worries Uzbekistan, which apparently fears at the same time water scarcity and flooding due to dam failure. This could have negative effects on its cotton fields and industry, which is particularly demanding of water. In Turkmenistan, the Karakum canal brings water from the Amu Darya all the way to Ashgabat and beyond; the Altyn Asyr lake is being filled by drainage waters through the main drainage canal of the Golden Age that runs across the country from Turkmenabat and through the Karakum desert; water is also brought to the depression in the northwestern part of the country, significantly extending the river basin.

In general, all downstream countries are greatly concerned by the overexploitation of water resources, which is causing the disappearance of the Aral Sea and which is having serious consequences for the livelihoods of millions of Kazakhs and Uzbeks. As the UN Secretary General put it after visiting the area in 2010: "During my flight over the Aral Sea, from Uzbekistan, I was particularly shocked by what I saw. A sea that was once the fourth largest inland bodies of water in the world has shrunk by nearly 90 percent [. . .]. Clearly, this is a collective problem requiring collective effort – not just from regional leaders, but the entire international community."[1] All countries understand that international cooperation is needed to support efforts at the global, regional, national, and local scale to mitigate the causes and to adapt to the consequences of the so-called "tragedy" of the Aral Sea. It must be noted, however, that the Aral Sea is not the only transboundary lake in the area. There is also the large and shallow Sarygamysh Lake, which finds itself in a depression between Turkmenistan and Uzbekistan and which consists mainly of agriculture drainage waters originating from the Amu Darya. For this reason, its level and the quality of its water is also an issue of concern for the two countries.

Given that drainage water for the Altyn Asyr lake originates from the Amu Darya, this contribution focuses mainly on this catchment area. In order to fully understand the geopolitical situation, it is however fundamental to have a clear picture of all transboundary waters of Turkmenistan, starting from the Murgab river basin. From the mountains of Afghanistan, the river extends itself to the Turkmen city of Mary, where it mixes with the Karakum canal and north of which it ends up in the desert. The Tejen represents another significant transboundary river for Turkmenistan. It also originates from the reliefs of Afghanistan, flows westward to Herat and northward along the border with Iran, defining it, before disappearing in the Karakum desert. In 2004, Iran and Turkmenistan inaugurated the Doosti dam,

[1] Quoted from his briefing to the Security Council of April 15, 2010, on the Secretary General's visit to Central Asia.

also known as the Iran–Turkmenistan Friendship Dam, which finds itself on the Tejen river. Located on the border between the two countries and very close to the border with Afghanistan, its reservoir plays an important role in stabilizing water supply to avoid social, economic, and environmental catastrophes such as the ten month drought of 2000. Through a pumped scheme, the reservoir also provides drinking water to the distant city of Mashhad, which is the second largest city in Iran. This greatly extends the scope and import of the Tejen river basin.

Finally, the Atrek river flows from the Iranian to the Turkmen side west of the Kopet Dag mountains. With a changing riverbed, its waters are used mainly for irrigation and reach the Caspian Sea only in flood season. The latter is a transboundary water body itself, shared with Azerbaijan, Iran, Kazakhstan, and Russia. It is of great importance for Turkmenistan because of its coastal and underwater oil and natural gas reserves, because of its influence on regional climate and environmental change, and because it is the natural habitat of a Turkmen staple produce such as sturgeon, from which caviar is derived. The government is also investing a lot in tourist infrastructure through the Avaza development project. As in-depth analysis of this water body goes well beyond the economy of this chapter, the reader is invited to refer to other titles of this series by the same editors, for further information [22, 23].

3 Historical Perspective

In order to fully understand the context, it is important to at least provide an overview of the kinds of interactions experienced in the lifetime of individuals that are now at senior positions of government structures. We all know that the Soviet period was characterized by massive investment in water infrastructure and ambitious projects that greatly impacted nature, society, and the economy. Cooperation among Soviet republics was mediated and sometimes enforced by Moscow and it resulted in a system where – by greatly simplifying it – the downstream Kazakh, Turkmen, and Uzbek SSRs were providing oil and natural gas, in which they are rich, to the upstream Kyrgyz and Tajik SSRs in exchange for water. Research, surveying, and design were carried on mainly by the branch of the "Hydroproject" Institute in Tashkent, including the planning of dams and canals. This is perhaps the reason for the popular claim reported by Erika Weinthal that the Uzbeks are the "water people" or *vodniki* of Central Asia [24, 25]. A series of agreements was in place with Iran and Afghanistan to manage transboundary rivers.

By the early 1980s, the situation of the Aral Sea was catastrophic. On top of that, the Soviet invasion of Afghanistan and the Iranian revolution of 1979 greatly contributed to the destabilization of the whole region. The Soviet response to the situation was the launching of a large-scale planning effort to save the Aral Sea. In 1982, a Water Resources Master Plan for the Amu Darya and Syr Darya river basins adopted the principles of limiting water extraction per hectare of irrigated land and of sharing available water among the riparian SSRs. By the mid-1980s, detailed regulations were issued for the operationalization of these plans and two river basin organizations were created for the management of the Amu Darya and the Syr Darya, respectively. According to most witnesses, in this period, the role of Central Asian SSRs, including the Turkmen SSR, was rather passive, as major decisions were taken in Moscow.

In the early 1990s, the fall and dissolution of the Soviet Union created a vacuum. This meant the need to establish a new mechanism at least capable of mediating disputes among the newly independent states. This did not come from the Community of Independent States (CIS), but in the form of the Interstate Commission for Water Coordination (ICWC), which was created as a regional intergovernmental arrangement, where all states of the region are equally represented. The two river basin organizations were restructured as joint companies and an ICWC Scientific Information Centre (ICWC SIC) was established in Tashkent to exploit synergies with the "Hydroproject" Institute. Over time, the ICWC SIC has become a key resource for water information in Central Asia. The fall of the Soviet Union also brought along a shift from planning to programming. An International Fund for Saving the Aral Sea (IFAS) was established to finance projects to mitigate the causes and to adapt to the consequences of the situation in the Aral Sea basin. Riparian and donor countries have pledged and invested hundreds of millions of dollars through this mechanism. Without the mediating role of Moscow, some consistency was lost in transition, despite the best efforts at coordination by many partners. Moreover, the sudden absence of an authority capable of arbitrating problems and enforcing solutions meant the emergence of disputes in the long term.

For Turkmenistan as for most newly independent states, these were eventful years of hope and enthusiasm under the leadership of Saparmurat Niyazov, also known as Turkmenbashi. Few individuals knew how to run a fully independent country. Few knew exactly what they were doing: on one hand, there was the tendency of welcoming all initiatives coming from abroad; on the other hand, path dependency from Soviet structures can be observed alongside the desire to renew everything, such as in the case of the Aral Sea. As it can be noticed comparing the list of participants to international meetings and the number of treaties signed and ratified since independence, Turkmenistan was participating actively in international processes until the mid-1990s. The number of projects implemented with the assistance of donor countries, international organizations, and financial institutions was also quite high.

It was not until the decision taken in 1995 to strictly adhere to permanent neutrality in its foreign policy that Turkmenistan started to progressively withdraw from the international scene, refusing to participate in international meetings and projects and, of course, to become member of new organizations or party to new conventions, with few exceptions [26]. There are many hypotheses about this empty chair policy: some believe, for instance, that it was a reaction to overexposure and negative experiences in the early period, while others argue that it was an explicit foreign policy choice. The fact is that, for the following decade, Turkmenistan effectively closed itself to international cooperation, even refusing development aid. The only exception was a general support to the United Nations because of its universality and neutrality. A significant gesture was calling a national holiday after the Turkmen proverb "a drop of water is a grain of gold," which is still celebrated on the first Sunday of April and is an opportunity to take stock of what the country does in the water sector. Another example is the Framework Convention for the Protection of the Environment and Sustainable Development in Central Asia, which was proposed in 2006 and which includes provisions for the management of transboundary waters.[2]

Since late 2006, the arrival to power of Berdimuhamedov and his policy of reform and increasing openness brought along a new wave of hope for international cooperation. With regard to foreign policy, the reform process started from improved relations with neighboring countries, from Afghanistan, which receives humanitarian and development aid from Turkmenistan, and Iran to Kazakhstan and Uzbekistan. This represents a welcome development for improved management of transboundary waters and has already been reflected in a more active stance in IFAS, which is the only true regional arrangement that is truly functioning at present time, as ICWC and other processes are captive of either the rivalry for leadership in the region between Uzbekistan and Kazakhstan or the conflicting interests of upstream and downstream countries.

Ashgabat slightly distanced itself from Russia, considering that dependence on gas exports to Moscow was excessive, and carefully balanced its relations with all major powers, from the USA to the EU and from India to China, particularly through economic policy. The pursuit of positive neutrality is possible also because of the relative wealth of a country, which is considered medium income by global levels and which allows it to act as a donor more than a recipient country. In 2010, development aid to Turkmenistan accounted for only about 16 million USD, according to UNDP. At the multilateral level, Ashgabat increased its participation in United Nations projects and processes and relaunched relations with development banks. Under Berdimuhamedov's leadership, it now pursues the adoption of international standards. To do so, it has partnered with international institutions such as the European Bank for Reconstruction and Development (EBRD) and the UNECE, which is a standard setting organization in fields ranging from water to the environment and from housing to transport, especially under the leadership of its former head Ján Kubiš.

In this framework, Turkmenistan often volunteers to act as chair of multilateral processes and to host international conferences in the magnificent buildings completed over the last few years in Ashgabat. Moreover, because of its economic resources, it often proposes to host international centers, such as the new Regional Center for Climate Change mentioned above or the United Nations Regional Center for Preventive Diplomacy for Central Asia, which was launched in 2007 and whose presence in Ashgabat is a reflection of the opening and neutrality of the country. Water and the environment is also one of the three priority areas of the center[3] [18, 19]. While the country is still young and developing, it is too early to make a balance of foreign policy under Berdimuhamedov.

---

[2] See article 9. On November 26, 2006, the framework convention was signed only by Kyrgyzstan, Tajikistan, and Turkmenistan.

[3] See its program of actions for 2009–2011 and, most recently, for 2012–2014.

4 Main Issues

After having looked briefly at transboundary water bodies and having provided a short historical overview, we will now analyze the three major drivers of cooperation in the water sector. These are regular supply of water from upstream countries, sharing water with neighboring countries, and sufficient supply of water to the Aral Sea. These are also the main issues for the Altyn Asyr lake and, more generally, the Amu Darya basin. Other issues such as climate change or risk management, which may be of great import at the global or regional scale, but that, for the geographical configuration of the country or for other contingencies are not currently at the very top of the agenda, will also be mentioned. It can be noted that main drivers are relatively short term, while the latter issues are more long term. This tension is frequent in all kinds of decision making – not only in Turkmenistan – and is a major concern for the sustainability of any given policy. Because of the sensitive nature of the first set of issues, most international partners have no choice but to work on questions that are currently not at the top of the government agenda, while they aspire to contribute to more critical issues such as solving the problem of the Aral Sea.

The regular supply of water from upstream countries is of great import for downstream countries such as Turkmenistan and Uzbekistan. Of course, it also important that water is sufficient to meet the needs of downstream countries, but this responsibility is shared by both upstream and downstream countries and will be considered from the perspective of the whole basin. The emphasis here is on regularity because, besides natural variability, such as seasons, there is also human generated variability. This depends mostly on interventions upstream, such as the construction of a new dam or the operations of existing ones or the launching of large irrigation schemes. Of course, upstream countries can and have the right to do so and downstream countries can and have the right to be concerned about undesirable effects such as the extremes of draught and flooding. With international agreements and judicial decisions, international law provides principles, instruments, and examples of how to solve these issues and international partners are working closely with the governments of the region to achieve peaceful solutions [15, 27]. Particularly in the latter period, Turkmenistan has consistently highlighted the need to avoid confrontation, military and otherwise, which would be detrimental to all. It has insisted on the importance of reaching a "mutually beneficial" agreement for the "rational use" of water resources.

Another key issue for Ashgabat is the sharing of water with neighbors. Turkmenistan is downstream with regard to Iran and Afghanistan and is both upstream (middle part of the Amu Darya) and downstream (lower and upper part of the same river) for Uzbekistan. Here, we mean active sharing alone, i.e., the water flow that is left for downstream countries, given that passive sharing or the water that is received from upstream, has been and, at the same time, will be dealt with in the previous and following paragraph. For the Amu Darya basin, this is still regulated by the 1992 Almaty Agreement, which allocated 43% of the water drainage of the Amu Darya to Turkmenistan and which, by the way, also created the ICWC. Other issues, such as infrastructure maintenance, are regulated by a more specific agreement. Governments hold bilateral and sometimes multilateral meetings on these topics and there is regular collaboration between operators on both sides, such as in the case of the Qarshi pumping stations shared by the two countries [21]. As the country is currently under the agreed share and given that exchange with Uzbekistan is generally positive, this would be no great issue, if it was not that the country seems to be aware of the fact that, if current trends are confirmed, according to frequently quoted government sources,[4] Turkmenistan risks running out of water by 2020. This is one of the reasons why, in recent years, the government has been putting so much emphasis on saving water and, more generally, on the rational use of water. As it can be seen from the data below, this is especially pressing in the Amu Darya basin, also considering that the amount of water used by Uzbekistan is comparable.

	How much of the share agreed in 1992 of the Amu Darya is actually used by Turkmenistan?[5]	How much water of the Amu Darya is used by Turkmenistan for nonirrigation purposes?[6]
1990	76%	1%
1997	70%	2%
2010	79%	9%

There is no need, however, to wait until 2020 to be concerned about water consumption in the whole Amu Darya basin. In Soviet times, ambitious projects and irrigation practices resulted in the excessive use of the river's water, which caused in turn water supply to the Aral Sea to become insufficient. Moreover, while the 1992 Almaty Agreement provides grounds for distributional justice among the five Central Asian republics, it theoretically allows them to withdraw 100% of the water of the Amu Darya, if you sum the share of each state. This leaves nothing for the Aral Sea or, to put it differently, puts the responsibility solely in the hands of riparian states and their capacity and good will to ensure that sufficient amounts of water end up in the Aral Sea, which is clearly not the case. In this regard, Turkmenistan is often criticized in international contexts for its large and liberal consumption of water. Its majestic fountains are often cited as example, even if consumption for nonirrigation purposes is relatively small compared to the agricultural sector, including cotton, as it can be seen in the figures above. In order to solve the issue of excessive water use, IFAS has been implementing large programs funded by the countries of the region and by donors, frequently with the assistance

[4] This must be at the national scale. I am not aware of exactly how the projection was calculated.

[5] According to simple calculations of the author based on data produced by the Joint Company "Amu Darya" and published on the web portal CAWATERinfo. The total mean annual flow of all rivers in the Amu Darya basin is assumed to be constant and estimated at around 74.22 km^3 (without the Zeravshan).

[6] Based on the same data as above. UNECE reports that the 1997 figures are actual water uses, while the 2010 figures are prospective water requirements [20].

of international organizations. However, there is no binding agreement and no way for the international community to force riparian countries to keep water use to sustainable levels and to manage the Amu Darya so to ensure that it receives sufficient amounts of water. The environmental pressure caused by the social and economic damage along the shores of the Aral Sea is serious, but seems, at this stage, to be considered less important than the losses that would derive from reducing water consumption in certain areas and for specific activities. This is a conscious and explicit political choice. There is, moreover, the fear that if a given country went ahead with large water saving plans, other countries would not do the same, which would result in a comparative disadvantage for the virtuous country. This is a typical cooperation dilemma. This does not mean, however, that the perception and understanding of the situation or the actual situation might not change in the future and that countries could not find the right incentives and political will to limit water use to sustainable levels.

Another important issue but one that has not reached the very top of the government's agenda yet is that of climate change. As one of the editors of this volume has pointed out in several occasions [28], Central Asia and Turkmenistan are among the parts of the world that have already experienced the highest increases of temperature and that are expected to suffer the highest increases in the coming future, which is cause of particular concern given the relative scarcity of water, that most of the country is desert and that many settlements are located in mountain areas. Most international organizations and development partners are somehow involved in trying to bring the attention of the government on this issue. The leadership of the country proved to be sensitive to the issue, launching high visibility initiatives such as the already mentioned regional center. It is likely that changing climate will result in more extreme events such as draughts and floods, which are already cause of concern. Risk management needs not only sustainable land and water management but also early warning and alert systems that sometimes need an international reach, such as in the case of most transboundary rivers. The governments of the region, which is prone to natural disasters such as earthquakes, are in regular contact on the matter and many international partners are also involved strengthening the capacity of the governments to respond.

Other issues of concern with an international dimension are environmental impact assessment of transboundary projects, such as dams and new irrigation schemes, prevention of and fighting against transboundary pollution, and transboundary effects of industrial accidents that can contaminate water. Access to environmental information, such as water quality in specific areas by the public, is another issue often raised by international partners. Nontraditional issues such as payments for ecosystem services, such as water sanitation performed by certain ecosystems, or more generally the so-called "green economy" discussed at recent international conferences, including ecotourism in wetlands, are relatively new. Another nontraditional approach to water management is including trade in agricultural products also in the water balance of countries. It is often said that exporting one tomato is like exporting four gallons of water. As far as I know, this approach is new to the region. Finally, technology exchange must also be mentioned. Turkmenistan actively uses international conferences and scholarly exchanges to invite experts from the USA, Israel, as well as other countries, to introduce new technologies and innovative techniques in the country.

5 Formal and Informal Interactions

International cooperation is not limited to formal interactions. An issue that is often overlooked in the various analyses is the co-presence of different types of interactions. There is in fact a wide range of informal activities going from general monitoring to the daily running of irrigation schemes and the cleaning up of riverbanks. These practical activities are usually performed at the level of operators and local governments. While there can be local rivalries and misunderstandings, in most cases cooperation with the other side of the river seems to be regular and positive, especially on environmental issues, cemented as it is by the sharing of common resources and by many years of living side by side. Many individuals working on two sides of the same border have studied together in Soviet institutes and have developed links of friendship. This seems to hold true with all neighboring countries, from Uzbekistan to Afghanistan. These activities are usually performed below the radar of officialdom as there is no need to have formal meetings and exchanges. Of course, the official level regularly monitors the situation and provides inputs, ultimately exercising control, if needed. On the Amu Darya, government authorities are more vigilant toward the Afghan than the Uzbek border for obvious security reasons.

Turkmenistan, however, distinguishes itself for its high level of formality both internally and externally. This is a reflection of its Soviet past and of the huge role that the public sector and government structures play in the national economy, as well as perhaps a cultural trait. Level of formality is an important trait in Turkmen domestic and foreign policy and is often used as a way to prioritize, also in the water sector. Huge importance is given, for example, to high level foreign guests. Their participation in official celebrations, such as the national holiday "a drop of water is a grain of gold," tends to have positive effects on relations between Turkmenistan and international partners, from donor countries to international organizations. Respect for elders and generous hospitality are traditional values in Turkmenistan, as well as in the greater region. There are many kinds of formal interactions and they range for the most formal, such as agreements and commissions, to the less formal, such as meetings and programs. The general trend is toward less formality, but there is still – and I have recently argued that there should be [29] – room for formal frameworks, including legal agreements and institutional structures. This is particularly true in the case of Turkmenistan also in the water sector.

6 Bilateral and Multilateral Relations

Two main types of relations can be identified in international cooperation. On one hand, there are bilateral relations between Turkmenistan and other countries individually. In the water sector, these countries can be classified in neighbors, donors, and others. On the other hand, there are multilateral relations at different scales: subregional, regional, and global. Multilateral relations often take place in the framework of international agreements or organizations. The following paragraphs are going to present the main bilateral and multilateral relations of Turkmenistan relevant for the water sector. This will allow to complete an overview of the status of international cooperation of the country in this issue area.

6.1 Bilateral Relations

The most important partner of Turkmenistan in the water sector is Uzbekistan, as the two countries share a significant part of the Amu Darya. From the trinational border shared also with Afghanistan, the river moves northwest well into Turkmen territory. It then defines the border with Uzbekistan north of Turkmenabat before fully entering Uzbek territory south of Urgench. As with all its neighbors, Ashgabat cultivates friendly relations with Tashkent. Meetings are frequent both at formal and informal levels. The countries jointly operate irrigation schemes such as the Qarshi pumping stations. These are regulated by the Agreement between Turkmenistan and the Republic of Uzbekistan on Cooperation on Water Management Issues, signed in Turkmenabat on January 16, 1996. This agreement includes some provisions for dispute resolution and is still in force. Also, the joint management of the large and shallow Sarygamysh Lake should not be forgotten. Its level and the quality of its water are of vital importance for the inhabitants of the surrounding area. While it is true that the two countries have some basic interests in common due to their geographical position and share many positions, it is unfair to equate the foreign policy of the two countries as far as water is concerned. Turkmenistan is very careful at maintaining its neutral stance and good neighborly relations, while Uzbekistan generally favors a bilateral approach. For this reason, it is an exaggeration to characterize them as a downstream block against upstream countries.

The second most important partner for water management is Afghanistan. Three important rivers originate from there: the Amu Darya, the Murgab, and the Tejen. For decades, relations have been complicated by the Soviet invasion, civil war, the Taliban regime, and the current war. In recent years, Turkmenistan has tried to foster good neighborly relations, also extending humanitarian assistance and development aid. In Soviet times, attempts were made to establish a shared water monitoring system, but war got in the way. In case of floods upstream, alert mechanisms for downstream countries are weak. The situation makes it difficult to know exactly what happens on the Afghan side and Turkmen experts are eager to learn more about it, especially considering the importance of transboundary rivers. Moreover, for security reasons, it is difficult for foreigners, including official representatives of international organizations, to obtain permission to visit and assess the situation on the border. Peace and prosperity in Afghanistan will definitely have consequences for water use and for downstream countries. The potential for collaboration between experts and administrations of the two countries is great and neutral frameworks such as the UN Special Programme for the Economies of Central Asia (SPECA), where both countries are full members and whose chairmanship is currently held by Turkmenistan, are ideally positioned to facilitate these contacts.

The third most important partner is Iran. The two countries share two rivers: the Tejen and the Atrek. Only the latter originates in Iran. We have seen that the Tejen flows from Afghanistan, defines first part of the Afghan–Iranian border and part of the Turkmen–Iranian border, where it feeds the Doosti dam, also known as the Iran–Turkmenistan Friendship Dam. The management of the dam is regulated by an agreement signed in 2004, which is similar to the bilateral agreement with Uzbekistan. With the exception of the Doosti dam, where a joint coordination commission was created, it must be noted that Turkmenistan has not established bilateral commission for the management of transboundary rivers, despite the fact that they are an instrument chosen by many countries and enshrined by several agreements and conventions [30]. The dam is of vital importance for the Iranian city of Mashhad, so trinational cooperation among Iran, Turkmenistan, and Afghanistan is essential, considering that the Tejen river also flows through the Afghan city of Herat. The joint management of the Atrek river and the Caspian Sea are also important but are more distantly related to the focus of this book. Relations between the two countries are friendly and meetings are regular.

Another important country in the region is Tajikistan. With 80% of the Amu Darya's run-off originating from there, the country is a "water superpower."[7] Of course, any change in water use in Tajikistan has consequences for Uzbekistan and Turkmenistan and the Aral Sea. The first and most immediate issue of concern is the presence of adequate mechanisms to alert downstream countries in case of flood upstream. A second issue is the potential consequences of the failure of existing or planned hydropower plants for downstream countries. A third issue is the integrated management of the river system, particularly with regard to finding a balance – and mechanisms to regulate it – between hydropower production, land irrigation, and clean water, on one hand, and guaranteeing that a sufficient amount of water ends up in the Aral Sea, on the other hand. For these reasons, relations between the two countries have known moments of tension in the past, but these have been much lower than the levels experienced with Uzbekistan on these issues. Turkmenistan tends to deal with the situation through regional platforms such as IFAS, SPECA, and UNRCCA. Diplomatic relations between the two countries are normal. These instances will be discussed in more detail below together with other multilateral processes.

[7] Calculation of the author based on data published by the IFAS Executive Committee.

Finally, relations with donor countries in the water sector, such as the USA and the European Union, are cordial, but conditioned by the double reluctance of donors to fund activities in a middle-income country and of Turkmenistan to receive financial assistance it has not requested. There is a general agreement that the country needs technical assistance and capacity building, as many experts left the country after the collapse of the Soviet Union. This is a kind of assistance that donors are normally happy to provide, as it allows them to give a competitive edge to their own experts, particularly in the case of Germany, France, and other European countries. Turkmenistan generally welcomes such assistance, such as in the case of the TACIS program, as long as it remains technical assistance and does not come with a hidden agenda. Spontaneously or as a result of technical assistance, Turkmenistan also collaborates with research institutes or individual experts from Russia, Israel, the USA, as well as other parts of the world, particularly for the implementation of its water projects. International conferences organized in Turkmenistan are usually the occasion to foster these collaborations.

6.2 Multilateral Relations

Moving to multilateral relations, there are at least three distinctions to be made: first, between formal and informal groupings of countries; second, based on sector or functions; and third, according to scale. With regard to the first distinction, we will focus on formal processes. As we have already discussed above, Turkmenistan keeps a neutral stance and tends to favor formal interactions. Concerning the distinction among the various sectors, the most important difference is between development banks and other international organizations Turkmenistan is member of. In fact, while countries usually find themselves in a position where they request the assistance of development banks, such as the World Bank, the Asian Development Bank (ADB), and the EBRD, to finance various initiatives, the relationship with other international organizations is normally the opposite. Organizations such as the United Nations often make proposals, but they rarely have resources themselves. They need to partner with donor countries to obtain these resources and they need to obtain the agreement of recipient countries to implement projects. This puts countries such as Turkmenistan decidedly in the driver's seat. Now, the relative wealth of the country puts it in a position where its need for funding from development banks in the water sector is limited, so this distinction is also not fully relevant to our case. Therefore, we chose scale as the main organizing principle for the concluding paragraphs, distinguishing between the subregional (Central Asia) and regional (Europe or Asia) scale, on the hand, and the global scale, on the other hand. We will see that the position and relative weight of Turkmenistan with regard to other countries at the different scales makes a significant difference for its attitude toward various platforms.

At the subregional scale, the three main platforms are ICWC–ICSD–IFAS and UNRCCA. As it has already been mentioned above, ICWC was created in 1992 to act as secretariat for the Almaty Agreement. Turkmenistan played an important role for its creation in the early 1990s. Under the ICWC, there is also an Interstate Commission for Sustainable Development (ICSD), which takes care of more specifically environmental issues, with a good degree of success. The environment is the most advanced sector in terms of subregional cooperation and Turkmenistan played an important role for the creation of this body as well. A SIC was established under the ICWC in Tashkent and acts as the main source of information for the water sector in Central Asia. There is a branch of the SIC in each member state, including Turkmenistan, and Turkmen authorities regularly share relevant information with the SIC. Turkmenistan strives to maintain positive working relations with all members of the ICWC, the ICSD, and the SIC. The greatest efforts are made, however, with regard to IFAS, which is arguably the only fully functioning autonomous subregional arrangement in Central Asia.

Like the five other member states, Turkmenistan also has a permanent representative in the IFAS Executive Committee, whose headquarters change on a rotating basis. This makes sure that the interests of all countries are taken into consideration. This is one of the advantages of IFAS, which resulted in the approval of the Aral Sea Basin Program (ASBP), which has already reached its third cycle. Supported by donors, the ASBP is basically a project container that is the result of a careful balance between the position of both upstream and downstream countries as requested by the presidential summit of 2009. There are in fact projects to support adaptation to the consequences of environmental change along the shores of the Aral Sea, as well as projects to promote mitigation of its causes in all riparian countries. To be fully adopted, the ASBP needs however to be approved at the national level by all member states. At the time of writing, Turkmenistan is about to join Kazakhstan and Uzbekistan, who have already approved it. It is hoped that the fact that no upstream country has approved it yet is not a sign of politicization of the program, which is mutually beneficial and quite neutral, which is in turn one of the prerequisites of Turkmenistan to support it. It is worth noticing that a significant part of the funding for the ASBP will come from Central Asian republics themselves, particularly for projects at the national level. Donor support was requested for regional initiatives, particularly from Germany and ADB.

Another relevant platform is represented by the UNRCCA, considering that water and the environment is one of its three priority areas. Mission created in 2007 through the UN Secretary Council, the UNRCCA constantly engages in political dialogue with all Central Asian republics to prevent conflict, also in the water sector.[8] This comes in the form of good offices of the UN Secretary General, who visited the region in 2009, and of his special representative, Miroslav Jenča, whose office is hosted in Ashgabat. This comes also in the form of regular consultations at the highest political level, of meetings, seminars, and trainings on the general situation, but more frequently on specific issues such as the joint

[8] See the letter dated May 7, 2007, from the Secretary General to the President of the Security Council (S/2007/279).

management of transboundary waters. The UNRCCA currently manages a project sponsored by the government of the USA to promote dialogue and a mutually beneficial agreement on water resources management. In this manner, it supports the work of IFAS and it builds capacity about international law, mediation of potential disputes on transboundary waters, and for the creation of an early warning mechanism for transboundary water issues, with the support also of France. The idea of such a center in Central Asia has been in the air for several years, but the offer of Turkmenistan to host it in Ashgabat once again proved fundamental for the actual opening of the center.

At the regional level, important frameworks of reference for water issues are the two UN Regional Commissions, the UNECE, which is based in Geneva, and the Economic and Social Commission for Asia and the Pacific (ESCAP), which is based in Bangkok. Together they manage and service SPECA. All Central Asian republics are at the same time members of both regional commissions, which are essentially standard setting organizations also in the water and environment sectors. The general neutrality of the United Nations and the technical nature but political leadership of the regional commissions – the Executive Secretary is traditionally a former minister of foreign affairs and an Under-Secretary General (USG) – makes them ideal platforms to advance cooperation in the water sector in the region. The last UNECE "Environment for Europe" Ministerial Conference, held in Astana in 2011, focused on water and the green economy and provided an opportunity for the countries of the region to discuss issues of common interest and to prepare for Rio + 20.

It must be noted, that the UNECE, in particular, has developed and services the 1992 Water Convention, which enshrines most generally accepted principles for the management of transboundary waters. In the framework of the EU Water Initiative's (EUWI) National Policy Dialogue (NPD) on Integrated Water Resources Management (IWRM), the UNECE is supporting the government of Turkmenistan in the accession process, which is expected to take place shortly. In particular, the UNECE is supporting a working group of national experts that are assisting the government in the preparation of the technical documents and draft legislation needed for accession. The EU has partnered with the UNECE to support the NPDs in Central Asia. The 2002 EU Water Framework Directive and the 1992 UNECE Water Convention are the two main frameworks of reference. With several non-UNECE member states that expressed interest in joining the convention, including Iran and Afghanistan, it must be noted that the 1992 Water Convention, on the hand, is evolving from a regional to a global convention, on the other hand, was caught in the dispute between upstream and downstream countries and politicized, even if the letter of the convention merely reflects general principles that are commonly accepted in many other subregions.[9] Some countries proposed to develop a water convention specific to Central Asia. Turkmenistan itself had

[9] The 1997 New York Convention, which was developed by the International Law Commission of the UN General Assembly and was supposed to be the global convention, has not managed to enter into force yet because of some controversial provisions.

presented the Framework Convention for the Protection of the Environment and Sustainable Development in Central Asia in 2006, but these approaches have not gathered consensus from all interested countries yet.

Other frameworks active in the water sector at the regional level are a development bank such as the Islamic Development Bank (ISDB) and an international organization such as the Organization for Security and Cooperation in Europe (OSCE). While the former is providing loans for rural water supply infrastructure, the latter is implementing small projects focusing on capacity building for the sustainable management of land and water to fight against soil degradation. On the side of development banks, the absence of the EBRD and the ADB from the water sector is significant if compared to other countries in the region. Again, the availability of financial resources for water projects gives Turkmenistan a high degree of autonomy in this regard. A specific feature of these regional arrangements is that they are sometimes dominated or have a strong imprinting from a large country or group of states in the broader region. While this is not necessarily a problem, this may clash with the strict neutrality of Turkmenistan. The OSCE, for instance, is perceived to be dominated by Western European countries and their values, the ADB by China, the EBRD by the United Kingdom, the ISDB by Saudi Arabia, etc.

Finally, we move to the global level, where somehow Turkmenistan, because of its foreign policy, feels more comfortable, particularly in the framework of the United Nations. Because of their neutral platform, the United Nations are in a position to collaborate with Turkmenistan much more closely than other international partners. However, water being a territorial resource, global initiatives in the water sector are struggling. We already saw how a regional agreement, such as the 1992 Water Convention, is de facto becoming a global standard. It is interesting to see how Target C to "halve, by 2015, the proportion of the population without sustainable access to safe drinking water and basic sanitation" of Millennium Development Goal (MDG) 7 to "ensure environmental sustainability" is generally being pursued at the national scale. There are traces of this in the United Nations Development Assistance Framework (UNDAF) negotiated between the United Nations and the government, but there are no projects currently being implemented specifically about water supply and sanitation, as far as I know. The United Nations Development Programme (UNDP), which is the only international organization present in Turkmenistan capable of implementing large projects, has instead obtained funding from the new Adaptation Fund to implement a two million dollar project to address climate change risks to farming systems at national and community level, with particular attention to the water sector. Concretely, this means that some analysis, support to the revision of the Water Code (in collaboration with the UNECE), and plenty of activities at the farmer, communal, and water users association level will be implemented. Again, while the government focuses on core functions such as water supply and sanitation, international partners try to promote forward looking issues such as climate adaptation. This resonates well with the government, which we saw promoting high visibility initiatives, such as that of launching a Regional Center for Climate Change in Ashgabat. Often, these international initiatives in the environment sector are also supported by UNDP, which is implementing projects to prepare the countries of Central Asia, including Turkmenistan, for their participation in large international conferences such as Rio + 20.

A peculiar case is that of the World Bank. In the late 1990s, it had approved a thirty million dollar project to improve water supply and sanitation in the northern region of Dashoguz in the framework of the ASBP. In the early 2000s, it had also performed a study on integrated water resource management at the subbasin level, where the need is particularly acute because of the presence of the Sarygamysh Lake and the proximity of the Aral Sea. Besides national and local authorities, some of these activities were implemented in collaboration with the United Nations Children's Fund (UNICEF). For reasons that are described in the final report of the project, which is published online on the bank's website [31], the project was not completed and its implementation was considered unsatisfactory, I assume by both the bank and the government. This was followed by a long period when the bank did not grant any loan to the country, which coincided with the closing up of the country until the mid-2000s. In recent years, relations with Turkmenistan have normalized and the World Bank is once again making investments. As far as I know, no loan has been granted in the water sector yet, but this may come in the future. In the framework of the NPD, the Ministry of Water Economy has recently expressed some interest in launching a pilot project of integrated water resource management at the subbasin level in the Dashoguz region, which may be an opportunity to build upon the work of the World Bank in the early 2000s.

Finally, we must not forget more traditionally environmental initiatives in the water sector such as the sites designated under the 1971 Wetlands Convention and the UNESCO Biosphere Reserves. These are purely scientific initiatives, where cooperation is relatively easier and which receive strong support from the government. In the case of the former, Turkmenistan rejoined the convention, which focuses on the protection of wetlands and of the migratory birds that inhabit them, in 2009 (its territory had been under the convention until the fall of the Soviet Union). The only Ramsar site in Turkmenistan is the Hazar State Nature Reserve on the Caspian Sea coast south of Turkmenbashi. The site is being supported by the UNDP with generous funding from the Global Environment Facility (GEF). Together with four other natural sites, including the Amu Darya State Nature Reserve, the site is now also on the national tentative list of Turkmenistan to enter the UNESCO World Heritage List. No natural property is currently located in Turkmenistan. The inscription of a site on the list would be not only a great recognition for Turkmen heritage but would also be an excellent manner to ensure continuous monitoring of the protection and sustainability of these sites, also in terms of tourism development. Another site on the national tentative list, the Repetek Biosphere State Reserve, is also a UNESCO Biosphere Reserve, the only one in the country. This is another tool to ensure continuous monitoring of sights, as well as a way to transform them in living labs to improve our understanding of coupled human–environment systems [32, 33].

To conclude our overview of bilateral and multilateral relations of Turkmenistan, it is important to mention a peculiar platform, the Environment and Security Initiative (ENVSEC), which brings together six global and regional partners – the United Nations Environment Programme (UNEP), UNDP, UNECE, OSCE, the Regional Environmental Center (REC), and the North Atlantic Treaty Organization (NATO) – to fight against environmental threats to reduce the risk of conflict. Interagency coordination is well known to be an arduous exercise, but this one has been more successful than others. Moreover, because of its many water and environmental issues, Central Asia is certainly one of the key areas for this initiative, which has recently produced an analysis of the situation in the Amu Darya River Basin [21].

Of course, this quick perusal does not include all aspects and certainly some international partners and cooperation activities of Turkmenistan in the water sector have not found their place here. The objective of this chapter was to describe, to provide a conceptual framework to analyze the situation, and to highlight major elements, so the reader can understand the overall picture and possess the elements to deepen specific issues. Even if there is no intention to evaluate the foreign policy of Turkmenistan in the water sector, the picture emerging from this analysis is that of a country principled in its relations, selective about its partners, in good terms with its neighbors, with a solid, balanced, and expanding network of international connections. In this manner, Turkmenistan is contributing to developing institutions capable of managing transboundary waters in times of increasing environmental pressure.

References

1. Haas PM (1990) Saving the Mediterranean: the politics of international environmental cooperation. Columbia University Press, New York
2. UNECE (2012) Environmental performance review of Turkmenistan. Draft presented at the Expert Group on Environmental Performance, Ashgabat, March 13–15, 2012. UNECE, Geneva
3. Stanchin I, Lerman Z (2010) Water in Turkmenistan. In: Arsel M, Spoor M (eds) Water, environmental security and sustainable rural development: conflict and cooperation in Central Eurasia. Routledge, Abingdon, pp 250–266
4. Berdiyev A (2007) International and national water law: Turkmenistan, 2nd edn. TC ICWC, Tashkent
5. Babaev AG, Kolodin MV (1995) The water resources of Turkmenistan. Problems of desert development. J Eurasian Stud 4:18–23
6. UNDP (2010) Assessment of water sector in Turkmenistan. UNDP, Ashgabat
7. Kostianoy AG, Kosarev AN (eds) (2010) The Aral Sea environment, vol 7. Springer, Berlin
8. Létolle R (2008) La mer d'Aral: entre désastre écologique et renaissance. Harmattan, Paris
9. Zonn IS, Glantz M, Kostianoy AG, Kosarev AN (eds) (2009) The Aral Sea encyclopedia. Springer, Berlin
10. Dukhovnyĭ VA, Schutter JD (2011) Water in Central Asia: past, present, future. CRC/Balkema, Boca Raton
11. Amansaryýew B (2009) Türkmenistanyň tebigaty täze galkynyş eýýamynda. Türkmen Döwlet Neşirýat Gullugy, Ashgabat

12. Turkmenistan (2008) "Suw hakynda" Türkmenistanyň bitewi kanuny: 2008-nji ýylyň Oguz aýynyň 30-na çenli girizilen üýtgetmeler we goşmaçalar bilen berildi. Türkmen Döwlet Neşirýat Gullugy, Ashgabat
13. Turkmenistan (1999) The state of environment of Turkmenistan. Ministry of Nature Protection, Ashgabat
14. Kurbanov PK, Turkmenistan. Ministry of Nature Protection, Portal to Asian Internet Resources Project (2003) The state of environment: Turkmenistan. Retrieved July 5, 2012, from http://enrin.grida.no/htmls/turkmen/soe2/index.htm
15. Grzybowski A, Mandell L, Menaker A, Paisley R (eds) (2011) Best practices and international experience with transboundary water dispute resolution: final report of the seminar held in Almaty, on December 6–7, 2010. UNRCCA, Ashgabat
16. Libert B, Trombitcaia I, Enderlein R, Vykhryst S, Steklov Y (2011) Strengthening water management and transboundary water cooperation in Central Asia: the role of UNECE environmental conventions. UNECE, Geneva
17. Nepesova MG (2006) The water resources of Turkmenistan and their transboundary aspects. In: UNECE (ed) Transboundary water cooperation: trends in the newly independent states. United Nations, New York, pp 85–88
18. Pupols A, Rahmanova B (eds) (2009) International seminar "Central Asia and global challenges," Ashgabat, March 10–11, 2009. UNRCCA, Ashgabat
19. Pupols A, Rahmanova B (eds) (2010) International seminar "Security and stability in Central Asia: interaction with international and regional organizations," Ashgabat, April 21–22, 2010. UNRCCA, Ashgabat
20. UNECE (2011) Second assessment of transboundary rivers, lakes and groundwaters. UNECE, Geneva
21. UNEP, GRID Arendal, Zoi Environment Network (2011) Environment and security in the Amu Darya River Basin. ENVSEC, Geneva
22. Kosarev AN, Kostianoy AG (eds) (2005) The Caspian Sea environment. Springer, Berlin
23. Zonn IS, Kostianoy AG, Kosarev AN, Glantz MH (2010) The Caspian Sea encyclopedia. Springer, Berlin
24. Jones Luong P, Weinthal E (2010) Oil is not a curse: ownership structure and institutions in Soviet successor states. Cambridge studies in comparative politics. Cambridge University Press, Cambridge
25. Weinthal E (2001) State making and environmental cooperation: linking domestic and international politics in Central Asia. MIT, Cambridge
26. Turkmenistan (2005) The permanent neutrality of Turkmenistan. Türkmen Döwlet Neşirýat Gullugy, Ashgabat
27. Wouters P, Vinogradov S (2003) Analysing the ECE Water Convention: what lessons for the regional management of transboudary water resources. In: Stokke OS, Thommessen ØB (eds) Yearbook of international co-operation on environment and development 2003/2004. Earthscan, London, pp 55–63
28. Kostianoy AG, Lebedev SA, Solovyov DM (2011) Satellite monitoring of water resources in Turkmenistan. Int Water Technol J 1(1):4–13
29. Church JM (2012) Recent trends in regional cooperation for the implementation of the concept of sustainable development. Paper presented at the International seminar "Sustainable development: key factor for stability and peace in Central Asia," Tashkent, March 1–2
30. UNECE (2009) River basin commissions and other institutions for transboundary water cooperation: capacity for water cooperation in Eastern Europe, Caucasus and Central Asia. UNECE, Geneva
31. World Bank (2004) Implementation completion report (SCL-41710 TF-25244 PPFB-P3120) on a loan in the amount of US $30.3 million to the Government of Turkmenistan for the Turkmenistan Water Supply and Sanitation Project. World Bank, Washington, DC
32. Turner BL, Clark WC, Kates RW, Richards J, Mathews JT, Meyer W (eds) (1993) The Earth as transformed by human action: global and regional changes in the biosphere over the past 300 years, 2nd edn. Cambridge University Press, Cambridge
33. Price MF (1995) Man and the Biosphere (MAB) Project 6 in Europe and the Former USSR. Mt Res Dev 15(3):267–282

Conclusions

Igor S. Zonn and Andrey G. Kostianoy

Abstract This book presents a brief description of the environment and water resources in Turkmenistan. The focus is made on the water bodies of Turkmenistan – the Caspian Sea, Kara-Bogaz-Gol Bay, Sarykamysh Lake, Amu Darya River, and the Karakum Canal. The information on the landscape-geographical features, the Karakum Desert, biodiversity (birds and fishes) and ecosystems, as well as the regional climate change is given. Special attention is paid to the "Altyn Asyr Lake" water reclamation project, the morphometric characteristics of the Karashor Depression, and four-years-long satellite monitoring of the area of construction in the vicinity of the Karashor Depression. The publication is based on observational data, scientific literature mainly published in Russian editions, and long-standing experience of authors from six countries in the scientific research of the environment in Turkmenistan. This is the first book on the Altyn Asyr Project published in a Western edition. The book is addressed to the specialists working in various fields of environmental problems and ecology, water resources and management, land reclamation and agriculture, regional climate change, and international cooperation in the water sector in Turkmenistan and Central Asia.

Keywords Altyn Asyr Lake, Central Asia, Environment, Turkmenistan, Water Resources

---

I.S. Zonn
Engineering Research Production Center for Water Management, Land Reclamation and Ecology "Soyuzvodproject", 43/1, Baumanskaya ul., Moscow 105005, Russia

A.G. Kostianoy (✉)
P.P. Shirshov Institute of Oceanology, Russian Academy of Sciences, 36, Nakhimovsky Pr., Moscow 117997, Russia
e-mail: kostianoy@gmail.com

I.S. Zonn and A.G. Kostianoy (eds.), *The Turkmen Lake Altyn Asyr and Water Resources in Turkmenistan*, Hdb Env Chem (2014) 28: 311–320, DOI 10.1007/698_2013_239,

In 2010, at the suggestion of Tajikistan, the UN General Assembly announced the year 2013 as the International Year of cooperation in the field of water resources. On 11 February 2013, at the headquarters of UNESCO in Paris, the official launching ceremony of this International Year took place. The goal of the International Year is to draw attention to the various aspects of cooperation in the field of water resources in terms of sustainable economic development, climate change, food security, and gender equality. Today, at least 145 countries in the world share water resources with their neighbors. Examples of close cooperation in this area are much more than those of conflicts. Evidence of this is the project for the equitable distribution of socioeconomic benefits from the cooperative use of water in the Nile Basin; an agreement between Argentina, Brazil, Paraguay, and Uruguay on the protection of the Guarani Aquifer; the program of comanagement of the Mekong River basin; the cooperation agreement recently signed between Moldova and Ukraine on the protection and sustainable development of Dniester River; and others. In a recorded message for the 20th World Water Day on 22 March 2013, UN Secretary-General Ban Ki-moon said: "Water is central to the wellbeing of people and the planet. We must work together to protect and carefully manage this fragile, finite resource." This year World Water Day highlighted spread of cooperation and security issues.

President of Turkmenistan Gurbanguly Berdymukhamedov at the conference "Environmental Cooperation of Turkmenistan with Major International Organizations: Achievements and Success," held in Ashgabat on 21–22 November 2011, said: "Putting forward new proposals and initiatives on environmental issues, Turkmenistan stands ready for intensifying and promoting positive cooperation on global scale." One of the objectives of that conference was to discuss the proposal to establish the Caspian Environment Council and to create the Regional Center for Climate Change in Ashkhabad [1]. We hope that our initiative in preparation of the present international book project on water resources in Turkmenistan and the Altyn Asyr Lake project fits well in the stream of activities of the 2013 International Year of cooperation in the field of water resources, as well as meets the expectations of the President of Turkmenistan.

Turkmenistan is a Central Asian country of about 490,000 km^2, the fourth largest by area in the former Soviet Union after Russia, Kazakhstan, and Ukraine. Over 80% of the country is covered by the Karakum Desert, one of the largest and driest sand deserts in the world [2, 3]. Some regions in the country have an average annual precipitation of only 12 mm. The highest temperature recorded in Turkmenistan was 51.7°C in July 1983. The habitable area is strictly limited, and this huge country has a small population of about six million people, which rapidly grows. Turkmenistan possesses the world's third-largest proven reserves (8.7%) of natural gas after Russia and Iran, but land and water are the two scarcest and most precious resources in this country. Turkmenistan, like all other Central Asian countries, is critically dependent on water because of its arid desert climate, which is becoming warmer and drier [4]. Amu Darya River, flowing from the Pamir and Tien-Shan Mountains along the entire length of the northeastern border with Uzbekistan to the tragically dying Aral Sea, is the main source (84%) of water for all agricultural and nonagricultural uses in Turkmenistan [5]. Water intake from Amu Darya River is supplemented with surface runoff from three other rivers – Murghab, Tedjen, and Atrek, as well as minor quantities from small rivers and springs. Groundwater plays a marginal role in Turkmenistan's water resources [6]. Water has become the principal strategic resource that determines the region's economic development.

Freshwater resources in Turkmenistan should be discussed with other water resources like the Caspian Sea, Kara-Bogaz-Gol Bay [7], and Sarykamysh Lake [8] which play a very important role in different sectors of the economy of the country. This is done in the present book. Turkmenistan is connected with the other Caspian Sea countries by shipping routes. Offshore, there are oil and gas fields, which are developed by national and foreign companies. The sea is rich in bioresources, and fishery is a part of the economy of the country. Kara-Bogaz-Gol Bay during the twentieth century played a key role in the chemical industry of the Turkmen Soviet Republic and today in Turkmenistan. The national tourist and recreation zone "Avaza" near Turkmenbashi town is progressively developed at the coast of the warm Southern Caspian Sea. The Caspian Sea water (after desalination) is an immense potential source of potable and technical water for the country, living in the desert conditions [7].

The history of irrigation on the territory of the present Turkmenistan goes back to the late fourth to early third millennia BC, when people used mountain springs in the piedmont areas of the Kopetdag Mountains bordering the Karakum Desert for cultivation of cereals. Herodotus, Polybios, and Pliny (third to first century BC) wrote in their treatises about extensive application of land irrigation in Southern Turkmenistan based on construction of large canals and irrigation facilities. The idea to use the water of Amu Darya River for irrigation of the Karakum Desert was shaped in the eighteenth century and it was partially realized during the tsarist time in Russia. But only in the 1950s the Karakum Canal (1950–1988), the world's largest hydraulic engineering project, was designed and constructed in the USSR. Its length is 1,380 km, and the head structures take out annually about 13.5 km^3 of water from Amu Darya River [9].

After Turkmenistan became an independent state this canal was renamed into Karakum River. The artificial Karakum River connects Amu Darya, Murghab, and Tedzhen Rivers into a single water system making the basis for economic development of the country. This artificial river permitted to extend the irrigated lands for growing cotton, fodder crops, vegetables, and melon crops; to create fishery farms; to water desert pastures and, accordingly, stimulate development of distant-range grazing of cattle; and to develop local navigation and use the waters of this river in industry and power engineering. In 2010 in the canal area over 900,000 ha were irrigated, almost four times more than before its construction. In the future, by means of the canal up to 1 million ha of fertile lands will be irrigated. After completion of construction, the canal will reach the western edges of the Kopetdag and come to the coast of the Caspian Sea [9].

Continuous massive irrigation works in Turkmenistan during 60 years required construction of the appropriate system of collectors and other drainage facilities intended for the removal of excess water from soil. Without proper drainage, soil may become waterlogged due to the rising water table and its salinity may increase to levels detrimental to crop growing. Expansion of irrigated areas naturally requires expansion of the collector-drainage network, whose inadequacy is reflected in severe deterioration of soil quality. By the beginning of 1982 the collector-drainage system was built on an area of 800,000 ha. This network allowed to divert annually 11.45 km^3 of collector-drainage water (CDW) from the irrigated areas by a system of canals to the Karakum Desert, Sarykamysh Lake, and Amu Darya River. As a result, mineralized CDW deteriorates water quality characteristics in receiving water reservoirs, and in the beginning of 1982 about 65,000–75,000 ha of pasture lands in the Karakums were flooded. In the flooded and underflooded zones, an active process of salt accumulation, waterlogging, salinization of soils, and deterioration (salinization) of water in the distant pasture wells is going on. Furthermore, transport communications are damaged; operation of gas pipelines and some of gas fields becomes difficult.

The Altyn Asyr (Golden Age) Lake Project is a new approach to disposal of CDW from irrigation [10, 11]. Following a decision adopted in August 2000 by the President of Turkmenistan, the country is constructing a huge artificial lake in the middle of the Karakum Desert, on the site of the natural Karashor Depression. The lake is on the border between Balkan and Dashoguz velayats, some 300 km north of the capital Ashkhabad. The lake will be filled with CDW via the Dashoguz branch (northern route) and a new Great Turkmen Collector from the south with a combined length of about 1,200 km. The total length of the main and supplying collectors will reach 2,654 km. According to the Project, Altyn Asyr Lake will be 103 km long, 18.6 km wide, and 69 m deep. It will have a surface of about 1,915.8 km^2 and a volume of 132 km^3 [10]. It is planned that it will receive annually 10 km^3 of CDW; thus it will be filled during about 15 years. When accumulation of water is completed, the lake will look similar to the modern western part of the Large Aral Sea.

The forecast of the Altyn Asyr Lake water and salt budget should take into account the long-term experience with the behavior of Sarykamysh Lake [8]. Sarykamysh Lake is one of about 2,500 artificial lake-collectors of drainage water in Central Asia. The lake is also located in a natural depression in the northwestern part of Turkmenistan, and it receives irrigation surpluses and CDW from Dashoguz and Khoresm oases. The area and volume of the lake have grown from 12 km^2 and 0.6 km^3 in 1962 to 3,955 km^2 and 68.56 km^3 in 2006 [8]. If CDW is partially diverted from Sarykamysh Lake to fill Altyn Asyr Lake, its water and salt budget will be changed. In the book we show the results of the study aimed to model the possible scenarios in the development of the lake following a change of inflow [8]. This research deals with the retrospective study of the parameters of the lake in the past 40 years using GIS and remote sensing methods in order to suggest a forecast of these parameters. The forecasted parameters will enable the mitigation of the negative regional impacts of the lake's changes. A three-dimensional model of the Sarykamysh Depression was built using the 1940s' topographic maps. TOPEX/Poseidon altimeter data, early Corona satellite images, and time-series of the Landsat satellite images were applied to the digital elevation model (DEM) together with ground measurements of the parameters of the lake and meteorological data. The model was calibrated and validated, and the water balance of the lake was calculated, enabling to suggest an optimal future inflow with higher accuracy [8].

The forecast of the Altyn Asyr Lake characteristics and behavior should take into account the characteristics of the regional climate change in Central Asia, and in Turkmenistan in particular [4]. According to the Intergovernmental Panel on Climate Change's (IPCC) Fourth Assessment Report (AR4), developing nations and arid regions of the world are particularly vulnerable to climate change and climate variability. Climate change and variability affect arid ecosystems and their productivity through the changing patterns in temperature and precipitation, droughts, floods, heavy winds, and other extreme events. Meteorological data reveal an increase of annual and winter temperatures in Turkmenistan since the beginning of the past century. The mean annual temperature has increased by 0.6°C in the northern part of the country and by 0.4°C in the south since 1931. At the same time the number of days with temperature higher than 40°C has increased since 1983 [4]. The precipitation has decreased in the area between the Caspian and Aral seas (i.e., in the location of Altyn Asyr Lake) since 1960, which coincided with desiccation of the Aral Sea and Kara-Bogaz-Gol Bay (caused by the construction of a dam in 1980 [7]). This has also caused significant changes in albedo, hydrological cycle, and mesoclimatic changes throughout western parts of Kazakhstan, Uzbekistan, and Turkmenistan [4].

Climate models predict a temperature increase in Turkmenistan by 1°C by 2025, 2–4°C by 2050, and by more than 5°C by the end of the century. Precipitation projections are highly uncertain but given the existing aridity in Turkmenistan, even a slight temperature increase is likely to deepen the existing water stress in the region [4]. Thus, Turkmenistan is projected to become warmer and drier during the coming decades. Aridity is expected to increase in all republics of Central Asia, but especially in the western part of Turkmenistan. This regional climate change is likely to have a profound impact on agriculture, particularly in western Turkmenistan and Uzbekistan, where frequent droughts affect cotton, cereals, and forage production, lead to extremely high water demands for irrigation, aggravate the already existing water crisis, and accelerate desertification [4].

In the book there is a description of the DEM of the Karashor Depression that was constructed on the base of digital surface topography of the area with the spatial resolution of 90 m and the vertical resolution of 1 m, provided by the Shuttle Radar Topography Mission (SRTM) [12]. The absolute (sea) level, surface, volume, and maximum depth of Altyn Asyr Lake were calculated with a step of 1 m. This information was given in a table and different graphical forms, which would be useful for scientific and practical purposes related to the Altyn Asyr Project. The results showed that the maximum depth of the lake will be about 60 m (mean depth – 37 m), surface – about 1,325 km^2, and volume – about 50 km^3. Two scenarios of the lake filling showed that a decrease of the rate from 10 to 5 km^3/year will increase the total time of the lake filling from about 6 to 17 years due to enormous loss of water for evaporation. When the lake is filled out, only 2 km^3/year will be required to compensate yearly evaporation and keep Altyn Asyr Lake at a constant level [12]. The obtained morphometric characteristics of the lake will help decision-makers in Turkmenistan to take correct decisions at the final stage of the Altyn Asyr Lake construction.

Authors of this research are planning to include modeling of water budget of Altyn Asyr Lake basing on the analysis of the potential water supply, precipitation, and evaporation from the changing lake surface, taking into account typical seasonal meteo conditions (air and water temperature, wind speed, and humidity) [12, 13]. This will allow to provide a more accurate relationship between the amount of water supply to the lake and its level (surface and volume) at different stages of its filling. This will also allow a correct forecast of the time required to fill the lake by CDW.

The knowledge of the present morphometric, hydrological, and chemical characteristics of canals and the future Altyn Asyr Lake allowed to forecast the probable fish and bird biodiversity in the area of the lake [14, 15]. The results of the Sarykamysh Lake studies showed that water salinity is the dominating factor for fish population development and evolution. Fish population will reach the lake via the collector and drainage systems from the lower and middle reaches of Amu Darya River, the Karakum Canal, Tedjen and Murghab river basins, and irrigation and drainage network of the Dashoguz velajat. Considering the volumes and salinity of the diverted CDW it may be assumed that the ichthyofauna of the lake will consist of the semi-anadromous fish, aboriginal fish of the Aral-Amu Darya fauna, and Chinese introducents adapted in the region. Ichthyofauna specialists are expecting to find in the lake 25–30 fish species, including carp, breams (eastern), sabrefish, Aral roach, Aral shemaya, Aral asp, Aral (Caspian) barbell, khramulya, (European) catfish, European pike-perch, and under favorable conditions also goldfish, silver carp and bighead, grass (Chinese) carp, white Amur bream, (northern) pike, and snakehead being of commercial significance [14]. The fish species not very demanding to external factors may also reveal mass development, such as stripped bystranka, Korean sawbelly, stone morocco, Aral stickleback, Chinese goby, and others being of no fishery significance [14]. The water salinity level of 4–6 g/L will be favorable for reproduction and fattening of many of the above-mentioned fish species. It should be noted that for most of these species the salinity level of 10–12 g/L will be a barrier [14]. The fish productivity in the lake will depend on the measures on fishery regulation, protection, and farming, which in their turn strongly depend on the special scientific research which has to be conducted in the lake and canals [14].

The Karashor Depression is already considered one of the important bird areas (IBA) in Turkmenistan and in Central Asia. 23 out of 50 IBAs of the Turkmen wetlands are located in the hydrographical network of the Altyn Asyr Project [15]. Currently, in the IBA Karashor we can find 10 species of resident birds, 20 – migrating-breeding, 8 – migrating-wintering, and 85 – migrating. In the dry Karashor Depression there prevails a complex of plain-chink species, of the so-called desert type of avifauna. The following resident birds can be observed: Saker Falcon, Golden Eagle, Long-legged Buzzard, Chukar, Rock Pigeon, Eurasian Eagle-Owl, Little Owl, Brown-necked Raven, Streaked Scrub-Warbler, etc.; among migratory-nesting species – Egyptian Vulture, Short-toed Snake-eagle, Common Kestrel, Greater Sand Plover, Alpine Swift, Finsch's Wheatear, Eurasian Hoopoe, etc. [15]. During filling of the lake a complex of waterbirds will be formed. The abundance of aquatic vegetation will contribute to the great diversity and biomass of aquatic invertebrates, which will serve as a corresponding food base for vertebrates and, above all, fish. This will provide food resources for such fish-eating birds as Gulls, Terns, Pelicans, Cormorants, and Herons, and among prey species – for White-tailed Eagle, Eurasian Marsh Harrier, and Osprey. Thus, in the future, the water surface of Altyn Asyr Lake and its dense riparian vegetation should become a shelter for a lot of migratory and nesting birds and in warm winters for wintering wetland birds [15].

Satellite monitoring of water resources and land is of great importance for Turkmenistan and other Central Asian countries, located in the arid zone, especially now when significant changes in the regional climate are observed [16, 17]. Today, an integral part of any modern environmental monitoring of land, sea, lakes, and rivers is satellite-based monitoring, which has great additional features and advantages over ground-based. An opportunity to survey huge water and land areas in a short period as well as a possibility of repeated observations of the same region with a short interval of time (1 day) make remote sensing the cheapest, fast, and objective method for environmental monitoring. Turkmenistan has supplementary advantages for satellite monitoring, because the country is located in the area with the number of cloudy free days which varies between 240 and 300 and sunshine of 3,100 h/year. These are the highest values in the Former Soviet Union.

The book shows numerous examples of the analysis of different multisensor satellite data for the Southeastern Caspian Sea, Kara-Bogaz-Gol Bay, Sarykamysh Lake, area near the Karashor Depression, and Amu Darya River. Special attention was paid to the drainage lakes Sarykamysh and Altyn Asyr, which we started to monitor since the beginning of its construction in July 2009 [16, 17]. High-resolution satellite imagery looks very effective in monitoring of the Amu Darya River area. Satellite remote sensing can be applied also to monitoring of other smaller water reservoirs, lakes, and rivers, and for agriculture by the analysis of the index of vegetation (NDVI) and vegetation liquid water content (NDWI).

Further sea level, wind and waves, sea surface temperature, suspended matter and chlorophyll concentration, and oil pollution monitoring at various points of the Caspian Sea and Kara-Bogaz-Gol Bay with the use of satellite remote sensing and other observations should allow us to follow the ecological state and future changes which are extremely important for designing, constructing, and operating industrial installations and infrastructure in the Caspian Sea and on its coasts, and first of all for providing ecological security for economic activities in the Caspian Sea region [16, 17].

In June 2011, we proposed to organize in Ashkhabad the National Center for Satellite Monitoring and Regional Climate Change in Turkmenistan with the support of the European Space Agency (ESA) and the Committee on Space Research (COSPAR) [17]. This proposal corresponds well to the initiative of the President of Turkmenistan in organization of the National Space Agency, the Caspian Environment Council, and the Regional Center for Climate Change in Ashkhabad. This Center could effectively monitor the ecological state, oil pollution, and sea level of the Caspian Sea waters of Turkmenistan, especially in such sensitive areas as the National tourist zone "Avaza," as well as in the areas of offshore oil and gas production, and in the areas of new offshore projects, such as the Trans-Caspian gas pipeline. The Center could monitor the state of Sarykamysh Lake and the filling of the new Altyn Asyr drainage lake, Amu Darya River, as well as other water bodies and land areas. The regional climate change in Turkmenistan and Central Asia should be an important task for the research in the Center. Satellite monitoring and climate research performed by the Center could provide very useful information for agriculture, water management, and other sectors of economy, academic and applied science, and education in Turkmenistan.

References

1. Church JM (2013) International cooperation of Turkmenistan in the water sector. In: Zonn IS, Kostianoy AG (eds) The Turkmen Lake Altyn Asyr and water resources in Turkmenistan. Springer, Berlin
2. Zonn IS, Essenov PE (2013) The Karakum Desert. In: Zonn IS, Kostianoy AG (eds) The Turkmen Lake Altyn Asyr and water resources in Turkmenistan. Springer, Berlin
3. Rustamov EA (2013) Turkmenistan: landscape-geographical features, biodiversity and ecosystems. In: Zonn IS, Kostianoy AG (eds) The Turkmen Lake Altyn Asyr and water resources in Turkmenistan. Springer, Berlin
4. Lioubimtseva E, Kariyeva J, Henebry GM (2013) Climate change in Turkmenistan. In: Zonn IS, Kostianoy AG (eds) The Turkmen Lake Altyn Asyr and water resources in Turkmenistan. Springer, Berlin
5. Zonn IS (2013) Water resources of Turkmenistan. In: Zonn IS, Kostianoy AG (eds) The Turkmen Lake Altyn Asyr and water resources in Turkmenistan. Springer, Berlin
6. Esenov PE (2013) Groundwaters and salinization of soils in Turkmenistan. In: Zonn IS, Kostianoy AG (eds) The Turkmen Lake Altyn Asyr and water resources in Turkmenistan. Springer, Berlin
7. Kosarev AN, Kostianoy AG, Zonn IS, Zhiltsov SS (2013) The Caspian Sea and Kara-Bogaz-Gol Bay. In: Zonn IS, Kostianoy AG (eds) The Turkmen Lake Altyn Asyr and water resources in Turkmenistan. Springer, Berlin
8. Orlovsky L, Matsrafi O, Orlovsky N, Kouznetsov M (2013) Sarykamysh Lake – collector of drainage water: the past, the present and the future. In: Zonn IS, Kostianoy AG (eds) The Turkmen Lake Altyn Asyr and water resources in Turkmenistan. Springer, Berlin
9. Zonn IS (2013) The Karakum Canal. In: Zonn IS, Kostianoy AG (eds) The Turkmen Lake Altyn Asyr and water resources in Turkmenistan. Springer, Berlin
10. (2010) Turkmen Lake "Altyn Asyr", Ashgabat, 104 pp. (in Turkmen, Russian and English)
11. Zonn IS, Kostianoy AG (2013) The Turkmen Lake Altyn Asyr. In: Zonn IS, Kostianoy AG (eds) The Turkmen Lake Altyn Asyr and water resources in Turkmenistan. Springer, Berlin
12. Kostianoy AG, Lebedev SA (2013) Three-dimensional digital elevation model of Karashor Depression and Altyn Asyr Lake. In: Zonn IS, Kostianoy AG (eds) The Turkmen Lake Altyn Asyr and water resources in Turkmenistan. Springer, Berlin
13. Pyagay ET (2013) Forecast of the water-salt balance in the Turkmen Lake Altyn-Asyr. In: Zonn IS, Kostianoy AG (eds) The Turkmen Lake Altyn Asyr and water resources in Turkmenistan. Springer, Berlin

14. Shakirova FM (2013) Present-day condition of ichthyofauna in reservoirs of Turkmenistan. In: Zonn IS, Kostianoy AG (eds) The Turkmen Lake Altyn Asyr and water resources in Turkmenistan. Springer, Berlin
15. Rustamov EA, Belousova AV (2013) Waterbirds of hydrographic network of Altyn Asyr. In: Zonn IS, Kostianoy AG (eds) The Turkmen Lake Altyn Asyr and water resources in Turkmenistan. Springer, Berlin
16. Kostianoy AG, Lebedev SA, Solovyov DM (2013) Satellite monitoring of the Caspian Sea, Kara-Bogaz-Gol Bay, Sarykamysh and Altyn Asyr Lakes, and Amu Darya River. In: Zonn IS, Kostianoy AG (eds) The Turkmen Lake Altyn Asyr and water resources in Turkmenistan. Springer, Berlin
17. Kostianoy AG, Lebedev SA, Zonn IS, Lavrova OYu, Solovyov DM (2011) Satellite monitoring of Turkmenistan. Signal, Moscow, 16pp

Index

I.S. Zonn and A.G. Kostianoy (eds.), *The Turkmen Lake Altyn Asyr and Water Resources in Turkmenistan*, Hdb Env Chem (2014) 28: 321–324, DOI 10.1007/698_2013,

Printed by Printforce, the Netherlands